Migration Ecology of Marine Fishes

Migration Ecology of Marine Fishes

David Hallock Secor

JOHNS HOPKINS UNIVERSITY PRESS   *Baltimore*

9 8 7 6 5 4 3 2 1

Johns Hopkins University Press
2715 North Charles Street
Baltimore, Maryland 21218-4363
www.press.jhu.edu

Library of Congress Cataloging-in-Publication Data
Secor, David H., 1960–
  Migration ecology of marine fishes / David Hallock Secor.
    pages cm
  Includes bibliographical references and index.
  ISBN 978-1-4214-1612-0 (hardcover : alk. paper) —
ISBN 978-1-4214-1613-7 (electronic) — ISBN 1-4214-1612-3
(hardcover : alk. paper) — ISBN 1-4214-1613-1 (electronic)
1. Marine fishes—Migration. 2. Fishes—Migration. I. Title.
  QL639.5.S43 2015
  597.156'8—dc23   2014018367

A catalog record for this book is available from
the British Library.

Special discounts are available for bulk purchases of this book.
For more information, please contact Special Sales at 410-516-
6936 or specialsales@press.jhu.edu.

Johns Hopkins University Press uses environmentally
friendly book materials, including recycled text paper that
is composed of at least 30 percent post-consumer waste,
whenever possible.

To my dad, and his example of
contemplative scholarship

Contents

Acknowledgments

Vince Burke, executive editor at Johns Hopkins University Press, asked me several years ago, "Why not write a book?" Years earlier, my then-5-year-old son posed a similar question: "Mom's running marathons. Why don't you?" Unfortunately, endurance running and writing cannot be sustained by the "why not" motivation. My motivation to write this book was largely selfish—a chance to sit in quiet scholarship and try to make sense of an era of discovery on fish migration. My career (1983–present) spans the information age, which has generated a flood of empirical and modeling advances during a period of strong public and private support for marine science. We now find ourselves in the happy dilemma of being deluged in discovery. My goal was to sample recent discoveries and place them into common themes related to life cycle diversity. The inspiration came from my dog-eared volume of *Fish Migration*, a thesis on the life cycles of marine fishes written by F. R. Harden Jones more than 40 years ago. Could I build on where Harden Jones left off?

I wrote most of this book at the James P. Muldoon River Center at St. Mary's College of Maryland, hosted by Bob Paul and the Department of Biology. Chesapeake Biological Laboratory Directors Margaret Palmer and Tom Miller and the University of Maryland Center for Environmental Science President Don Boesch were quite supportive in accommodating periodic absences to work on the book. I am indebted to the members of my laboratory over the past several years, who have endured my frequent absences. In particular, Alex Atkinson, Ben Gahagan, Mike O'Brien, and Matt Siskey have helped hold down the fort. Librarian Kathy Heil kept me out of hoc with late-returned books, and Mike O'Brien obligingly put together most of the maps appearing in the volume. Sara Rains provided fish and bird artwork appearing in chapter 2. I am grateful for initial review of material by Joe Zydlewski (chap. 1), Robert Humston and Helen Bailey (chap. 2), Sue Sogard (chap. 3), Ryan Woodland (chap. 4), Edwin Niklitschek (chap. 5), Lisa Kerr (chap. 6), and Pierre Petitgas (chap. 7).

Lots of contemplative quiet hours at St. Mary's College may sound a bit lonely, but in fact there was a constant din of voices from colleagues as I visited (or revisited) their work. I am privileged to work amidst an incredibly active and prolific cohort of scientists, whose enthusiasm for unraveling the hidden lives of marine fishes is inspiring. I have tried to listen to what they have had to say and hope that I have given their voices due reflection. Most influential, however, were my own former students: Lisa Kerr, Richard Kraus, Edwin Niklitschek, and Ryan Woodland. The daily interactions of "supervising" them as past PhD students have opened doors, heightened my awareness, and helped me witness new discoveries that I would have otherwise missed. Likewise, many friends and colleagues have helped shape this book. Foremost in motivating me were book writers Ken Able and John Waldman, who provided large footsteps to follow. Influential conversations (albeit during sometimes brief interactions) shaped this book, including those with Ken Able, Andy Bakun, Steve Berkeley, Barbara Block, Denise Breitburg, Deirdre

Brophy, Steve Cadrin, Greg Cailliet, Steve Campana, John Casselman, Dave Conover, Ad Corten, Chuck Coutant, Bob Cowen, John Mark Dean, Julian Dodson, Dewayne Fox, Jean-Marc Fromentin, John Graves, Mart Gross, Sharon Herzka, Robert Humston, Karin Limburg, Molly Lutcavage, John Manderson, Jim McCleave, Ian McQuinn, Tom Miller, John Neilson, John Olney, Pierre Petitgas, Adrian Rijnsdorp, Alexei Sharov, Mike Sinclair, Susan Sogard, Masaru Tanaka, Katsumi Tsukamoto, Bob Ulanowicz, John Waldman, Paul Webb, and Mike Wilberg. I am particularly grateful to my long-term collaborators Ed Houde and Jay Rooker for their friendships, unique takes on the world, and abiding support of my scientific pursuits.

Finally, I thank my nature-enthusiast mom and wife, Eriko, who understands all about endurance sports.

Migration Ecology of Marine Fishes

1 Introduction

The Hidden Lives of Marine Fish

Animal migration is a phenomenon far grander and more patterned than animal movement. It represents collective travel with long-deferred rewards. It suggests premeditation and epic willfulness, codified as inherited instinct.
—David Quammen (2010)

Imagine the clandestine lives of marine fishes —their migrations, affinity for habitat, social interactions, and pursuit of prey—all obscured below the sea's oft-opaque surface. The lives of marine fishes have long inspired frustration, wonder, and creative impulse among naturalists, fishermen, scientists, and writers. Two generations ago, visionaries Jacques Cousteau, Rachel Carson, and Eugenie Clark dramatically expanded our view of the dynamic seascapes and the fish that inhabit them. Their technical and literary breakthroughs, passion and creativity generated a wave of public enthusiasm for undersea exploration and catalyzed a generation of marine scientists dedicated to making seascapes and the lives of marine fishes ever more overt. The result is an unprecedented period of discovery arising from advances in ocean observing systems, in methods to track individual fish movements, and in computing and telecommunications systems designed to detect, summarize, and simulate fish migrations. The Census of Marine Life is but one, albeit grand, example: a 10-y program (1990–2000) of exploration resulting in hundreds of expeditions, >6,000 new species descriptions, and countless newly described fish migrations and life cycles (Snelgrove 2010).

This book was motivated by the recent period of exploration and discovery, but also by Harden Jones's 1968 thesis *Fish Migration*, which presented a general concept of marine fish migration—his emblematic if overextended migration triangle (fig. 1.1). Harden Jones theorized that directed adult migrations compensated for high dispersal during the larval stage, which is common for most marine fishes. Spawning migrations led to life cycles that were closed, leading to population integrity. Natal homing (return to place of origin) and breeding philopatry (multigenerational return to place of origin) were well known for certain vertebrates, particularly birds and salmon. Harden Jones proposed that these migrations were general to marine fishes, drawing from exhaustive case studies on Atlantic herring, Atlantic cod, North Sea plaice, and Atlantic eels. But there is much more to revisit in Harden Jones's *Fish Migration*: a more nuanced view of the diverse nature of migrations and life cycles, particularly the role that minority behaviors play in population and species persistence (Secor 2002a). This book takes up on a "loose end" of previous treatments (Harden Jones 1968; Sinclair 1988); namely, what happens to individual or groups of fish that break away from the migration triangle and do not follow dominant or expected patterns of migration.

After Harden Jones's classic work, other books on fish migration followed: reviews (Smith 1984; McKeown 1985; Lucas and Baras 2001; Quinn 2005), edited compilations (McCleave et al. 1984; Dadswell et al. 1987; Haro et al. 2009; Metcalfe 2012), and additional conceptual theses (Sinclair 1988; MacCall 1990). This book is of the last category—a treatment of the causes and consequences of diversity as it applies to marine fish migrations and related life cycles. Harden Jones opted for in-depth case studies on relatively few marine fishes

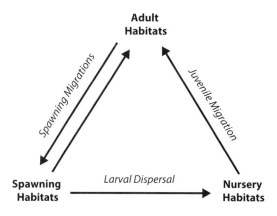

Figure 1.1. Harden Jones's classic migration triangle showing a fish population's lifetime migration track. Spawning migrations lead adults to spawn at particular times and places that favor larval dispersal away from these areas into nursery habitats. Juveniles then enter adult migration circuits. The entire track is driven by natal homing, the capacity of individuals to spawn where they themselves originated. Many fishes reproduce multiple times throughout their lives, exhibiting spawning site fidelity, shown here as the to-and-fro migrations between adult and spawning habitats. Adapted from Harden Jones 1968.

but lamented his omission of tuna and other species, confessing that his "knowledge is too small." *Migration Ecology of Marine Fishes* takes a moderately more ambitious tack: all marine fishes (teleosts and elasmobranchs) are fair game. Still, this volume is necessarily abbreviated, representing an incomplete treatment on fish migration in that its objective is to explore old and new concepts rather than to catalog migration behaviors. Further, although freshwater fishes provide important evidence underlying the concepts of fish migration developed in the book, freshwater ecosystems are not specifically considered. On this subject the reader can turn to Lucas and Baras (2001), who have produced an excellent synthesis on fish migration in freshwater ecosystems, including reviews on navigation and movement ecology, methods, climate, and applied aspects as well as a comprehensive taxonomic review of migration behaviors.

The purpose of this book is to examine how migration and life cycles of fishes confer production, stability, and resilience to populations, metapopulations, and species. In it I seek to portray migrations and life cycles as complex

adaptive systems, which govern population and metapopulation responses to environmental change. New empirical approaches have produced a flood of evidence for diverse migration behaviors that merit a review of our current definitions and explanations. The goal here is to broaden the causal framework of migration, borrowing from avian and systems ecology and expanding the current disciplinary emphasis on individual capacities and motivations (movement ecology) to one that also emphasizes collective agency and population outcomes (migration ecology).

The remainder of this introduction covers (1) past efforts to classify fish migration, (2) definitions of migration that emphasize individual and collective behaviors, (3) causal frameworks in understanding migration, and (4) the book's organization.

Classifying Migration

Types of Fish Migration

While some scholars have struggled for a general definition of migration across diverse animal taxa (e.g., sponges, insects, wildebeests;

Baker 1978; Dingle 1996), others have argued that such generality would render the term meaningless (Kennedy 1985). This exceptionalist view, whether deliberate or unintentional, has contributed to a legacy of idiosyncratic terminology in marine fishes and other taxa (Holyoak et al. 2008; Secor and Kerr 2009). Fish migration terminology, perhaps best exemplified in ichthyology textbooks but widely adopted in the scientific literature (see p. 22), is most often applied to how fish move among marine, estuarine, and freshwater environments (e.g., anadromy, catadromy, marine residents, estuarine stragglers, etc.; see appendix). Such terms are useful in describing broad patterns of habitat use and can aid in species and habitat conservation efforts. For instance, knowing that an anadromous sturgeon requires unimpeded access to large rivers for reproduction leads to an emphasis on dam removal in their conservation. From a broader view, though, why emphasize transiting through salinity boundaries? No doubt an epitome comes to our attention: the remarkable migrations of Pacific salmon that take them from distant ocean habitats into freshwater foothill and montane ecosystems to reproduce (chap. 5, Homing on a Fixed Itinerary: Fraser River Sockeye Salmon). But one could argue that this is a peculiar life history among marine fishes. Alternatively, salinity-based typologies may have arisen because marine fishes are most easily studied in coastal regions and salinity represents a robust means to delineate coastal zones and habitats, particularly in estuaries. But in considering all marine taxa, of which diadromous fish (those that cross salinity boundaries as part of their life cycle) represent but a small fraction (<1%; McDowall 2009), are the homing migrations of Atlantic herring or bluefin tuna across seas and oceans any less remarkable than those of anadromous sturgeon, shad, or smelt? Similar arguments apply to fish migrations within freshwater ecosystems (Lucas and Baras 2001). Salinity boundaries are too narrow a lens in considering marine fish migration. As Holyoak et al. (2008) admonishes, "previous taxonomies of movement have encouraged us to assume

that we understand the patterns, mechanisms, and motivation for movement once movement has been named." Here these classifications are set aside in favor of broader process-oriented terms (table 1.1).

Fisheries Science: The Unit Stock

Fish migration figures only implicitly in how we manage marine fishes. Modern fisheries science has been motivated by the need to understand how fishing, climate, habitat loss, pollution, and hatchery augmentation—forces that are bounded geographically—influence the internal dynamics of certain groups of fish. Such understanding requires two central inquiries: (1) which group of individuals is subject to these forces and, having identified the affected group, (2) how are its internal dynamics affected? Early 20th-century oceanographic sampling, age and growth studies, and statistics showed that exploited fishes such as Atlantic cod and herring were regionally structured, composed of populations with unique internal demographic dynamics. Later, mark-recapture and transplant experiments (particularly on salmons) showed that fish, like birds, were capable of high rates of natal homing. But unlike birds, fish larvae and young juveniles cannot "tag along," learning seasonal and natal migrations on the fly from their parents. The uncertain dispersal fates of fish larvae led to alternative theories on natal homing that continue to fuel scientific debate (Secor 2004).

That migration leads to self-renewal through natal homing is a principal tenet in fisheries stock assessment. Here so-called "unit stocks" are defined so that their internal dynamics can be efficiently assessed against fishing rates (Cadrin and Secor 2009). Generations of scientists have measured migration and other indices of separation (e.g., molecular tracers related to genetic lineage) to test population integrity and classify unit stocks—biological entities that are self-renewable against certain rates of fishing. Stocks are expected to exhibit separate patterns of migration and habitat use; most importantly, their breeding areas should be unique in space and time. A popular theory

Table 1.1. Glossary of terms used to describe the migration ecology of marine fishes

Adopted Migration: Movements that are acquired through social transmission. Behaviors are typically transmitted from experienced (e.g., adults) to naïve (e.g., juveniles) individuals through their interactions, but in some instances movement behaviors may be transmitted through numerical dominance (Huse et al. 2002; Petitgas et al. 2007).

Complex Life Cycle: Life cycle that entails passage between two or more ecologically distinct life history phases (Istock 1967).

Connectivity: Degree of demographic exchange between geographically discrete populations or subpopulation groupings (Cowen and Sponaugle 2009). For marine fishes, connectivity often emphasizes larval dispersal, termed *larval connectivity*, but connectivity can occur through the migrations of juveniles and adults as well. Consideration of the entire life cycle, termed *population connectivity*, specifies reproductive exchange among populations (Pineda et al. 2007).

Contingent: Subpopulation group whose members exhibit a similar migration behavior over major life history phases or over an entire generation (Secor 1999).

Differential Migration: Change in migration behaviors that are associated with demographic attributes age, size, and sex (Terrill and Able 1988).

Dispersal: Collective displacement of individuals that occurs chiefly through external drivers such as advection and diffusion. Dispersal can include active swimming behaviors or control of vertical position, as occurs for many teleost larvae. Dispersal as an outcome also includes losses that occur during collective displacement (Pineda et al. 2007).

Habitat: Region of sustained occupancy separated by ranging behaviors or phases of migration.

Homing: Directed movements to a previous habitat. Seasonal homing often occurs to feeding, wintering, and breeding habitats for fishes. High-frequency (hours to weeks) recurrent visits between the same set of habitats has been termed *commuting* (Dingle 1996). *Vertical migration*—hourly, diel, tidal, or otherwise frequent (< 1 wk) changes in vertical position—can occur as either a commuting or ranging (short-phase exploratory) behavior. The latter term is preferred absent evidence of obligate and constant recurrence.

Intermediate agencies: System operations that bridge mechanistic (proximate) and global (ultimate) causes. Structural flows of information and energy result in such agencies, including positive and negative feedbacks, thresholds, carryover, latency, and overhead, which are relevant at ecological scales of inference (Ulanowicz 1997). *Collective agencies* are intermediate agencies that pertain to group behaviors (Sumpter 2006).

Irruptive migration: Coincident expansion in a group's abundance and spatial distribution. Irruptive migration is often associated with mass movement by a population into novel habitats and ecosystems.

Metapopulation: A set of populations or subpopulations that exchange individuals through migration. Traditionally, metapopulation structure emphasized genetic exchange, but in marine population studies, ecological (demographic) exchange is emphasized in what are termed *patchy metapopulations* (Harrison and Taylor 1997).

Migration: Collective movement of individuals that occurs chiefly through motivated behaviors, resulting in changed ecological status. Related (paraphrased) definitions include:

- mass directional movements of large numbers of a species from one location to another (Begon et al. 1990),
- directional movements made by individuals or (more commonly) groups during specific times within a generation (Johnson and Gaines 1990),
- broadscale movements of populations (Wiens 1997),
- movement of individuals and populations from one well-defined habitat to another, usually on a cyclical basis (Metcalfe et al. 2002), and
- synchronized movements of all or a large part of a population between two or more separate habitats (Brönmark et al. 2010).

Migration Syndrome: Program of specialized behaviors by individuals involving directed movements, neural stimulus-suppression feedbacks, and schedules of physiological activity and energy acquisition that control whether an individual will move away from, or remain local within, its current habitat (Dingle 1996).

Movement: Basic unit of spatial displacement for an individual. *Movement path* is operationally defined by the empirical approach used. *Movement phases* comprise sets or subsegments of movement paths that correspond to goal-oriented behaviors such as foraging or homing (Nathan et al. 2008).

Natal Homing: Within-generation return of an individual to its place of birth for the purpose of reproducing.

Partial Migration: Coexistence of two or more life cycles within the same population. Partial migration traditionally considers concurrence of migratory and sedentary life cycles (Lack 1944).

Philopatry: Multiple-generation return of a population to a breeding site that leads to reproductive isolation.

Population: Group of interbreeding individuals that are reproductively isolated from other groups of the same species.

School: Mobile fish aggregation. *Shoal* is broader term used to describe all types of aggregations (Pitcher and Parrish 1993).

Spawning Site Fidelity: Repeated return by an adult to the same breeding habitat. Note that spawning site fidelity can occur for the fish's natal breeding habitat (natal homing) or to another breeding habitat (straying).

Straying: Breeding outside of the natal breeding habitat.

---

holds that fish populations remain intact owing to "life cycle selection," where natural selection leads to population-specific imprinting, navigation, migrations, and reproductive behaviors (Sinclair 1988; chap. 5, Hydrographic Containment: Member-Vagrant Hypothesis). Although marine fish stocks are often defined as reproductively isolated populations, in many instances, group entities other than populations are most influenced by fishing and other environmental factors, leading to a call for more flexible definitions of stock that consider other collective entities (Petitgas et al. 2007; Cadrin and Secor 2009; chap. 5, Stock Biocomplexity and Spatially Explicit Stock Models).

An emphasis on lifetime migrations, population structure, and reproduction is not unique to marine fishes and fisheries science. In Baker's (1978) encyclopedic review of animal migration, seasonal and lifetime migrations were conceived as a series of habitat transitions, each governed by improvements to overall reproductive success. Avian ecology in particular has emphasized the roles of seasonal migrations and natal homing in breeding success (Greenberg and Marra 2005).

What Causes Fish to Migrate?

Movement Ecology: A Series of Individual Movement Phases

Movements are motivated by the trade-off of staying put versus moving to a new location. In many ecosystems, habitat conditions change seasonally, and through seasonal movement an individual can control (reduce) the amount of environmental change it experiences over a year (Leggett 1985). External drivers (e.g., prey, predators, and physicochemical factors, mating opportunities) and internal motivations (e.g., energetic status, size, experience, navigational capacity, maturity level) influence the costs and benefits of movement. Underlying fish migration, then, is what drives an individual to relocate—its movement ecology (Nathan et al. 2008). Organismal design and the capacity to orient, navigate, move through fluids, and store and generate energy will motivate an individual to take flight or stay put when environmental conditions or its internal physiological state changes (fig. 1.2).

Movement ecology as a subdiscipline is driven by the large technical inroads made in tracking individuals in nature (from seeds and butterflies to bears and albatrosses). The motion paths of individuals are remotely telemetered or archived in electronic tags that are conveyed by the animal for years at time, recorded in body chemistry and genomics or observed by widely arrayed receivers (satellite, radio, and acoustic). Data processing and simulation modeling have kept pace with the volumetric growth of the exquisitely detailed records of stops, starts, distances, directions, and changed locations. Still, without a conceptual framework, the millions of measured movement paths present the dilemma of losing sight of the forest for the trees. In their call for a new paradigm in move-

ment ecology, Nathan et al. (2008) lament "that we still lack a general framework for studying why, how, where and when organisms move."

The ascendance of the individual in movement ecology (Nathan et al. 2008) and other ecological subdisciplines (DeAngelis and Gross 1992) is in keeping with an emphasis on natural selection of individual traits as a central agency. As defined by Nathan and his colleagues, movement ecology combines or partitions movement paths (i.e., observations) into functional movement phases, which can be scaled up further into entire lifetime tracks (fig. 1.3). Central to their paradigm is classifying movement phases as goal-oriented behaviors, such as foraging, predator evasion, or territoriality (Nathan 2008). Consistent with earlier concepts on why individuals change habitat (Istock 1967; Baker 1978; Werner and Hall 1988), Darwinian fitness criteria drive movement phases. Within this construct, feedbacks center on how relocation affects the organism's internal state (fig. 1.2), but flexibility exists in assigning relative priority to individual attributes such as navigation or swimming capacity.

Migration as a Collective Behavior

Similar to the popular view expressed so eloquently in this chapter's opening epigraph, migration is defined here as *the collective movement of individuals resulting in a change to their ecological status*. Doing so takes a deliberate turn from the recent emphasis in understanding causes of migration at the individual level (Baker 1978; Kennedy 1985; Dingle 1996; Åkesson and Hedenström 2007; Nathan et al. 2008) and restores a central endeavor within marine fisheries ecology—that is, to understand population-level behaviors. Here individuals partake in a migration as members of aggregations, schools, contingents, populations, metapopulations, species, and mixed-species assemblages. Migration is not a defined quantity (e.g., a movement path) but requires integrating group status (growth, death, recruitment, social interactions, and trophic and reproductive outcomes) with that group's spatial displacement (fig. 1.4). In movement ecology the

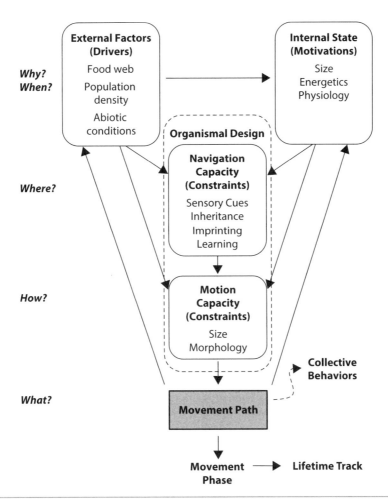

Figure 1.2. Drivers, constraints, motivations, and feedbacks influencing the movements of individuals in Nathan et al.'s (2008) concept of movement ecology. External factors such as changes in forage, predation risk, and seasonal changes in abiotic conditions influence an individual's motivation to move (*why* and *when*). The internal state of the individual—e.g., its size, stored energy, and neurohormonal status—will determine that individual's readiness to move. Navigation and motion capacities represent constraints on the individual's wherewithal to undertake a directed movement (*where* and *how*). A resultant movement path (*what*) can alter the individual's environment and internal state, influencing an individual's decision to continue to move or to undertake a sedentary behavior. The movement ecology "paradigm" posits that processes influencing the movement paths of individuals can be broadened to larger-scale movement phases and lifetime tracks of individuals and populations. Redrawn from Nathan et al. 2008.

question of "why move?" is planted at the individual's feet, wings, or fins, driven by the environment and constrained by organism design and status. In migration ecology the question of "why migrate?" is related to emergent properties of complex systems that are indeed influenced by the movement of individuals but also relate to "intermediate" structural agencies that operate at social, food web, and ecosystem scales (fig. 1.4; table 1.2).

Stipulating "ecological status" in the definition emphasizes the outcome of what migration

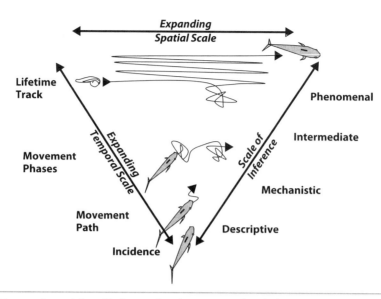

Figure 1.3. Temporal, spatial, and inferential scales associated with the movement of an individual fish. In this conceptual depiction, axes are logarithmically scaled. Movement path is defined by a short sequence of spatial displacements of an individual. Movement phases are composed of movement paths and entail an ecological "goal" or consequence; for instance, diffusion (random walk behavior), directed search behavior, or territorial defense. Lifetime tracks are the complete sequence of movement paths from birth to death, here depicted by rapid dispersal during the larval period, relatively sedentary behavior by the juvenile, and repeated natal homing during the adult period.

achieves (Rogers 1983) but is admittedly vague. The intent is to specify migrations that substantially alter a group's status (e.g., survivorship, feeding, growth, reproduction, range, and metapopulation dynamics). Substantial alterations require time, so migration typically references seasonal, annual, or lifetime (ontogenetic) behaviors. Vertical migration—ascents and descents in the water column that occur in hourly to daily time steps—is not considered to be migration per se, but rather a ranging behavior that relates to short-term exploratory movements associated with transport, habitat, and niche. (Still, vertical migrations underlie important ontogenetic and seasonal migrations that are reviewed further in chaps. 3 and 6.) The lack of specificity on ecological status in the definition is decidedly agnostic (e.g., group status can either improve or degrade) and contrasts with more purposeful goal-oriented terminology of the past. A view posited in this book is that in many instances migra-

tion is not an optimal strategy, nor is it implemented exclusively through natural selection, but rather that migration and associated life cycles are complex systems emerging through multiple processes that operate at individual, population, food web, and ecosystem levels. Natural selection has historically operated to constrain and influence migration ecology, but over ecological timescales the actions of phenotypic plasticity, cultural and ecological inheritance, and intermediate structural agencies are often of greater consequence than the genetic inheritance of adaptive traits (fig. 1.5). Of particular interest in this book are the latencies, carryover effects, nonlinear feedbacks, overheads, and redundancies that influence how migration and life cycles respond to ecosystem change (table 1.2).

Dingle (1996) warned against moving beyond individuals: "attempts to define migration ecologically [would] . . . suffer because they focus on outcomes rather than the behavior that produces those outcomes." As a prime example, he introduced difficulties with the term "dispersal," which is vague because, as an outcome, dispersal represents an unknown mix of passive and active processes. True enough were we only interested in individual movements, but a central tenet of this book is that migration is not the mere sum of individual physiologies and behaviors. Schooling, for instance, represents a central collective behavior occurring in the majority of marine fishes, yet this seminal behavior cannot be deduced without integrating behaviors across individuals (Couzin et al. 2002; chap. 2, Dynamic Aggregations).

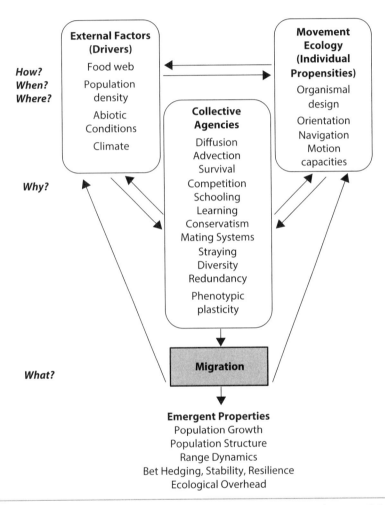

Figure 1.4. Drivers, propensities, and feedbacks underlying the migration of marine fishes. External factors affect individual behaviors through their movement ecology, which influences the timing and degree of displacement that groups of animals will undertake (*how*, *when*, and *where*). But feedbacks will govern how groups will respond to their physical and social environments (*why*). The resultant displacement of that group (migration; *what*) will put it into a novel ecological setting, resulting in a feedback between migration and the propensity to continue to migrate. The migration ecology "paradigm" posits that environmental and social agencies operate differently at individual and collective scales, and that emergent properties associated with migration cannot be understood from the outcome of individual movements alone.

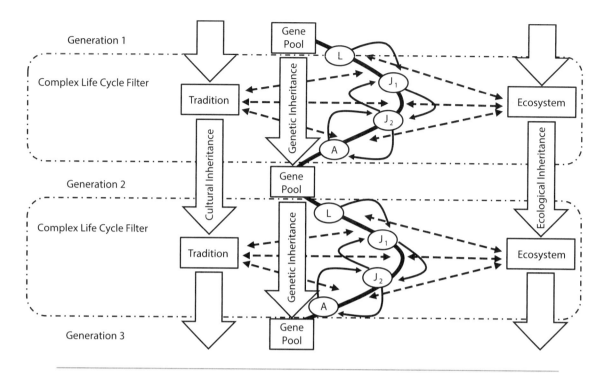

Figure 1.5. Systems of inheritance that influence how migration behaviors are transmitted from one generation to the next. The gene pool of a population interacts with the environment to effect migration traits within each generation specific to different stages: larvae (L), juvenile (J_1, J_2), and adult (A). Each generation and constituent stage has the potential to modify its environment, causing changes to the environmental selection regime that can convey to the next generation (ecological inheritance). Similarly, conserved and "new" traditions convey from one generation to the next through social transmission (cultural inheritance). The information conserved in gene pools, ecosystems, and traditions is thus altered through genetic, ecological, and cultural inheritance pathways and their feedbacks. Complex life cycle "filters" represent carryover effects, latencies, and feedbacks between life history stages that can promote or retard exchange between systems of inheritance. Illustration concept from Odling-Smee et al. 2003.

Certainly, individual movements are critical to collective behaviors. For instance, movements are often accomplished outside of schools and flocks as solo flights. But whether moving in isolation or collectively, it is the integrated outcome of these movements, influenced by both individual and collective agencies, that defines whether a population grows, expands its range, builds resiliency, or goes extinct. In this endeavor, we cannot separate biological hierarchies to understand cause. If we are interested in migration as an ecological process, a focus on individual operations alone is too narrow.

Migration: A Trait, Syndrome, or Complex System?

Directed animal movements are rarely if ever controlled by a single trait or gene; rather, they require an integrated set of morphological, physiological, and behavioral attributes. Dingle's (1996) widely adopted definition of migration specifies it as a specialized behavior, the result of a syndrome, a program of hormonal, metabolic, and neural control that preconditions an individual to migrate away from—or remain local within—its current habitat. As a

Table 1.2. Collective and other structural agencies that operate on migration

Collective agency	Action	Examples
Individual Integrity	The basic unit of collective behaviors defined by the individual's experience and genetic, physiological, and behavioral propensities to move and interact with others	*Chapter 2* Moving in fluids: physical and energetic constraints Random walks and area-restricted search behaviors Association rules: avoidance, alignment, and attraction Inherited orientation compasses Imprinting to specific habitats Strays
	Individual integrity generates group variability, error, bias, stochasticity, leadership, and other collective responses	*Chapter 3* Sperm competition: courtship behaviors Egg provisioning Constraints on early movements owing to viscous forces Ontogenetic development, metamorphosis
		Chapter 4 Foraging behaviors, prey selection Settlement, nursery habitat selection Tidal stream transport
		Chapter 5 Natal homing, spawning fidelity, and breeding philopatry Density-dependent habitat selection Roamers Behavioral syndromes (personalities)
		Chapter 6 Switching partial migration Long-distance roaming behaviors by dogfish, cod, black drum, and goliath groupers Strays
Positive Feedback	Amplification of migration behavior through positive reinforcement Positive feedbacks are de-stabilizing and can lead to geometric increases or cata-strophic loss in population states	*Chapter 2* Cues for seasonal migrations (e.g., Dingle's migration syndrome) Information transfer in the accelerated growth of schools and flocks Frequency-dependent adoption of oriented movements Reinforced navigation errors associated with overshoot, mirror, and reverse migrations
		Chapter 3 Early larval feeding and locomotion Early larval feeding and growth and development Increased aggregation and vulnerability to fishing with decreased abundance

(cont'd)

Collective agency	Action	Examples
		Artificial selection for certain spawning behaviors through fishing or hatchery operations
		Chapter 4 Size-dependent survival and growth in larvae and young juveniles (surfing the size spectrum) Predator swamping and predator breakout by pelagic forage fish Population and range dynamics in opportunistic life history strategists
		Chapter 5 Adopted migration hypothesis (size-, age-, or frequency-dependent adoption of established migrations) Pheromone hypothesis (sea lampreys) Self-recruitment type of larval dispersal kernel
		Chapter 6 Genetic accommodation of migration thresholds in novel environments Evacuations
		Chapter 7 Shifts between stable states (regimes) apparent in long-term oscillations of anchovies, sardines, and bluefin tuna Colonization events catalyzed by habitat expansions (equatorial Atlantic bluefin tuna, Icelandic monkfish) Depensation from increased aggregation in overfished northern cod Resilience in Atlantic striped bass following period of overfishing as a result of the storage effect
Negative Feedback	Regulation or inhibition of migration behavior leading to stabilization	*Chapter 2* Evolutionary convergence of body form and oscillating appendages in swimming and flying ecomorphologies Body size (allometric) constraints in flying fish and swimming birds Daily recurrence of herring schools Stable aggregation states (swam, torus, school), dependent on association rules Guess-and-go and pinball navigation self-correction by Pacific salmon
		Chapter 3 Frequency-dependent mating success Trade-off between offspring size and number, conditioned by adult body size

Collective agency	Action	Examples
		Control of buoyancy and vertical migration in larval fishes
		Chapter 4 Balance of forage reward and predation risk associated with ontogenetic habitat shifts Trade-off in producing more numerous small progeny versus fewer larger progeny Selection for variable spawning times to match offspring with favorable growth and survival conditions Predator regulation of school abundance states (predator pit)
		Chapter 5 Life cycle selection (selection against spawning behaviors that cause advective loss)
		Chapter 6 Frequency-dependent (nonnatal) partial migration
		Chapter 7 Variance dampening through the insurance effect
Response Thresholds	Changed migration behavior as a result of an environmental or social threshold	*Chapter 2* For schools and flocks, threshold changes between stable aggregation states resulting from small changes to zone of alignment Threshold response between average and single-vector orientations owing to the number of leaders
		Chapter 3 Improved locomotion as larvae transition from viscous to inertial media Loss of behavioral components of populations through fishing
		Chapter 4 Ontogenetic habitat shifts Predator breakout and irruptions by pelagic forage fish Rapid contraction following unsustainable irruption by pelagic forage fish
		Chapter 5 Barriers to ideal free distribution (the basin model)
		Chapter 6 Discontinuous phenotypic response to the reaction norm for partial migration (threshold model) Latent partial migration in invasive and colonizing species

(cont'd)

Collective agency	Action	Examples
		Chapter 7 See negative feedback examples above
Leadership and Consensus	Degree of individual influence on migration behaviors; leadership tends to result in positive feedback and increased polarity, whereas consensus results in distributed information and stable, less variable, collective outcomes	*Chapter 2* A minority of individuals exerts strong influence on school or flock polarity Learned seasonal migration behaviors *Chapter 4* Exploratory individual or school behaviors may catalyze population breakouts by pelagic forage fish *Chapter 5* Adopted migration: experienced individuals convey migrating circuits to naïve individuals
Redundancy and Central Limit Theorem	Degree of repetition in the movement behaviors between individuals. Results in increased sampling of resources and more stable collective outcomes	*Chapter 2* Lévy flight and summed movement distributions State-space models of aggregate movements and distributions Collective compass orientation (many wrongs principle) Redundant navigation abilities *Chapter 3* Increased patchiness with development of oriented behavior in larvae Larval dispersal: diffusion and advection outcomes *Chapter 4* Oversampling of patchily distributed resources by abundant cohorts of small larvae Variance in time of spawning associated with transient prey resources for offspring (match-mismatch hypothesis) Periodic life history strategies, iteroparity Fractal nature of school nuclei, schools, and school cluster Storage effect *Chapter 5* Ideal free distribution Propagule rain larval dispersal kernel Patchy metapopulations *Chapter 6* Nonnatal and natal partial migration *Chapter 7* Response diversity and cross-scale diversity Rivet and passenger and driver models

Collective agency	Action	Examples
Synchronization	Adjustments of individual movements toward mean group behaviors	*Chapter 2* Alignment in schooling and shoaling Recurring school types Anonymity of individuals within schools *Chapter 3* Courtship behaviors Spawning migrations, spawning cues *Chapter 4* Spawning waves timed for periods of peak prey production for offspring (match-mismatch hypothesis) *Chapter 5* Spawning runs Recurrent seasonal migrations *Chapter 7* Cohort resonance
Carryover (Path Dependency)	Sensitivity to initial or intermediate conditions; in concert with other collective agencies—e.g., positive feedback—carryover can lead to asymmetric responses in group behaviors	*Chapter 2* Allometric constraints on flying fish Path dependency in the effect of alignment spacing on stable aggregation states *Chapter 3* Influence of spawning behaviors on larval dispersal *Chapter 4* Influence of spawning behavior on larval production (surfing the food web) *Chapter 5* Source-sink larval dispersal kernel *Chapter 6* Carryover effects of larval growth and development on lifetime migrations in systems of partial migration Differential migration *Chapter 7* Regime shifts Cross-scale diversity
Ecological Overhead (Response Diversity)	Oversampling in indeterminate systems; the degree of redundancy and disordered responses in migration behaviors (modified from Ulanowicz 1997; Elmqist et al. 2003)	*Chapter 2* Collective compass orientation (many wrongs principle) *Chapter 3* Polygamous mating systems Copious sperm

(cont'd)

Collective agency	Action	Examples
		Within-species diversity in male mating behaviors
		Variance in larval duration influencing transport outcomes
		Multiple sources of larvae (spawning grounds) and transport fates
		High production of larvae offsetting transport losses
		Multiple larval retention and dispersal behaviors
		"Skipped" spawning behaviors
		Chapter 4
		High production and mortality of larvae, offsetting losses in size structured marine food webs
		Chapter 5
		Population connectivity
		Straying as complement of natal homing
		Metapopulation effect/spillover
		Chapter 6
		Phenotypic plasticity
		Genetic assimilation
		Biocomplexity (diversifying selection at ecological timescales)
		Chapter 7
		Storage and portfolio effects

Note: Terms and definitions are modified and amended from Sumpter (2006).

prime example, birds in preparation for migration store fat, enlarge flight muscles, increase oxygen transport, and become increasingly restless. Similarly for Pacific salmon, cessation of the spawning migration triggers suites of traits related to reproduction and senescence. This integrated program of physiology and behavior is viewed as purposeful. Once initiated, the syndrome drives an animal to target habitats and causes the individual to remain unresponsive to other habitats en route, regardless of their suitability. Migration as a specialized behavior fits well with the strongly oriented seasonal migrations of many vertebrates and the intergenerational dispersals of insects such as those of the monarch butterfly.

As a behavior, set of traits, or syndrome, migration is commonly explained through genet-ic inheritance. Despite convincing evidence drawn from several insect model species, Dingle (1996) was appropriately circumspect: "What remains still largely an open question is how much of the variation we see in migration syndromes is the result of genetic variation." Evidence for genetic inheritance of migration traits mostly comes from those insects, salmons, and birds that exhibit partial migration (for a definition, see table 1.2). Lineage influences the fraction of migratory versus resident animals within a population. For instance, populations of Arctic charr *Salvelinus alpinus* and blackcap warblers *Sylvia atricapilla* range between fully migratory to fully resident, and crossing experiments between migratory and resident strains produce offspring with traits associated with intermediate levels of migration (Nordeng 1983;

Helbig 1991; Berthold 1996; chap. 6, Threshold Model).

Outside of controlled crossing experiments, genetic inheritance is supported by several lines of evidence: (1) groups that are discrete in their life cycles (sets of migrations) also exhibit genetic separation (Leggett and Carscadden 1978; Hendry et al. 2000; Beacham et al. 2004); (2) imprinting and navigation behaviors are developmentally ordered in a manner consistent with a genetic program (Cury 1994; Dingle 1996); and (3) sets of traits that are genetically correlated also covary with migration behaviors (Taylor 1990; Dingle 1996; Hutchings and Jones 1998; Cooke et al. 2004). A rich literature for insects, birds, and salmons upholds these premises. Migration syndromes are particularly well studied in insects, for which some species exhibit obvious wing polymorphisms (e.g., winged and wingless forms) associated with flight and migration. In the laboratory, wing traits were shown to be responsive to selection regimes for wing size or flight duration, and in some instances reproductive traits were also responsive to these selection regimes (Dingle 1996). In the field, factors that affected wing traits included insect density and habitat stability. Although genetic inheritance of traits associated with migration is well supported for some birds and insects, it remains largely unstudied for marine fishes.

Views on genetic inheritance have changed substantially during the last several decades. The classic mechanistic view of selection shaping intergenerational genotypes through rules of heredity (e.g., mutation rates, recombination, and polygenic inheritance according to the Hardy-Weinberg equilibrium; Falconer 1981) has been substantially broadened, with increased opportunities for change through environmental–genetic interactions (Mariani and Bekkevold 2013). Complete elucidation of genomes has captured a legacy of genetic duplication and introgression of gene inserts from other species (e.g., hybridization and horizontal genetic transfer from pathogens). Thus slow-acting mutation is not the single source of genetic variation upon which selection acts,

nor does genotypic change exclusively occur through descent (Amores et al. 1998; Zhang 2003; Thornton et al. 2007). Environments that adults encounter alter gene expression of their offspring through epigenetic feedbacks (Dingle 1996; Balon 2002; Youngson and Whitelaw 2008). Environments can also influence large sets of traits through their action on specific control regions (i.e., pleiotropy and reaction norms), resulting in threshold phenotypic responses to environmental change (Dingle 1996; Colosimo et al. 2004; Piche et al. 2008; chap. 6, Threshold Model). Thus genetic inheritance has come to resemble a complex system, with multiple, redundant, homeostatic, and threshold environmental–genetic interactions that are unlikely to be fully understood through simple rules of trait selection and genetic inheritance (Davies 2012).

In addition to epigenetic effects, the theory of niche construction has challenged the traditional view that environmental–genetic interactions principally occur in a single direction (fig. 1.5). This idea builds on the keystone species concept, wherein species and communities can fundamentally structure their environments, leading to altered or "constructed" food webs, habitats, and ecosystems. When a species alters an environment in such a way, then the selection regime changes as well (Laland et al. 1999; Kinnison and Hairston 2007), leading to a feedback to the keystone species and its descendants, known as ecological inheritance (Odling-Smee et al. 2003). The best-known example is the role of Precambrian photosynthetic species in collectively oxygenated earth's atmosphere. As a more relevant example, consider alewife *Alosa pseudoharengus*, a coastal fish species that seasonally invades freshwater lakes from coastal waters for spawning and foraging but occasionally becomes trapped in lakes by natural impoundments (e.g., beaver dams) and adopts a resident life cycle. Alewives feed selectively on larger zooplankton in lakes, causing the zooplankton's size distribution to change, which in turn results in morphological changes to the alewife's feeding structures to allow more efficient feeding on the changed zooplankton community (Post and Palkovacs 2009). These changes are conveyed to subsequent generations of alewives. Because marine fishes are well known to alter food webs through their trophic interactions (chap. 4, Effect of Migration on Size Spectra), it seems feasible that ecological inheritance could influence migration-associated traits. Such "eco-evolutionary" feedbacks can lead to rapid evolution of traits at ecologically relevant timescales (Kinnison and Hairston 2007; Post and Palkovacs 2009).

Cultural inheritance is yet another means by which migration behaviors can be modified (Laland et al. 2011). Although one typically attributes intergenerational transmission of learned behaviors to vertebrates of greater cognitive abilities—for instance, the linked migration and foraging behaviors of orcas and dolphins that are transmitted across generations—marine fishes also conserve migration behaviors that are apparently learned (Dodson 1988; Corten 2001; Petitgas et al. 2010). In general, cultural inheritance should lead to establishment of traditions that are resistant to genetic and environmental influences. Depending on how behaviors are transmitted, cultural inheritance can result in rapid migration changes that may or may not be coupled with changed environments (ten Cate 2000; Corten 2001). One example that will be detailed in chapter 4 (Schooling through Food Webs) is when behaviors are acquired in a density-dependent manner, which can lead to abrupt colonization of new food webs and ecosystems (Bakun and Cury 1999; Huse et al. 2002).

Layered within genetic, ecological, and cultural inheritance systems are the complex life cycles of fish where what happens earlier in life carries over to later behaviors and traits (Balon 2002). Natural selection, ecological feedbacks, or behavioral interactions at one stage can influence an entire life cycle or migration syndrome (Pechenik 2006; Podolsky and Moran 2006). Latencies and feedbacks attend complex life cycles and confound simple deductions on proximate (immediate) causes for movements and migrations (Laland et al. 2011). Partial migration and differential migration represent two prime examples where migration outcomes are largely determined by what happens early in life (chap. 6, Partial Migration Writ Large).

Emerging from multiple systems of inheritance and multipartite life cycles are complex systems (fig. 1.5)—a network of interactions, redundancies, feedbacks, latencies, and carryover that cannot be understood by simple mechanisms alone (Laland et al. 2011). This explanatory framework is far from complete. Further, because concepts underlying the three types of inheritance rely largely on circumstantial evidence, there is no way to ascertain their relative influence. Although this causal framework is tenuous, the goal here is to illustrate that migrations as we observe them are unlikely to be solely the result of natural selection on coupled traits. There are simply too many intervening structural agencies related to (1) latent ontogenetic effects; (2) collective behaviors; and (3) reciprocity in systems of genetic, ecological, and cultural inheritance. This conclusion does not exclude propensities in evolving systems to acquire increased fitness; rather, the suggestion is that natural selection is an embedded process, one of several that can account for the collective behaviors that we observe as migrations.

The difficulty of fully understanding constituent processes that underlie migration leads to the pragmatic corollary that emergent property will continue to supply an important basis for inferences related to fish migration. As defined by Salt (1979), "an emergent property of an ecological unit is only discernible by observation of that unit itself." Emergent properties are particularly important in understanding complex systems where nonlinear and variable processes interact across multiple scales (Ulanowicz 1997). As shown throughout this book, both empirical and modeling frameworks can be deployed to evaluate emergent properties associated with complex life cycles, such as (1) how spatial distributions and migrations are structured and (2) whether they show covariance, synchrony, modalities, and patterns of self-organization in association with physical and ecological attributes of marine ecosystems. These questions align with the emerging subdisciplines of population connectivity and metapopulation ecology. In this book, I adopt the pragmatic view by Sumpter (2006)—a call for an increased (but not exclusive) focus on collective agencies in natural history, experiments, and simulation as an explanatory framework for migration.

Book Organization: Migration Propensities

Migration is conditional: the propensity to migrate depends on organismal and phylogenetic constraints, mating systems, larval dispersal, systems of inheritance, habitat resources, climate and ocean forcing, and the influence of humans through fishing and changes to marine habitats. These processes and circumstances will be examined in this book, moving from lower to higher levels of ecological organization. Although this structure is logical and convenient, it could countervail efforts to demonstrate that migration emerges from multiple ecological scales, and that it is not possible to separate biological cause and ecological consequence (Mayr 1961; Laland et al. 2011). Therefore efforts are made throughout the book to emphasize intermediate-scaled agencies that bridge individual and collective properties. To aid with the flow, each chapter ends with summary and segue sections. Most chapters also provide an example of how migrations and life cycles are considered in fisheries management and habitat conservation.

Following this first chapter on historical concepts and definition of fish migration, chapter 2 provides a treatment on what constrains individual movements of fishes. Some of this is necessary background—for instance, how animals move in fluid environments. More central to the chapter, however, is an attempt to capture the explosion of digital-age discoveries on fish movements, including (1) the capacity to track individual movement behaviors at regional and global scales, (2) computationally intensive models that emulate complex collective behaviors, and (3) insights into vertebrate

geomagnetic reception. These discoveries are reinforced when we include birds together with marine fishes in a comparative analysis. Fishes and birds are similarly tagged, telemetered, and otherwise observed during their migrations. But recent empirical advances have resulted in a common challenge, which is to extract an integrated picture from the millions of movement paths now available. Fishes and birds follow similar rules of aggregation leading to self-organizing behaviors in schools and flocks. Their migrations depend on common vertebrate navigation capabilities. Despite remarkable geomagnetic and other navigational abilities, birds and fish go astray. Multiple orientation compasses and self-correcting mechanisms mitigate against navigation errors. Anomalous migrations in birds are commonly observed, and theory applied to avian migration has been accommodating of this source of diversity, more so than for marine fishes.

In chapter 3, we review mating systems and larval dispersal, focusing on how collective agencies counteract an inherently dispersive aquatic environment that tends to separate sperm from egg and cause dispersion of small fish larvae. Complex mating behaviors include schooling, gamete provisioning and overproduction, courtship, parental assurance systems, and spawning site fidelity. We dissect what happens inside the gamete clouds left in the wake of spawners—how mating systems influence fertilization and the dispersal of eggs, embryos, and larvae. The alignment between mating systems and larval dispersal is viewed through the lens of oceanographic numerical models. These models, together with new empirical estimates of larval swimming capacities, indicate that larval dispersal is not a fully passive process. Indeed, larvae of some species are as active in controlling their spatial behaviors as later-stage juveniles. Despite the wondrous diversity and adaptations in systems of mating and larval dispersal, we learn that in most instances the fate of larvae is highly uncertain. This ecological overhead results in extremely high losses, but it can also cause substantial gains for the cohort of larvae that is well placed, a topic emphasized in chapter 4. Finally, we look at how targeted harvest of spawners can influence mating systems, larval dispersal, and population sustainability.

Chapter 4 examines how the life cycles of fishes are adapted to marine food webs. First, marine food webs are contrasted with terrestrial ones, emphasizing biomass structure among trophic levels. We then move on to size spectrum theory, why many fish start life at such a small size, multistage (complex) life cycles, ontogenetic habitat shifts, and exploitation by fish of transient and patchy prey resources. Schooling is introduced as a prevalent means by which fish encounter marine food webs. As a self-organizing system, schooling can result in rapid nonlinear changes to a population's spatial distribution and migration behavior. How life cycles persist given high variability in early survival is evaluated through a broad examination of life history traits among species. The periodic life history mode (late maturity, high fecundity, and low early survival) is dominant among teleost fishes, for which early losses are offset by storing gametes in relatively large and invulnerable adults. In contrast, elasmobranchs exhibit a different life history mode (late maturity, low fecundity, and relatively high early survival), but here too life cycles must be considered against a trophic seascape where "big things eat little things." Periodic "strategists" are among the most migratory of marine fishes. Because traits and spatial behaviors of periodic strategists vary with age, conservation of age structure in fisheries management will favor resilience and stability.

For marine fishes, common challenges of reproducing in a dispersive environment, advection of young, and transiting through size-structured food webs are met through life cycles within populations that are both closed and open to emigration, the topic of chapter 5. Natal homing remains a central aspect of life cycles bolstered by recent discoveries that point to the remarkable abilities of marine fishes to find their way back home. Alternate explanations for life cycle closure—imprinting, genetic and cultural transmission, and eco-

logical inheritance (the pheromone hypothesis)—are reviewed. Concepts of open populations, applicable over shorter phases of life, are explored with simple null models of dispersion and more elaborate density-dependent models. Reef-associated fish populations have attracted substantial interest in spatial management strategies that rely on patterns of connectivity. Some reef fish show surprising degrees of natal homing, termed "self-recruitment" by reef ecologists. Studies of larval connectivity track the oceanographic fates of larvae depending on where they originated, leading to emergent population and metapopulation structures. The dual open and closed nature of populations of marine fishes is well accommodated by metapopulation models, which explicitly recognize the role of straying in persistence. On the other hand, this dual state confounds a central simplifying assumption in fisheries stock assessment that attempts to define populations as unit stocks. Examples of misspecified stock structure are highlighted as well as advances in stock assessment approaches that incorporate more complex population structures into management advice.

In chapter 6, we review partial migration, the simultaneous incidence of resident and migratory components (a.k.a. contingents) within the same population, a ubiquitous feature in marine fishes. Partial migration theory links early threshold responses to environmental conditions to modes of behaviors that carry over to entire life cycles. Migratory and resident contingents are maintained through genetic inheritance of thresholds, which cause individuals to adopt one mode or the other, conditional on prevailing environmental or social conditions. Further, populations that are fully resident or migratory can rapidly evolve alternate behaviors because these behaviors remain latent through genetic assimilation. The pattern of latency and repeated emergence of partial migration is evident in Arctic charr, threespine stickleback, temperate eels, and invasive species. This same mechanism can also promote the spread of populations through colonization and speciation. An exciting new field, loosely termed "biocomplexity," uses functional molecular markers to examine how contingent structure may be shaped by recent environmental change. General classes of partial migration are reviewed that cause population members to (1) diverge in seasonal feeding/wintering migrations, (2) stray, (3) undertake size- and sex-specific migrations, (4) "skip" spawning, (5) evacuate and irrupt, and (6) diverge in vertical behaviors. A goal of chapter 6 is to show how partial migration accommodates the large volume of discoveries on the diverse, nonuniform migration behaviors, now well documented in marine fish populations.

A recapitulation is amended to chapter 6, which revisits central issues in the book: (1) migration as an adaptation to abundance, (2) migration as a means for ensuring population integrity, (3) migration as a sum of individual movement behaviors, (4) collective agencies that operationally influence migration, and (5) migration as a contingency plan.

The final chapter examines migration and population structure as systems of adaptation and resilience to fishing and environmental change. Resilience (stable state) theory predicts that fishing and forcing variables can sometimes lead to alternate population states through discontinuous threshold changes. The long-term depression of some stocks following periods of overexploitation is suggestive of such shifted states, but regime shifts are also associated with multiyear climate oscillations. Overshadowing these regime changes is an overall vector of climate change and the prospect for novel hybrid ecosystems. Fisheries and ecosystem stewardship goals will therefore need to accommodate nonstationary ecosystems through adaptive management frameworks and broadened reference points, those that relate to population resilience, stability, and persistence. The discovery of life cycle diversity over the past several decades provides important opportunities and challenges in conserving yield and ecological functions of marine fish populations through fisheries regulations, spatial management strategies, and implementation of ecosystem-based management.

Appendix   Traditional Migration Classifications for Marine Fishes

Term	Definition	Number of citations
Migration typology		
Potamodramous	Truly migratory fishes that live and migrate wholly within freshwater (Myers 1949; McDowall 1988)	18
Diadromous	Truly migratory fishes that migrate between the sea and freshwater (Myers 1949; McDowall 1988)	143
Anadromous	Diadromous fishes that spend most of their lives in the sea and migrate to breed in freshwater (Myers 1949; McDowall 1988)	985
Semianadromous	Diadromous fishes that spend most of their lives in saline water and migrate to, or almost to, freshwater for spawning (Cronin and Mansueti 1971)	7
Catradromous	Diadromous fishes that spend most of their lives in freshwater and migrate to the sea to breed (Myers 1949; McDowall 1988)	71
Semicatadromous	Diadromous fishes that spend most of their lives in freshwater and migrate to the estuary to breed (Whitfield 1999)	0
Amphidromous	Diadromous fishes whose migration from freshwater to the sea, and vice versa, is not for the purpose of breeding (Myers 1949; McDowell 2007)	122
Habitat typology		
Oceanodromous (marine resident)	Species that utilizes ocean habitat all life (Myers 1949; Tsukamoto and Arai 2001)	4
Marine straggler (alternates)	Marine fish that occasionally occur in estuaries (Whitfield 1999)	7
Marine migrant (alternates)	Marine breeding fish that commonly occur in estuaries (Whitfield 1999)	19
Estuarine resident (alternates)	Truly estuarine fish that live wholly in estuaries (Whitfield 1999)	27
Estuarine migrant	Estuarine breeding fish that occur also in marine or freshwater (Whitfield 1999)	1
Freshwater resident	Truly freshwater fish that live wholly in freshwater (Whitfield 1999)	40
Freshwater straggler (alternates)	Freshwater breeding fish that occasionally occur in estuaries (Whitfield 1999)	1
Freshwater migrant (alternates)	Freshwater breeding fish that require estuarine or marine habitats (Whitfield 1999)	0

Source: Adapted from Secor and Kerr (2009).

Notes: Classifications are according to broad classes of habitats used (habitat typologies) and migrations between these habitats (migration typology). "Alternates" indicate that a single term is given for several closely related ones used in the literature. The number of citations for each term was determined by a search of abstracts and titles in Cambridge Scientific Aquatic Sciences and Fisheries Abstracts for the period 1975-2006.

2 Bird and Fish Migration

Movement Ecology as a Comparative Framework

We see them come.
We see them go.
Some are fast.
And some are slow.
—Dr. Seuss (1960)

A common starting point in understanding migrations of fishes, birds, and other taxa is the individual—how it moves and adapts to its environment. Movement ecology emphasizes mechanistic processes that drive, motivate, and constrain an individual's propensity to change location. The emphasis on organismal design in movement ecology—an individual's ability to perceive and respond to its physical and social setting and its capacity to accumulate energy and information and to translate these stores into stochastic or oriented movements—makes it particularly amenable to comparisons among taxa (Holyoak et al. 2008). In this chapter, movement ecology is compared between birds and fishes, taxa that show broad similarity in locomotion, seasonal migrations, mating systems, and navigation. First, biomechanical constraints in vertebrate design are considered as they relate to locomotion in fluids (water or air). Second, parallel developments in measuring and analyzing the individual movements of fishes and birds are reviewed. Then models of individual motion pathways are examined for highly migratory seabirds and tuna. Simple rules of interaction between individuals are scaled up to explain patterns of aggregation, schooling, and flocking as well as to introduce collective agency. Navigation capacities and adaptations are then reviewed for juvenile birds and fishes undertaking their first migration. We wrap up with an examination of locomotion, aggregation rules, and navigation as they pertain to migration as an emergent property of collective interactions.

Movement in Fluids

As opposed to the earth's solid support of ambulatory animals, fishes and birds principally move in fluids. Water as a fluid is 830 times denser than air and resists forward progress. Gravity in water, however, acts with much less force, and by matching its specific gravity with that of the surrounding water, a fish can efficiently adjust its position in the water column. Birds move through a far less dense fluid, but because they move at higher speeds than most fishes, friction and inertial forces still play important roles in their flight performance. Further, all flying birds must overcome gravity through lift. Some fishes (e.g., sharks) also generate lift through their movements. Principal constraints opposing locomotion in birds and fishes have resulted in convergence in body form—streamlined trunks propelled by oscillating appendages. Energetically, locomotion is more efficient for fish in water than birds in air (Schmidt-Nielsen 1972; fig. 2.1). Still, birds typically undertake longer migrations than fishes, capitalizing on higher speeds, recuperative stopovers (refueling stations), and specialized tactics such as gliding and soaring. Short reviews of fish and bird locomotion follow. In comparative treatments, we then look at what happens when one taxon adopts locomotion

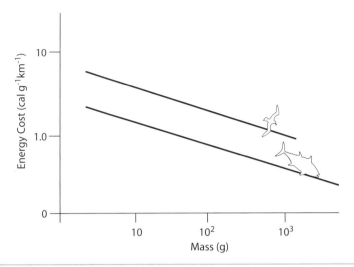

Figure 2.1. Cost of transport for birds and fishes, estimated as the energy required to move 1 g over 1 km. Travel costs are higher for birds because they must generate both lift and forward thrust. Adapted from Schmidt-Nielson 1972, with permission.

behaviors that typify the other: gliding fishes (whale sharks and flying fish) and swimming birds (guillemots).

Locomotion in Fishes

As a fish moves in water, mass, buoyancy, and hydrodynamic lift affect its vertical position; thrust and resistance influence its horizontal movement. Swimming relies on muscles that set up inertial forces in the surrounding water (Lindsey 1978). The produced thrust depends on the length of the fish as well as the frequency and amplitude of the tail (or fin) beat and the effect of streamlining. Thrust is counteracted by the imposition of inertia (density) and friction (viscosity). The relative importance of these two opposing factors (the Reynolds number, R) is dependent upon the fish's velocity and length according to

$$R = UL\left(\nu D^{-1}\right)^{-1},$$

where U is velocity, L is fish length, ν is viscosity, and D is specific gravity (Denny 1993). Note that as length or swimming velocity increase,

the relative contribution of viscosity in opposing locomotion (R) declines. At low speeds and small body size, the influence of viscosity dominates as the swimming fish carries along an enveloping boundary layer of water. Molecular forces within the boundary layer oppose forward progress.

With increased fish size or velocity, turbulence across the surface area of the fish disrupts laminar (streamlined) flow, and the boundary layer thins (Müller et al. 2000). The mass of the fish is increasingly opposed by the mass of the water, and inertial forces predominate. The manner in which acceleration is opposed in inertial environments is described by

$$\text{acceleration reaction} = \rho_b Va + C_a \rho_f Va,$$

where ρ_b is the density of the fish, V is the fish's volume ($\rho_b V$ is the fish's mass), ρ_f is the fluid's density, and a is acceleration. The second term of the equation expresses the mass of water that must be accelerated along with the fish, which depends upon C_a, a drag coefficient that relates to the shape of the fish (Denny 1993). This drag coefficient is proportional to the surface area exposed against forward progress: the fish's degree of hydrodynamic streamlining.

Most fishes start life quite small, at a size where viscous forces predominate (chap. 3, Embryo Dispersal). For small (<10 mm) and slow-swimming embryos and larvae, the strong resistive force of viscosity is countered by whole body flexions or undulations (Webb and Weihs 1986; Fuiman 2002). High-amplitude lateral movements cause the greatest deformation of the boundary layer, trading frictional drag for forward propulsion. Interestingly, small larvae can periodically escape viscous environments by swimming at high speeds (e.g., 10 body lengths s^{-1} for a 5-mm anchovy), although such rates are not sustainable (Webb and Weihs 1986). Swimming is energetically expensive in a viscous flow regime, resulting in lots of stops and starts. In viscous environments, swimming resistance exceeds friction during coasting, and a beat-and-glide pattern of locomotion is energetically favorable. Rapid growth in length will release the small larva from its viscous surroundings, improving its swimming performance (Webb and Weihs 1986; Müller et al. 2000).

For larger larvae, juveniles, and adults, body and fin flexions accelerate adjacent parcels of water, reducing the mass of water accelerated in lateral directions, resulting in increased swimming efficiency (Sfakiotakis et al. 1999). Swimming velocity and endurance are positively related to fish length owing to proportional (isometric) scaling between length, tail beat amplitude, and frequency (Webb et al. 1984; Videler and Wardle 1991). But species vary substantially in swimming performance. For instance, an epitome cruising fish flexes (oscillates) only the caudal portions of the body (thunniform or carangifom; fig. 2.2). Restricting lateral propulsive elements to the caudal region leads to anterior recoil, which in many fishes is reduced by deepening the trunk and median fins (Webb 1978, 1982). Swimming ecomorphologies classify the relative importance of linear versus angular acceleration to a fish's

ecology (Lindsey 1978). Epitome forms exist among cruisers, accelerators, and those that maneuver (Sfakiotakis et al. 1999). These ecological classifications are crossed with physical modes of swimming that are accomplished by propulsion of (1) the body and caudal fin or (2) median and paired fins (fig. 2.2). Over its life span, fish locomotion meets diverse demands of feeding, evasion, reproduction, and migration. Thus swimming epitomes (cruising, accelerating, and maneuvering) are incomplete descriptors of the ecological consequence for the large diversity of swimming modes in fishes.

Locomotion in Birds

In contrast to most fishes, birds must support the weight of their bodies aloft. Wings provide force against the bird's inertial environment (air flow) and provide lift:

$$\text{lift} = 0.5\, C_L \rho_f A v^2,$$

where C_L is a drag coefficient that relates to the shape of the wing, and A is the wings' area. Lift represents force deflected downward against gravity according to the angle, shape, and oscillation of the wing. Positive lift occurs as the extended wing deflects air downward resulting from its angle of attack, creating negative pressure opposed to gravity. In cross section, wings are similar to an idealized airfoil shape, which contributes to large pressure differences above and below the wing's surface. Lift depends on the fluid's density; as atmospheric pressure declines with altitude, birds must work harder to generate lift (Denny 1993). Formation flight, well known for geese and other water fowl, takes advantage of updrafts shed from wingtips of central leading birds, which contributes lift to those that follow (Newton 2008). Lift depends on speed; speed induces increased frictional and inertial drag, which is offset by streamlining of the bird's wings and overall body form (Hedenström 2002).

In flapping flight, the most distant portion of the articulated wing propels the bird forward as its tip oscillates in wide arcing down-

(A)

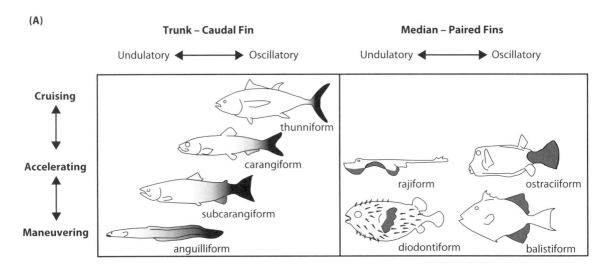

(B)

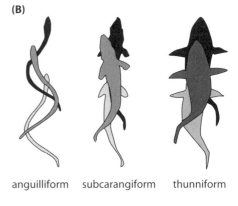

anguilliform   subcarangiform   thunniform

Figure 2.2. Swimming ecomorphologies of fishes classified on the basis of (A) undulatory versus oscillatory propulsive elements and (B) the degree of lateral recoil. The degree of tail and fin movements involved in propulsion is shown by shading. Physical mode of swimming is defined by the taxonomic family that epitomizes that mode. For instance, undulatory-style swimming in anguilliform, subcarangiform, and carangiform occur >50%, 30%–50%, and <30%, respectively, of the body's length. Oscillations are confined to the caudal peduncle and tail in thunniform swimming. Fish classifications are modified from Lindsey (1978) and Sfakiotakis et al. (1999). Modified from Sfakiotakis et al. 1999, with permission.

and upstrokes. The more fixed inner portion of the wing provides lift. In the downstroke, the wing's tip is accelerated downward and forward, causing proportional changes in air pressure that result in forward momentum. In the recovering upstroke, the wing is drawn toward the body to reduce inertial drag. At higher flight speeds, birds may opt to increase lift throughout the flapping cycle by reducing wing contraction toward the body during both strokes. Positive lift generated during the upstroke will oppose some of the bird's forward momentum (Hedrick et al. 2002). Wind has a strong impact on flight performance, and birds will often seek out favorable tail winds by flying during specif-

ic seasons or by selecting particular elevations or times of day (Dingle 1996). Upstream wind favors take-off and lift, but once in flight, countervailing winds will increase time and cost of travel, particularly for smaller birds.

Soaring flight exploits changes in air density caused by thermals: convection cells of rising air. Weather and landforms can cause seasonally predictable thermals that soaring birds such as hawks and condors exploit for foraging or long-distance movements. Soaring landbirds gain height, circling within these rising air masses, and then cover distance through gliding. Soaring seabirds ride over air pockets at low heights (<10 m), the result of wave action

(Newton 2008). Frigatebirds *Fregata magnificens* are not seabirds (i.e., they cannot land on the sea's surface) but are pelagic ocean predators, catching flying fish and squid on the fly. The surface area of their wings is extremely large in comparison to mass, allowing them to ride persistent tropical sea thermals at high elevation (100–2,500 m) in search of surface-leaping prey (Weimerskirch et al. 2006). Gliding flight is performed by most birds as they (1) descend from altitude, (2) recuperate from flapping flight, and (3) accelerate from height (e.g., to hunt or evade predation).

Birds, like fishes, have been classified according to morphological trade-offs and specializations associated with their flight performance (fig. 2.3; Rayner 1988; Norberg 2002). Wing shapes are specialized for gliding, soaring, acceleration, speed, and hovering. Interestingly, the aspect ratio—the thinness of the propulsive element (wings or fins)—indexes locomotion performance for both birds and fishes. More so than in fishes, mass in birds (indexed as wing loading = mass · gravity · wing area^{-1}) limits flight performance and style (Alerstam et al. 2007). Conveying extra stores of energy

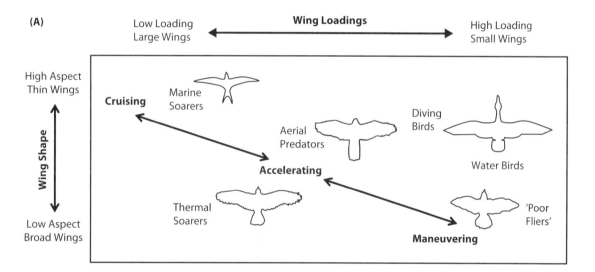

(A)

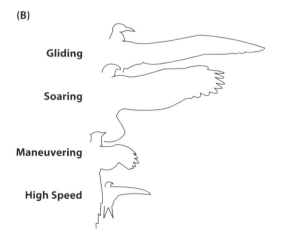

(B)

Figure 2.3. Flying ecomorphologies of birds classified on the basis of (A) wing loading and wing streamlining with (B) specific examples of wing form and flight performance. Marine soarers are epitomized by albatrosses and have very high aspect ratio wings. Water fowl have medium aspect ratios and high wing loading favorable for take-off and maneuvering. Aerial predators include swallows, terns, and some falcons with high aspect ratios and low wing loading. Thermal soarers include hawks and condors with low aspect ratios and low wing loading (owing to very large wing size). "Poor fliers" are birds such as turkeys and grouse that fly only short distances; these birds have very high wing loadings coupled with a low aspect ratio. Adapted from Lockwood et al. 1998, with permission.

(fat) during long migrations entails increased costs for birds, which must rely on refueling stopovers (Åkesson and Hedenström 2007). For flapping flight, travel ranges and speeds are reduced in larger birds, but these metrics scale positively with mass for soaring birds (Hedenström and Alerstam 1998). Flying species of birds range in size from 0.0015 kg (hummingbirds) to 15 kg (pelicans, condors, albatrosses; Newton 2008). Larger birds are flightless. Birds also show specializations in the types and arrangement of their feathers matched to flight performance.

Gliding Fishes

On planar surfaces (wings, fins, flattened bodies), gliding animals deflect their weight against the density of the surrounding medium. Through controlled descents, animals harness the potential energy of gravity. The flattened body and wide pectoral fins of the whale shark *Rhincodon typu*, a large pelagic planktivore, suggest a form adapted to gliding. Gleiss et al. (2011) evaluated the efficiency of controlled descents and ascents in terms of cost of transport: the energy required to move a certain distance (see fig. 2.1). Daily diary data loggers are short-term retrievable electronic tags designed to sense and record depth and accelerations relative to the fish's lateral movements. From these data, entire dive profiles can be reconstructed—ascent and descent phases, angles, distances, depths, and power (indexed by the shark's lateral acceleration). Dive profiles of nine whale sharks varied in shape and depth (typically <50 m) but showed the expected pattern of powered ascent followed by unpowered rigid body descent. Cost of transport was minimized at the lowest angles of pitch during the gliding descent stage. During ascent, cost of transport initially declined with increased ascent angle because a small level of positive pitch (10°–15°) favored lift of the negatively buoyant shark. Thereafter, greater pitch imposed increased power requirements.

From a strict optimality viewpoint, the cost of horizontal transport should favor neutral buoyancy without changed elevation. Recovery of gravitational energy entails cost. Yet, regardless of buoyancy attributes, gliding is increasingly recognized as an important mode of movement. Telemetry studies have shown that frequent diving ("yo-yo") behaviors are common in tunas, billfishes, sturgeons, and sharks (Holland et al. 1990; Erickson and Hightower 2007; Watanabe et al. 2008; Kraus et al. 2011; Nakamura et al. 2011). Similar to those performed on whale sharks, daily diaries deployed on Japanese flounder *Paralichthys olivaceus* recorded powered ascents followed by shallow angle glides favored by their planar body form (Kawabe et al. 2004). An interesting related discovery is that the laterally compressed bluefish *Pomatomus saltatrix* undertakes rigid-body glides by diagonally tilting its body's axis to expose a broader profile against gravity (Stehlik 2009).

Other ecological benefits of efficient gliding behaviors exist independent of distance traveled. Vertical dives allow predators to canvas a broader segment of the food web, particularly those that are vertically structured in the water column. For bigeye tuna *Thunnus obesus*, foraging occurs in cold deep waters, demanding thermal recovery in surface waters (Holland et al. 1992). Cues for navigation and orientation may require sensory inputs from multiple depths (Willis et al. 2009). Ascents to the surface by Pacific salmon have been linked to celestial navigation (Friedland et al. 2001). Sampling the seabed depth could be a form of station keeping by whale sharks maintaining themselves in shallow shelf habitats. Gleiss et al. (2011) also speculated that diving by whale sharks increases flow into the fish's mouth, favoring filter feeding and ram ventilation (i.e., while swimming, a gaped mouth causes continued flow of oxygenated water through the gills).

Several species of fishes (family Exocoetidae) take flight, breaking the water's inertial constraints and exploiting high-speed aerial gliding (up to 20 m s⁻¹) to evade underwater predators (Park and Choi 2010). Most spectacular are the flights of the four-winged flying fish that deploy enlarged and stiffened pectoral and pelvic fins in flight. As Davenport (1994) narrates, "fish swim towards the surface at very high speed [10–30 body lengths s⁻¹] with the lateral fins furled, leap through the water surface at a shallow angle, accelerate to take-off speed by taxiing with the lateral fins expanded and the tail beating in the water at up to 50 beats s⁻¹, and enter a free flight that may be prolonged by further taxiing." Flights range from meters to hundreds of meters, typically at altitudes <10 m (Hubbs 1918). Remarkable adaptations attend this flight: (1) the pectoral fins are greatly enlarged and exhibit a favorable aspect ratio for bearing the weight of the fish; (2) pectoral fins are supported by musculature and skeletal elements that cause a locked position while in flight; (3) pelvic fins stabilize against pitch and roll during flight; and (4) the stiffened and enlarged lower lobe of the caudal fin translates hydraulic propulsion into aerial flight during taxiing (Davenport 1994). Physics favors low-elevation gliding (<0.5 m) through the ground effect, where the sea's surface interrupts the turbulent flow generated by the wings, and drag is reduced (Davenport 2003; Park and Choi 2010). But operating in both air and water entails compromise. Starting life small imposes limits on the relationship between wing and animal size. While swimming, the undulating flying fish folds its wings along its trunk. Larvae and small juveniles <50 mm, despite their propensity to leap, do not fly: the surface tension of water is too strong to unfurl lateral fins (Davenport 1994). Later in life, at larger adult sizes, the constraint of folded wings during swimming imposes limits to the wing's size and load (Davenport 2003). In a comparison of aerodynamic properties, Park and Choi (2010) concluded that gliding performance by flying fish was similar to hawks and ducks and superior to that of flying insects.

Swimming Birds

Brünnich's guillemots Uria lomvia, diving seabirds in the auk family (~1 kg), use their wings to both fly and swim. As we have seen, the challenge in organismal design for a flying fish is generating lift against its own weight, or, alternatively stated, its negative buoyancy in air. In contrast, for breath-holding guillemots and other diving birds, the challenge is to oppose positive buoyancy (air-filled lung) in seawater through powerful downward wing thrusts as it dives. Interestingly, as one of deepest-diving (>100 m) seabirds, the guillemot's buoyancy actually decreases to nil at ~60-m depth, the result of lung compression at depth. Thus power requirements of wing strokes are expected to decline with depth during descent. Watanuki et al. (2003) attached daily diaries to the backs of three birds, which were subsequently recaptured several days later. Depth-gaining surges of each wingbeat were indeed higher in frequency in depths <20 m, frequently exceeding 3 surges s⁻¹, but once below 60 m, frequencies declined and swimming speed increased by ~40%. Midway in the ascent phase, the seabirds ceased their wing strokes and relied on increasing lung buoyancy to convey them to the surface.

In a comparison among breath-holding swimming taxa including whales, seals, turtles, and birds, Sato et al. (2007) found remarkable convergence in the relationship of body size to stroke frequency of the propulsive appendage (wing, tail, or flippers; measured again using daily diaries). Stroke frequency strongly scaled ($R^2 > 0.9$) to mass$^{-0.29}$ across masses that ranged from 0.1 to 10^5 kg. This conformed to expectations that appendage oscillation should scale directly to body length (length^{-1} or mass$^{-1/3}$). Sperm whales, seals, salmon, and seabirds all dived at ~1–2 m s⁻¹ regardless of their size. Guillemots were outliers on Sato et al.'s relationship. Wingbeats during diving were approximately fivefold less than those of other diving specialists. Sato et al. concluded that flying in both aerial and aquatic domains imposed compromises on the guillemot wing's hydrodynamic performance. The other swimming birds of note, penguins, do not fly and conformed to predictions of flipper stroke performance (Sato et al. 2007, 2010).

Analysis of Movements and Migration

Those who study fish migration have long en-
vied researchers of bird migration. Framed
against the sky, flying birds are often overt,
providing a long history of direct observation
by scientists, naturalists, and bird enthusiasts
(Newton 2008). Largely driven by commercial
fisheries and the need to manage them, scientists have developed efficient and sophisticated means to sample and model fish populations
on the move. Traditions of observing fish in
nets and birds against the sky have led to an
emphasis on population versus individual behaviors, respectively. Avian ecologists have
attached importance to both typical and anomalous bird sightings (Newton 2008), whereas
fisheries scientists have long discounted variant migrations in favor of predicting average
population behaviors (Cadrin and Secor 2009).
New approaches in observing movement and
migration emphasize approaches that serially follow individuals and aggregations over
ever-expanding periods and spatial ranges
(figs. 2.4 and 2.5). These are reviewed below as
(1) distributional snapshots, where observing
systems are used to detect aggregations over
curtailed periods of time, typically <1 week;
(2) conventional tagging and ringing, where
groups of individuals are marked and serially tracked over one to several periods of time;
(3) telemetry, where electronic tags on individuals transmit current and past positions and
environmental conditions; and (4) life cycle
tracers, where intrinsic biological and chemical tracers provide information on past environments over major phases of life history (fig.
2.5). Telemetry and life cycle tracers in particular have experienced rapid and parallel developments across birds, fishes, and other marine
taxa, allowing new understanding on patterns
of migrations for entire marine assemblages
(Block et al. 2011).

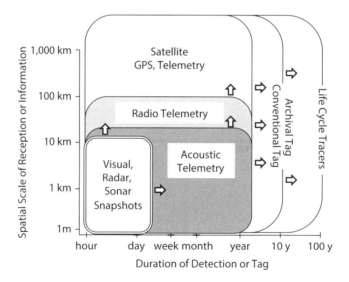

Figure 2.4. Spatiotemporal scales of detecting and analyzing individual movements. Note that although
the spatial scale of recovered information can be quite large for some types of tags and tracer approaches,
it can be obtained only by physical sampling or recovery of an animal, which can cause important biases
in the representativeness of recovered movement data. Arrows indicate that most technologies are expe-
riencing rapid gains in spatiotemporal scales.

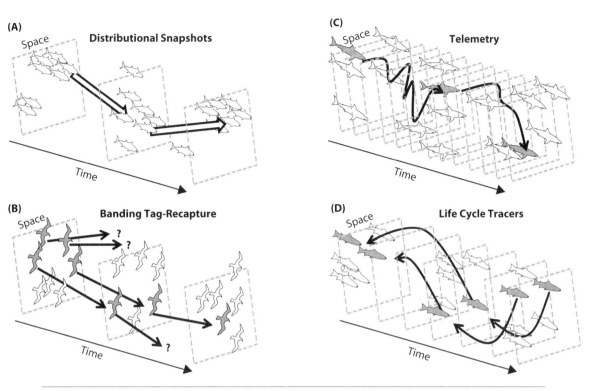

Figure 2.5. Scales of inference among approaches used to analyze movements of groups (white) and individual (gray) animals. (A) Distributional snapshots track animal distributions over short periods (typically <1 week), represented by each temporal pane (dashed square). Inferences on migration require assumptions related to the representativeness of distributions over sequences of snapshots (open arrows). (B) Tagging and banding approaches provide more definitive sequences of spatial histories but are typically limited by low probabilities of recapture and associated biases. Paths between recapture events are unknown (assumed as straight paths here). (C) Telemetry provides continuous data for individuals at multiple temporal scales (represented by numerous panes). Low sample sizes resulting from expense, inability to tag very small animals, and representativeness of tagged individuals in group dynamics are limits to this approach. (D) Life cycle tracer approaches are retrospective approaches that typically entail less precise information on spatial histories, and sampled individuals are not always representative of group migrations.

Distribution Snapshots

The comings and goings of fishes and birds were first understood by local residents and fishermen. Early communication networks allowed humans to follow seasonal movements. For instance, cod fishermen in the fjords along the Norwegian coast were well familiar with the pattern of increasing and decreasing catches as fish seasonally migrated from northern to southern fishing grounds. Similarly, the departure of birds from known breeding areas during fall and their subsequent migration along flyways became traditionally known among human communities (Newton 2008). The seasonal recurrence of birds and fish (principally salmon) in breeding areas led to an early belief that such individuals were homing to breeding habitats long before science supported such movements (Secor 2004; Newton 2008). Traditional knowledge about seasonal occurrences remains important in studying fish and bird distributions, now supported by diverse census and sampling techniques. Current approaches for birds include visual counts of birds at concentration points along flyways (Newton 2008) and nocturnal telescope counts of birds through "moon-watching" (Liechti et al. 1996) or through beams of artificial light directed skyward (Williams et al. 2001). Examples for fish include intercepting fish on the move in nets and traps (Olney and Hoenig 2001; Miller and Sadro 2003), analyzing harvest records from fisheries that intersect migration pathways (Ravier and Fromentin 2001), and monitoring fish with trawls and other sampling gear as they migrate seasonally (Rago et al. 1998).

Technologies that remotely sense animals have substantially enhanced the environmental settings over which fishes and birds can be observed. Bird and fish vocalizations are recorded and logged by acoustic receivers to uncover species incidence and behavioral interactions (Luczkovich et al. 2008; Goyette et al. 2011). Active transmissions from radio waves (radar; range ~10^2–10^8 m) and acoustic waves (sonar; range 10–10^3 m) from transducers on platforms (e.g., weather stations, moored buoys, airplanes, and boats) intercept flocks and schools as they migrate. Reflecting (echo) waves depend upon the direction and strength of the transmission and the target's (individual or aggregation) size, impedance, and distance. Importantly, active acoustic and radio transmissions and detections can occur in environments where direct observation is not possible because of light conditions, weather, visual obstructions (clouds, turbidity), visual range, altitude, and depth. Receptions are logged for data postprocessing, from which target sizes and densities of positional metrics are estimated and further analyzed to support measures of short-term and diurnal movements, including the speed, direction, and altitude (depth) of migration pathways. More focused transmissions (narrower beams) can permit improved resolution of individuals within aggregations, albeit over reduced spatial scales. Examples include

"pencil radar" that can track individual birds for short periods of time (Bruderer et al. 1995) and acoustic video cameras (e.g., DIDSON) capable of resolving three-dimensional images of moving fish under a range of environmental conditions (Holmes et al. 2006).

Remote sensing and imaging technologies have diversified rapidly during the past two decades. Airborne technologies for observing pelagic fish schools include video and light detection and ranging (lidar), the latter transmitting a focused green laser beam into the sea (~5 m at the surface) and detecting light backscatter from objects within the upper water column, typically <50 m deep (Brown et al. 2002). Doppler radar, used routinely to monitor weather, also can track densities and movements of birds during their migrations across hundreds of square kilometers (Dokter et al. 2011). A reflected microwave beam is altered by the target's motion relative to the radar transducer. Flocks of birds are a common component of reflected atmospheric spectra. The size of birds and their likely velocities allow them to be distinguished among other radar targets.

Acoustic and radar technologies do have important limitations. Species identification is often uncertain, an aggregation's structure can confound density estimates, some environmental settings are inaccessible to research platforms but not to the animals under study (e.g., polar and open ocean ecosystems), blind spots occlude portions of a receiver's range (e.g., regions immediately adjacent to the transducer and aggregations immediately adjacent to land or the seabed), and observation platforms can affect animal and distribution behaviors (e.g., vessel avoidance). Airborne detections of fish can be strongly biased owing to weather conditions that affect plane and laser performance, resulting in sometimes-poor correspondence in school density estimated by lidar versus underwater sonar (Brown et al. 2002). Still, vast areas can be sampled with airborne observing systems (200-km-long transects per hour of flight), far exceeding traditional seaborne platforms (Churnside and Wilson 2001; Newlands et al. 2006).

Distributional snapshots are well suited for short-term studies on the influence of environmental and social factors on the movement of groups of animals. Habitat and other environmental drivers can be simultaneously monitored so that the distribution of birds and fish can be transposed directly onto the environmental setting. Snapshots are particularly relevant to environmental variables that vary rapidly or are patchy, such as weather, prey resources, and oceanographic fronts. By looking across Doppler radar installations, Dokter et al. (2011) tested for short-term associations (~15-min resolution) between mesoscale weather conditions and bird movements (thrushes *Turdus* spp.) over a several-day period. Similarly, in the California Current ecosystem, Reese et al. (2011) simultaneously measured the distribution of pelagic fish schools and sea surface temperatures through airborne lidar and uncovered evidence for school affinity with thermal fronts. Thermal front formation and dissipation are associated with upwelling and are highly dynamic, as are the distributions of pelagic fishes; only airborne observations were expected to uncover these short-term affinities.

Detecting aggregations of birds and fish has rapidly advanced beyond instant snapshots, as technologies now allow wide-angle visualizations over hours to days (fig. 2.4). Videos of changing distributions now capture the short-term association with dynamic environmental drivers and serve to ground-truth models of how individual movements scale up to collective migrations (see Schooling and Flocking below). Still, animals cannot be tracked individually for days at a time with this class of methods, obscuring biological and ecological factors associated with movement pathways.

Tagging: Connecting the Dots

True, an aggregation can be tracked for short periods of time by radar, sonar, and other technologies, but at some point the aggregation moves out of range, dissipates, reorganizes, or merges, and we lose track of its original members. Under the assumption that animals continuously migrate together, spatially displaced aggregations can be connected to infer a general migration course. Examples include the observation of birds at stopovers along a flyway or the seasonal changes in harvest rates at fishing grounds arrayed along the migration route (fig. 2.5). But we also know that the routes taken between stopovers, and feeding and breeding habitats, are likely more complex and less certain than the pathways suggested by sets of distributional maps (Webster et al. 2002). For this reason, early scientists sought to connect the observed distributions of birds and fish by tagging (or ringing) and recapturing individual animals. By this means, at least two points—initial capture and recapture—along pathways were more definitively connected (fig. 2.5).

For well over a century, birds and fish have received external tags: metal leg bands and neck collars on nesting birds and disks and streaming tags secured to fin ray elements in fishes (fish tagging and marking methods are numerous and well cataloged in Parker et al. 1990). Tags are engraved with unique identification and contact information for reporting recaptured individuals. Early tagging studies were critical in refuting traditional beliefs on the comings and goings of breeding animals; for instance, the mystery of where fish and birds go after they depart known breeding areas. Early conjecture held that Norwegian cod left coastal fishing grounds for inaccessible feeding regions under the Arctic icecaps (known as the polar sea theory; Secor 2002a). In pioneering investigations, Hjort (1909) and his colleagues tagged spawning cod in fishing grounds near the Lofoten Islands and subsequently reported captures from fishing grounds hundreds of kilometers distant in the Barents Sea, showing that seasonal movements of the same cod connected these two fishing grounds. Many north temperate breeding birds were believed to hibernate and become cryptic during nonbreeding seasons. An unintentional early "tagging" study was a white stork conveying a West African spear, observed in Germany in 1822, offering evidence for a southern wintering migration (Newton 2008). Earlier still, practices of marking the birds on royal estates in Europe gave evidence of homing behaviors.

Instead of releasing animals at the site of initial capture, researchers have also deliberately displaced tagged individuals to investigate their migration behaviors, such as the ability to home back to a breeding site. As an extreme example, 18 Laysan albatross *Diomedea immutabilis* were captured on their nests (either incubating eggs or caring for nestlings) at Midway Atoll (tropical North Pacific Ocean) and carried to six sites around the margin of the North Pacific Ocean 2,100–6,600 km distant (Kenyon and Rice 1958). The US Navy Airbase at Midway facilitated the rapid transshipment of birds to other naval stations around the North Pacific, most of which occurred within the natural range of the species. The majority (14 adults) returned to the breeding site within 1 month of their release (range 9–72 d), with those incubating eggs returning more rapidly. Similarly, Snyder (1931) transplanted 15,000 juvenile Chinook salmon *Oncorhynchus tshawytscha* from the Sacramento River (California, USA) ~500 km north to the Klamath River. Ten were recaptured in ocean fisheries, 50 in the Klamath River, and none in the Sacramento River, providing early evidence that spawning adults returned to the stream of their early growth rather than to their parents' stream.

Hundreds of millions of fish and birds have been tagged by scientists, fishermen, and conservation groups (e.g., National Audubon Society, British Trust for Ornithology, American Littoral Society). Most fish recaptures depend on fishermen reporting them in their catch; recapture numbers thus depend on harvest and reporting rates. In contrast, avian mark-recapture studies depend principally upon the efforts of scientists and conservation societies (albeit, for waterfowl and other game birds, hunters often account for the majority of recaptures). Typical banding recovery rates are <1% (Newton 2008; Robinson et al. 2010), which curtails population-level inferences, particularly in weighting unexpected recapture locations and dates. Further, most recaptures occur

at the site of release owing to natal homing; observations can be exceedingly rare at stopover or wintering habitats. A large effort ringed over one million pied flycatchers *Ficedula hypoleuca* in northern Europe, yet only six were subsequently observed at African wintering grounds (reported in Webster et al. 2002). For fished populations, low survival contributes to poor recapture rates. Harden Jones (1968) generalized early studies on Pacific salmon: return rates from adults tagged as juveniles in natal streams ranged from 0% to 10%. But whether these low recapture rates indicated low survival after juveniles emigrated to the sea (predation, commercial fishing) or were the result of high navigational error during the return migrations (Jamon 1990) remains uncertain. Did salmon die, or were they were lost at sea?

In summary, for both birds and fishes, tagging provides fairly certain information on capture and recapture locations but requires careful consideration of biases associated with reporting systems, mortality, and tag loss (Nichols 1992). External marks can be efficiently and inexpensively applied, particularly at points where humans can gain access to animal aggregations such as banding stations or fisheries that target breeding grounds (Hobson and Norris 2008). Because large numbers of animals can be tagged, mark-recapture as a "quasi-longitudinal" approach represents a bridge between distributional snapshots and methods that focus on detailed movement pathways of individuals. The scope for inference is limited, however. We are typically connecting only two dots along a pathway, which is unlikely to be linear or representative of a collective migration (fig. 2.5). The most significant source of uncertainty is the fortuitous nature of how we observe recaptures. Without a global system of observation, recaptures will be biased by where and when they were originally tagged, and also by the inability to track the spatial and demographic fates of those tagged animals that are never seen again.

Telemetry: Distant and Transparent Conveyance of Information

Miniaturization of electronics after World War II—power sources, transducers, data processors, and sensors—resulted in new technologies to track animals. Instrumentation is now evolving so rapidly that any attempt to review state of the art in telemetry would be out of date within a year's time. Broadly, telemetry is defined as distant and remote detection of individual movements and includes (1) electronic tags that transmit position and other data and (2) tags that store such information. Of course there are tags that do both, transmitting archived data to remote receivers (e.g., satellites). Telemetry yields detailed positional records at multiple spatial and temporal scales (figs. 2.4 and 2.5). The same archival tag in a bluefin tuna that yields information on seasonal transatlantic migrations can relay hourly dive behaviors associated with spawning, bathymetry, and local oceanographic features (Walli et al. 2009).

Global capabilities of telemetry expanded rapidly through the French-American satellite system Argos, originally designed to provide complete GPS coverage in locating and retrieving data from thousands of drifting oceanographic buoys (Argos 1996). Doppler radar triangulates the geographic coordinates of sea-based radio transmitters, and that information together with other transmitted information (e.g., archived environmental data) is made available by Argos processing and service stations to investigators around the world. For birds, GPS records are continuously logged in data storage tags that are bodily attached; these data are then either retrieved from recaptured individuals (say, on breeding sites) or directly transmitted to Argos. GPS cannot locate fish underwater. Light levels (detected by a photo sensor) and date and time stamps for each record provide approximate latitude (day length) and longitude (time of local noon) coordinates. Analysis of position from light levels is substantially improved by incorporating information

on previous positions and encountered temperatures (Teo et al. 2004). In an alternative approach, tidal pressure changes are used to estimate bathymetric region for demersal fishes (Metcalfe and Arnold 1997). Tags can sense, record, and transmit a wide range information on environmental (depth, altitude, temperature, salinity), physiological (internal temperature, flight/swimming activity, heart rate), and even social variables (proximity to other tagged individuals). For students of animal movements, telemetry has brought an unsurpassed generation of discovery; three examples follow.

The gull-like sooty shearwaters *Puffinus griseus* perform remarkable pan-Pacific migrations as circuitous figure-eight pathways extending over 50,000 km (fig. 2.6A; Shaffer et al. 2006). Nesting birds were removed from breeding burrows in New Zealand and outfitted with small archival tags that recorded light level, pressure, and temperature at ~10-min intervals. A majority of these birds (*n* = 19/33) were subsequently recaptured in the same borrows but now with 7–10 months of data archived in their tags. Pressure data revealed undersea diving behaviors associated with periods of foraging during both austral and boreal summers. Rapid transequatorial transit between these foraging areas was accomplished by using Southern Hemisphere easterly trade winds during northwestward movements. Westerly trade winds attended their return southeastward transit to New Zealand, resulting in the overall figure-eight configuration (fig. 2.6A). The extreme migrations of sooty shearwaters place them in both south and north temperate latitudes during summertime peaks in those latitudes' production cycles. In summary, telemetry uniquely showed how sooty shearwater exploited "oceanic resources across the Pacific Ocean in an endless summer" (Shaffer et al. 2006).

Blue marlin *Makaira nigricans* is a large pelagic marine predator (maximum size >1,000 kg), distributed sparsely across tropical and subtropical oceans, and is presumed to be highly migratory (Ortiz et al. 2003). Conventional

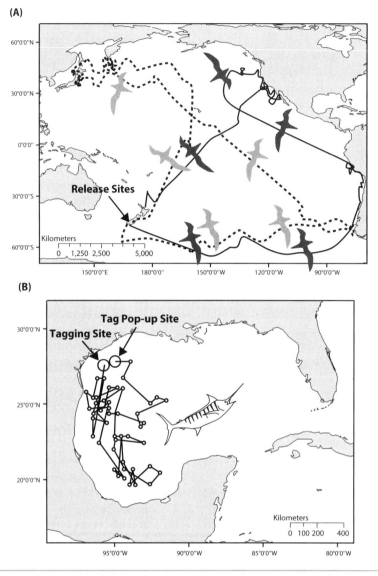

Figure 2.6. Telemetry tracks of highly migratory (A) sooty shearwater and (B) blue marlin. Two representative tracks of shearwater adults show their departure from New Zealand breeding colonies and their subsequent migrations. The 80-kg blue marlin was tagged and released in July 2008, and then released its pop-up tag nearly a year later (334 d). Its net displacement was 101 km. Redrawn from Shaffer et al. 2006 and Kraus et al. 2011, with permission.

tags suggested that blue marlin were "world travelers," moving across and between major oceans (Ortiz et al. 2003). For instance, a blue marlin that received a conventional tag in the northwest Atlantic Ocean (Delaware, USA) was recaptured 4 y later in the Indian Ocean (Mauritius). But we can only conjecture about the pathway and speed of such long-distance movements and whether transoceanic movements are common. A recent satellite telemetry study on Gulf of Mexico blue marlin suggests not (fig. 2.6B). Similar to the shearwater study, Kraus et al. (2011) bodily attached data storage tags to marlin (30–250 kg) that recorded light levels, temperature, and pressure to support daily position estimates. But in contrast to birds returning to nesting burrows, tagged marlins are unlikely to be recaptured, so information on the tags was retrieved by designing them to release and float to the surface after certain intervals (30, 90, 180 and 365 d). A continuing technical issue with these "pop-up" tags is early release, but 10 of the 34 tags designed to release at either 180 or 365 d remained attached for at least 200 d. By staggering tag deployments across seasons, Kraus et al. (2011) provided evidence for restricted year-round movements within the Gulf of Mexico Basin (fig. 2.6B), a fairly surprising result given the presumed capacity of marlin to undertake long oceanic migrations.

One of the world's largest fishes (maximum size ~12 m), the basking shark *Cetorhinus maximus* is distributed in boreal to subtropical shelf waters of the Atlantic Ocean and takes its name because it is often sighted in surface waters as it filter-feeds on plankton. Indeed, it was traditionally harvested by harpoon. Out of sight during winter months, it was believed that basking sharks curtailed their activity, hibernating in deeper but local waters (Sims et al. 2003). Through telemetry, researchers discovered that basking sharks are in fact quite active during fall and winter months, undertaking wide-ranging horizontal migrations to winter feeding habitats hundreds to thousands of kilometers distant. Further, basking sharks are not just surface oriented but in all seasons rely on vertical diving behaviors ranging up to 1,000 m as they prospect for zooplankton (Sims et al. 2003). Basking sharks tagged in shelf waters of New England (USA) entered tropical Caribbean and South American waters during winter, in regions thought to be "off limits" on the basis of previously observed distribution and expected thermal tolerance (Skomal et al. 2009).

More traditional telemetry approaches depend on locally based detection of radio or acoustic transmissions, technologies that have been available for over 50 y (Harden Jones 1968). In marine fish applications, radio waves dissipate too rapidly in salt water, so acoustic beacons have been exclusively used. In contrast, small radio tags are of greater advantage in avian applications, where small birds such as songbirds can be tagged and tracked. In endurance contests, investigators continuously track individuals so long as they are able to keep up with them in time and space (typically <3 d). These mobile tracking efforts have provided a wealth of information on the mechanisms of movement, such as the use of tides to facilitate transit in estuaries (Parker and McCleave 1997); in situ measures of metabolism, locomotion performance, and mating behaviors (Cooke et al. 2004); and habitat selection (Robinson et al. 2010). Remotely deployed receivers can also detect and log those detections over months and years. Arrays of receivers have been strategically deployed to investigate movements across broad regions (Welch et al. 2004; Grothues et al. 2005; Heupel et al. 2006).

By emphasizing individual movement behaviors, telemetry and particularly satellite telemetry have transformed our understanding of migration by uncovering previously hidden behaviors and pathways, and they have begun to show how the environment, physiology, and the repertoire of behaviors shape travel decisions (e.g., Cooke et al. 2008). The promise of the technology is "scale-independent tracking" (Webster et al. 2002), where movement phases are recorded from annual to subhourly intervals. But we must recognize certain limits (Hobson and Norris 2008; Hebblewhite and Haydon 2010). Telemetry tags are anything but transparent to the experimental animal. They are often large relative to the study organism, and care is needed to ensure that the tag itself does not bias behavior, causing anomalous movements, physiological stress, or mortality (e.g., Kerstetter et al. 2004). Satellite telemetry tags are larger than other types of tags, which limits their deployment to larger birds (>0.3 kg; Hobson and Norris 2008) and fishes (>5 kg; Jellyman and Tsukamoto 2005). Similar to conventional tagging and banding, the movement pathways we observe will depend upon where and when we tag individuals (fig. 2.5). If tagged individuals are drawn from a single aggregation, then our inferences are limited to that aggregation rather than to the population or species of interest. Further, owing to the expense of telemetry, investigators are less likely to representatively sample aggregations and populations. In their review of published literature on GPS terrestrial studies, Hebblewhite and Haydon (2010) reported a mean sample size of 18 study animals (range 4-82). In marine studies using pop-up tags, large tunas have received the most attention, and in a review of the literature (1980–2011; International Web of Science) the mean number of tags per study was 41 (range 2–98; n = 13). The Tagging of Pacific Predators Program, part of the Census of Marine Life, deployed over 1,000 satellite tags in 13 species of sharks, sea turtles, albatrosses, and pinnipeds (Block et al. 2011).

In their critical review of satellite (GPS) telemetry, Hebblewhite and Haydon (2010) concluded that "the most difficult problem facing ecologists using GPS data is how to scale up to the population consequences of movement." The issue here is to construct movement phases of ecological consequence from the many thousands of movement pathways that are measured on a nearly continuous basis. One means that will be reviewed below is to break up the data into seminal movement pathways and to reconstruct movement phases through simple mechanistic models of movement (e.g., random walk, Lévy flight, and state-space models). Still, from the standpoint of unintended consequences and current bioinformatics challenges, one could suggest that telemetry currently faces an embarrassment of riches. This challenge will likely be overcome as the technology becomes more accessible, samples of tagged fish become increasingly representative, and investigators continue to move from descriptive studies to ones focused on ecological questions guided by experimental design.

Life Cycle Tracers

Animals carry with them intrinsic tracers that can provide information on past environments

over entire or major portions of their life cycle, including their morphology, plumage, chemical composition, parasite loads, and genetic attributes (Harden Jones 1968; Hobson and Norris 2008). Most of these characteristics convey information that corresponds to seasonal, annual, and lifetime intervals. Still other chemical tracer approaches can resolve daily movement paths (Elsdon et al. 2008). This approach is thus broadly classified here as "life cycle tracers." Life cycle tracers are retrospective—we are inferring an animal's past history of movement between environments. An advantage of life cycle tracers is that an animal need only be sampled once, as opposed to traditional tagging studies, which require at least one recapture event. Further, the manner in which we sample does not influence the future behaviors of what we will observe nor their representativeness; rather, sampling constrains the range of past behaviors we are likely to see. Thus tracer approaches will give a biased depiction of movements for survivors only, which can be particularly important in fishes where early mortality rates are high (fig. 2.5). In contrast, for some conventional and electronic tagging applications, one can simultaneously track dispersal and demographic fates (Pine et al. 2003; Welch et al. 2008). One could feasibly do the same using retrospective tracer approaches, but it would require careful sampling design elements (e.g., Kerr and Secor 2012).

In this section, a single class of tracers— stable isotope composition—is reviewed briefly, leaving genetic markers and other tracers for later. Molecular approaches provide a diverse class of life cycle tracers that are undergoing rapid transition from methods used to track lineage and draw inferences on natal homing migrations and population mixing to those that evaluate selective processes that shape migration and population structure. Because genetic markers as life cycle tracers typically entail assumptions specific to population structure (Hedgecock et al. 2007; Mariani and Bekkevold 2013), a review of related applications is deferred until chapter 6 (Cod Ecomorphs and Biocomplexity). For review of other life cycle tracers, see Newton (2008), Cadrin et al. (2013), and Kerr and Campana (2013).

Ratios of stable isotopes represent changes in bedrock chemistry $\delta^{87}Sr$ (notation for $^{87}Sr:^{88}Sr$), precipitation (δ^2H, $\delta^{18}O$), autotroph nutrient sources ($\delta^{13}C$, $\delta^{34}S$, $\delta^{15}N$), marine-freshwater mixing ($\delta^{18}O$, $\delta^{13}C$, $\delta^{34}S$), and benthic-pelagic coupling ($\delta^{13}C$) and provide information on exposure to different watersheds, latitudinal ranges, ocean basins, food webs, and salinity (for more in-depth reviews on stable isotope environmental chemistry, see Hobson 1999, Inger and Bearhop 2008, Fry and Chumchal 2011, and McMahon et al. 2013). Birds and fishes take up stable isotopes through diet and respiration; the latter is of greater importance for fishes (Campana 1999). Stable isotopes are incorporated into soft tissue, where they are subject to metabolic turnover, or in hard parts such as feathers and otoliths (ear stones; see below), which experience no turnover. In soft tissues, stable isotopes convey information for the period of tissue synthesis—typically, weeks to months (Hobson 1999; Buchheister and Latour 2010). Differential turnover times between tissue types allow recent dispersal histories to be inferred. For instance, more rapid turnover in the liver than in muscle means that a change in $\delta^{13}C$ in liver but not in muscle tissue represents a recent change in the animal's food web (nutrient source) and associated habitat (Suzuki et al. 2008). Feathers and otoliths are metabolically inert and retain a temporal record of isotopic signatures associated with periods of feather and otolith growth. Stable isotope applications require (1) known and persistent differences in stable isotopes associated with location or food web; (2) evidence that stable isotopes in soft or hard tissues correspond to background geographic or food web source signatures; and (3) estimates of isotope incorporation (fractionation) and turnover rates in the animal (Post 2002; Elsdon et al. 2008; Wunder and Norris 2008).

Songbirds are too small (<10 g) to tag effectively, making them strong candidates for trac-er approaches (Hobson 1999). The black-throated blue warbler *Dendroica caerulescens* is an abundant songbird that nests in mixed deciduous forests of the Appalachian Mountains in the northeastern United States and southeastern Canada. It then overwinters in the Caribbean islands. Prior to their southern migration in autumn, the warblers molt and the new feathers incorporate $\delta^{13}C$ and δ^2H, at levels that are coarsely representative of breeding latitude (Rubenstein et al. 2002). Overwintering sites are arrayed across an east-west gradient between the islands of Puerto Rico, Hispaniola, Jamaica, and Cuba. Sampled feathers indicated that overwintering warblers in Puerto Rico and Hispaniola tended to originate from southern breeding latitudes (<43°N), whereas those in more distant Jamaica and Cuba came from more northern breeding latitudes. Although both $\delta^{13}C$ and δ^2H are influenced by temperature, elevation, and precipitation rates that vary with latitude, δ^2H in particular is commonly used to infer broad latitudinal environmental histories for birds that breed in temperate and boreal habitats (Hobson 1999; Rubenstein et al. 2002). In comparison, δ^2H has seen limited application in fishes (Whitledge et al. 2006).

Tracer studies have addressed important fishery management questions. Drake Passage, between Patagonia and the Antarctic Peninsula, is a critical navigation route for both humans and marine fishes. Of these, the southern blue whiting *Micromesistius australis*, a member of the cod family, supports South America's largest industrial fishery. At issue is whether individuals cross-navigate the passage and mix between fisheries that are centered on either side of Patagonia (40°S–60°S; southwest Atlantic and southeast Pacific Oceans). Seasonal aggregations south of Drake Passage (Scotian Sea) suggest that substantial mixing is possible. During their first year of life, juveniles originating on either side of the passage are exposed to ocean sources of $\delta^{13}C$ and $\delta^{18}O$ that substantially differ. In Pacific shelf regions off Chile, higher precipitation, greater freshwater inflow, and warmer temperatures contribute to a lower level of $\delta^{18}O$ (LeGrande and Schmidt 2006) and a higher expected level of $\delta^{13}C$ than in Atlantic shelf regions off Argentina. Otoliths were sampled from Pacific and Atlantic fisheries, and material corresponding to the first year of life was isolated and analyzed. Resultant stable isotope signatures gave evidence that jurisdictional fisheries were largely dependent upon local reproduction specific to the ocean of harvest (Niklitschek et al. 2010; this case study is given further detail in chap. 5, The Necessity of Straying).

Otoliths are timekeeping structures that have transformed our understanding of fish life cycles and migrations (Elsdon et al. 2008; Secor 2010). Concretions of calcium carbonate located in the hearing organs of teleost fishes, otoliths grow by daily and seasonal accretion, resulting in daily increments (rings) during the first several months of life, then seasonally deposited annuli (again, rings) for the remainder of life. For several chemical tracers, otoliths grow in isotopic equilibrium with surrounding water (Campana 1999; Kraus and Secor 2004a; Elsdon et al. 2010). Precise microprobe and milling procedures have been coupled with ever-improving sensitivity in chemical assays. Age-specific tracer analysis in fishes, termed profile analysis (Elsdon et al. 2008), is a particularly promising approach in linking (1) environmental factors to dispersal and migration, (2) spatial behaviors to demographic and physiological fates of individuals, and (3) demonstrating carryover effects from early dispersal behaviors to later ones (for examples, see reviews by Secor and Rooker 2000, Campana 2005, and Elsdon et al. 2008; chaps. 5 and 6). In avian studies, feathers also have important timekeeping attributes, but their resolution is principally seasonal and not age specific, dependent upon the molting cycle (Inger and Bearhop 2008).

The chief limitation of stable isotope and other chemical tracers is the degree to which the tracer represents spatial features relevant to movement phases (Wunder and Norris 2008). Tracers that represent large geographic features and gradients will necessarily limit applications to those that describe movement pathways, which are of similar scale (e.g., continental-scale autumn migrations of songbirds, interocean migrations of fish). Temporal stability is a critical issue that relates to underlying ecosystem dynamics. Thus tracers of bedrock are stable over geological epochs. On the other hand, tracers that relate to food web differences will depend on multiple dynamics—stability of the food web as well as feeding and growth dynamics of the study animal. Further, as stable isotope ratios propagate through a food web, metabolism causes disproportionate excretion of the lighter isotope and the ratio increases (heavier isotope proportionally increases), a process termed fractionation. Trophic fractionation varies across isotopes, tissue types, species, and systems (Elsdon et al. 2010), requiring some degree of calibration against lower trophic levels (Inger and Bearhop 2008; Woodland et al. 2012). Tracers associated with environment gradients such as the conservative mixing of strontium isotopes in estuaries will depend on the steepness and temporal stability of the mixing gradient (Fry 2002; Kraus and Secor 2004a). Further, most stable isotopes such as $\delta^{13}C$ can represent multiple environmental signals (e.g., salinity, precipitation, flow, nutrient source) for which it is not always possible to reconstruct a single environmental history. Use of multiple stable isotope tracers can help differentiate among multiple environmental sources (Inger and Bearhop 2008). Tracer uptake studies are infrequent in the literature, representing a second major source of uncertainty (Post 2002; Herzka 2005; Inger and Bearhop 2008; Elsdon et al. 2010). Such studies are on the rise and in general confirm expectations of (1) a positive exposure versus uptake relationship and (2) tissue-specific turnover rates (Herzka 2005; Buchheister and Latour 2010; Elsdon et al. 2010). Still, most published applications rely on untested assumptions related to temporal and spatial stability of the tracer in the environment, uptake and turnover rates, and metabolic fractionation (Wunder and Norris 2008).

Uncertainty and the Use of Multiple Approaches

Movement ecology has progressed rapidly, particularly with advances in telemetry, which has provided unprecedented details on the movement paths of individuals. The current challenge, well recognized in the field, is to representatively sample populations and scale up from individual movements to collective behaviors. In contrast, life cycle tracers are typically coarse in the reconstruction of past habitats and migration pathways, requiring that classification procedures be applied to samples rather than individuals (Webster et al. 2002), which in turn obscures individual movement behaviors (otolith profile analysis is an exception; chap. 6, Developmental Plasticity). Further, consideration of error and methods of calibration in assigning past geographic/oceanographic regions has received insufficient attention. Still, like telemetry, tracer applications have heralded important new discoveries on previously hidden life cycles during the past several decades (Hobson 2003; Secor 2010).

For distributional and telemetry studies, advances have increased the temporal and spatial scales of detection and inference (fig. 2.4). Arrays of radar and acoustic receivers can remotely detect aggregations and individuals at tens to hundreds of kilometers, with much greater coverage in store for the future as these receivers become ever more widespread. Coordinated arrays now track distributional changes during major seasonal migrations at subcontinental or ocean basin scales (Gauthreaux and Belser 2003; Dokter et al. 2011). The initial limitation of data storage tags was their physical recovery (birds on a breeding ground or fish in a net); this shortcoming has been overcome with tags that can transmit stored data to satellites. Intrinsic tracers and conventional tags remain the go-to approaches to provide longer-term information on movements, but with increasing battery lifetimes, electronic tags will soon be providing decadal-scale information.

An argument could be made that those who use telemetry are likely to frame migration inquiries differently than those using tracer approaches. Operationally, the surfeit of telemetry information on individual pathways invites questions on what motivates the individual to move—or movement ecology. For similar reasons, the statistical nature of life cycle tracer applications typically leads to a focus on group- or population-level questions—or migration ecology. We hope this is an overstated divide, because there is great advantage to combine approaches in addressing questions related to both individual and group movements. Increased resolution and accuracy can be attained through a mixture of distributional, tagging, and tracer approaches (Inger and Bearhop 2008; Kurota et al. 2009). For instance, initial tracer studies that uncovered broadly classified migration behaviors, such as dependence on upwelling regions by sooty shearwater (Minami and Ogi 1997) or lifetime freshwater resident behaviors by coastal striped bass (Secor and Piccoli 1996), were later followed up by telemetry studies that provided increased details on their specific movement pathways (Shaffer et al. 2006; Wingate and Secor 2007).

Maximum likelihood is a rigorous statistical framework to contrast competing models of spatial behaviors against empirical evidence drawn from multiple approaches (chap. 5, Stock Biocomplexity and Spatially Explicit Stock Models). Similarly, Bayesian likelihood frameworks can make use of multiple approaches and seemingly incompatible data sets by sequentially fitting probability functions to parameters of interest. For instance, on the basis of telemetry data alone, directional vector estimates were highly uncertain for a bird population owing to small sample size and large error, but vector parameters were improved by constraining these estimates to likely pathways obtained from a ringing database containing thousands of recaptured birds (van Wilgenburg and Hobson 2011). Integrating pop-up satellite, archival, and conventional tagging data sets for Atlantic bluefin tuna, Kurota et al. (2009) used a sequential Bayesian procedure to estimate >15 parameters related to transoceanic movement, mortality, and tag reporting rates. Their study is a window into the feasibility of drawing important population-level inferences from individual movement data. Still, despite the obvious advantage of combining distributional, telemetry, and tracer data sets, empirical and statistical studies that do so remain rare in the literature.

Rules of Aggregation

Can we explain patterns of movement simply by distance? On the surface, organismal design and ecological considerations would argue otherwise. Large animals have a greater capacity to store energy and move greater distances; longer movements are more frequent in seasonally variable high-latitude environments than stable tropical ones. Further, measures of discrete step pathways—distance, direction, turning angles, speed, depth, etc.—when examined across a sequence, suggest motion phases associated with ecological functions (Nathan et al. 2008; Patterson et al. 2008). For instance, a high turning rate might correspond to a foraging behavior, whereas a series of high velocities could correspond to predator evasion or a seasonal breeding migration.

Now, what if we dissected an individual's movement by discrete time steps and looked at the frequency of differing distances? Would we not expect shorter paths to be more frequent than longer ones, perhaps falling off exponentially as path length per interval increased? Interestingly, for diverse taxa the distributions of movement paths are often well fitted by simple statistical functions such as Gaussian (bell curve) distributions, despite expectations for environmental feedbacks, pathway latencies, or functions specific to organismal design or navigation capacity.

Lévy Flights

A modification of a simple random Gaussian distribution in movement lengths is the Lévy flight distribution: sequences of randomly oriented straight-line paths that start and stop at obvious reorientation points such as turns and dives (Reynolds and Rhodes 2009). In these distribution models, movement phases are typically ascribed to foraging or acquisition of other resources. In open systems, the spatial distribution of these resources is likely patchy and overdispersed (Bartumeus 2009). In comparison to the rare distribution of longer paths expected from a Gaussian distribution (a.k.a. a random walk), Lévy flights yield a fatter tailed distribution (fig. 2.7). These distributions are modeled as a power law where the probability of path length or other pathway metric $P(L)$ is estimated by

$$P(L) = L^{-\mu},$$

where exponent μ ranges from 1 to 3. For $\mu \geq 3$, the distribution becomes Gaussian, emulating diffusion or a random walk. More than a simple mathematical fitting exercise, Bartumeus (2007) and others hold that Lévy flights represent a biphasic search process—a local search pattern within a regional patch defined by an organism's range of perception, interrupted by occasional longer exploratory flights to more distant patches outside of the organism's immediate search area. The underlying theory is that without information or other movement drivers, a stochastic search for resources that includes a fractal (multiple-scale) element will increase the likelihood of successful searches. Lévy flights as an emergent feature have attracted attention as a general and possibly adaptive constraint underlying the movement behaviors of diverse taxa (Bartumeus 2007; Schick et al. 2008; Reynolds and Rhodes 2009).

Lévy flights are predicted for large oceanic predators, where prey are likely patchy, overdispersed, and often beyond the perception range of the predator (Sims et al. 2008). Such roving predators are epitomized by wandering albatross *Diomedea exulans* and bigeye tuna. Foraging search behaviors are critical to wandering albatross as they migrate hundreds of kilometers per day in pursuit of forage to support their large size and metabolism, and as they brood their young (Phalan et al. 2007). These seabirds undertake circumpolar migrations, foraging on

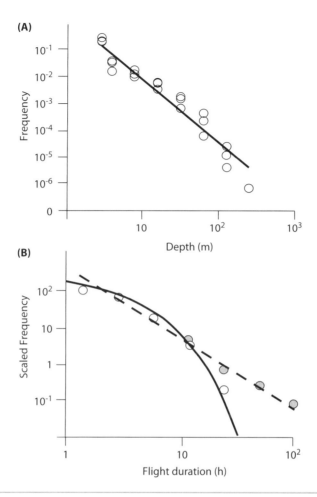

Figure 2.7. Lévy flight distributions for (A) bigeye tuna and (B) wandering albatross. (A) represents >200,000 movements for three individual tuna. (B) shows frequencies binned by flight durations for 19 foraging trips undertaken by five tagged albatross. Longer flight durations originally reported by Viswanathan et al. (1996) were later determined to be in error (gray circles). Redrawn from Edwards et al. 2007 and Sims et al. 2008, with permission.

squid, fish, crustaceans, and carrion in the surface layer of the water (<20-m depth). Repeated splashdowns and take-offs are energetically taxing (Lecomte et al. 2010). Bigeye tuna are large endothermic (heat-generating) fish found pan-globally in tropical and subtropical latitudes (≤40°). Their high energy demand is met through deep vertical foraging transits (200- to 400-m depth) into cold and sometimes hypoxic waters (Musyl et al. 2003). Bigeye tuna must frequently return to the surface layer to recover

their body temperature and respiratory function (see Gliding Fishes above).

Analysis of movement paths supported Lévy flight distributions for both these ocean predators (fig. 2.7). The seminal path metrics of search behaviors vary between these two species: for albatross, the duration of sustained flight between splashdowns; for bigeye tuna, the depth of each dive. Wandering albatrosses were outfitted with a seawater immersion logger at breeding sites and subsequently recovered upon their return many days later (Viswanathan et al. 1996). Bigeye tuna were implanted with archival tags containing a pressure sensor, and a portion of these fish was recaptured and returned by commercial fishermen. Frequency of periods of sustained flights and depth of dives showed the expected distribution with μ ranging between 2 and 3 (Viswanathan et al. 1996; Sims et al. 2008). Note that the longest flight intervals were well predicted by the function, exceeding 50 h (fig. 2.7). For bigeye tuna, there is a slight indication of a departure from the predicted equation at the highest depths, those exceeding 100 m. Later reanalysis of the albatross data uncovered an important error: the longest flight times were in fact time spent on the ground at the breeding site (Edwards et al. 2007). When these data were removed, a more rapid extinction rate in long flight times was observed than that predicted for a Lévy flight (fig. 2.7). In yet another examination of this data, together with new high-resolution GPS flight tracks, Humphries et al. (2012) found evidence for a truncated (bounded) Lévy flight distribution.

The emphasis on fitting distribution functions to motion paths is in part justified by the sheer volume of data generated by archival tags—the need to somehow make sense of it all. The search for scale-invariant movement rules for how animals interact with trophic and other environmental aspects of their seascape seems naturally suited for this type of data. For instance, the same archived data set of millions of individual movement records for dozens of tuna can contain multiple scales of information related to dive behavior and horizontal movements at temporal scales ranging from minutes to years (Walli et al. 2009). Further, comparison of distribution functions can be informative. The rapid extinction rate of flight durations above 10 h suggests threshold energetic constraints for albatross, whereas dive depth by bigeye may not be so severely constrained by physiology. In a recent comparative study of over 12 million path records, Humphries et al. (2010) compared ocean predators across ecosystems and found that the same predator exhibited random walk movement behav-

iors over continental shelves but Lévy flights in open ocean waters. They hypothesized that forage distributions and associated search path behaviors differed in kind between these two ecosystems.

State-Space Models

Movement ecology has moved rapidly to adopt a statistical method for testing simple process models of movement (such as those introduced in the previous section) against telemetry and other movement databases. State-space animal movement models predict a future spatial state at time step 1 on the basis of (1) the observed location at time step 1 and associated error, (2) the animal's past locations, and (3) an expected behavior constrained by (typically) a simple movement model (Patterson et al., 2008). The statistical approach explicitly considers error in both the observation and process model and as such provides a means of predicting movement behaviors while appropriately conveying uncertainty. The simultaneous fitting of several likelihood functions has thus far limited applications to fairly simple process models, but these are becoming more complex, with increased emphasis on applying multiple process models to the same individual track (e.g., switching between resident and transitory modes; Patterson et al. 2008; Pedersen et al. 2008; Schick et al. 2008). Among marine fishes, state-space models have been particularly useful in identifying hot spots (dense concentrations of aggregate activity) for foraging, reproduction, transit habitats, and oceanographic correlates to habitat use (Royer et al. 2005; Block et al. 2011; Abecassis et al. 2012). Such models remain firmly rooted in movement ecology, relying on the principle that migrations scale from individual capacities and motivations rather than collective agencies.

Schooling and Flocking

Dynamic Aggregations

At least half of all fishes and birds collectively migrate during some portion of their lives (Viscido et al. 2004). Within schools and flocks, individuals show ordered spacing and alignment, collectively coordinating their movements in propulsion, evading threats, orienting, and navigating. This coordination results in dynamic stability of the aggregation's size, velocity, and density. For the field observer—swimming, wading, flying, or walking in the midst of schools and flocks—the rapid synchronous response of dynamic aggregations is remarkable in its immediacy and coordination. Aggregations can suddenly bend, expand, split, and condense. Indeed, laboratory measures show that entire fish schools react to disturbance within 0.5 s (Partridge 1981). Schools and flocks are dynamic in membership, dimension, shape, and density. For instance, pelagic fishes are well known to aggregate during daytime and to disperse at night. In acoustic observations of large sardine schools (~10^5 individuals), Fréon et al. (1996) observed compact schools with densities >50 fish m^{-3} during the daytime. Then, beginning at dusk, the sardines gradually dispersed to midnight densities of ~4 fish m^{-3}; at dawn, individuals rapidly condensed back into schools. At smaller intervals, acoustic observations show schools of Atlantic herring *Clupea harengus* that seem in a constant state of flux, splitting and reassembling, dissipating or merging with other aggregations at subhour frequencies (Pitcher et al. 1996; Makris et al. 2009). Within daily periods, schools display a wide spectrum of conformations evoking descriptors such as tower, pole, compact, zig-zag, fluffy, hourglass, and funnel (Petitgas and Levenez 1996; Vabø and Skaret 2008).

Dynamic flocking behavior is epitomized by the massive flocks of European starling *Sturnus vulgaris* (~10^5 individuals), which pulsate against the sky like huge high-speed amebae rapidly expanding and condensing in tidal waves of dense bands stretching across the horizon, occasionally extending armlike "pseudopodia" (Hildenbrandt et al. 2010). Although not as obvious, schools can show similar dynamics. Vast schools of herring (>10^6 individuals) on Georges Bank were detected and imaged over an area ~1,000 km^2 using a new technology that propagates acoustic waves across the entire water column (Ocean Acoustic Waveguide Remote Sensing; Makris et al. 2006). At sunset, clusters of high-density shoals coalesced into large schools. The growth of the school occurred in serial waves of concentration bands (Makris et al. 2009). Interestingly, the speed of these waves (3–6 m s^{-1}) was greater than individual cruising velocities (<1 m s^{-1}). For starling and herring aggregations, information transmission operates at a collective scale, leading to dynamic conformational changes in flocks and schools, sometimes over several kilometers, which greatly exceed the spatial and temporal scales of individual movements.

Self-Organizing Systems

That the same individuals do not maintain their same position or persist in the same aggregation, and that schools and flocks dynamically reassemble over seconds, hours, or days, directs attention to the "anonymity" of the individual, which submerges its behavior within the aggregation (Partridge 1981; Sumpter 2006). The relationship between individual and collective movements has seen limited study in the laboratory. The position of marked saithe *Pollachius virens* within small schools (*n* = 20–30) was laboriously tracked through direct observation and videography as they swam in a circular raceway (Partridge 1981). Individuals were highly variable in their position within schools with no evidence of persistent leaders or followers, good or poor "schoolers," or differences in school position associated with size. Distances between nearest neighbors were maintained at about one body length but decreased with increasing swimming speed. Individuals tended to match velocity with their nearest neighbors but match their directional heading with the overall school's bearing. An individual tended to align with neighbors to the front and alongside rather than with those to the rear. This is consistent with the two sensory systems used for schooling: vision and the lateral line (the latter used for mechanoreception; Partridge and Pitcher 1980).

Visual and acoustic observations support

several classes of fish and bird aggregations: swarms, stationary and mobile aggregations, and a circulating pattern known as a torus (fig. 2.8). According to simulation studies, the manner in which these aggregation states develop and transition one to the other depends on a small set of individual behavioral rules (fig. 2.9). (1) Individuals are repulsed by each other at close distances, thereby avoiding collision and other negative interactions. (2) At intermediate distances, individuals orient to each other by aligning speed and direction. (3) At still greater range, individuals seek not to be alone, are attracted to distant individuals within their perception range, and will move to intersect their path. (4) When no compatriots are perceived, the individual will undertake nondirectional random movements until a neighboring fish is detected. These interaction rules can be summarized by a series of reactive distances that evoke changes in directional vector and velocity by the individuals (fig. 2.9). Limited laboratory and field observations support these rules (e.g., Partridge 1981). For instance, in overhead observations of floating flocks of surf scoters *Melanitta perspicillata*, Lukeman et al. (2010) found evidence for strong close-range repulsion and farther-range alignment and attraction among neighbors. Similar to Partridge's study on saithe, scoters tended to align themselves with neighbors to the front and alongside.

Simulation experiments represent a powerful means to scale up individual interaction rules to collective schooling and flocking behaviors (Parrish et al. 2002). Couzin et al. (2002) conducted numerical experiments allowing individuals to make vector changes at 1-s time steps (≤5,000 steps) and incrementally varied reactive distances for alignment Δr_o and attraction Δr_a. Two measures of group trajectory were modeled. Group polarity P_{group}, was the weighted average of individual directional vectors v_i over summed time steps t:

$$P_{group}(t) = \tfrac{1}{N} \left| \Sigma_{i=1}^N v_i(t) \right|.$$

Angular momentum, M_{group} indexed the degree of rotation of individual members' locations c_i around the group's center c_{group}, estimated as the centroid of c_i:

$$M_{group}(t) = \tfrac{1}{N} \left| \Sigma_{i=1}^N r_{ic}(t) \times v_i(t) \right|$$

and

$$r_{ic} = c_i - c_{group}.$$

From the numerical experiments, four classes of aggregation resulted (fig. 2.8). Swarms were characterized by low polarity and angular momentum and emerged when the zone of alignment was compressed relative to the zone of attraction. Swarms showed strong group cohesion through attraction, but high individual variance in direction resulted from inefficient information transfer. Maintaining a large Δr_a and slightly increasing Δr_o resulted in a torus (high M_{group}), but this occurred over a very narrow range of Δr_o, suggesting that it was not stable. Still, this cyclonic feature corresponds to the manner by which some schools of jacks, tunas, and barracudas mill in open water, particularly in association with physical structures (Couzin et al. 2002). Increasing Δr_o farther resulted in more polarized group behavior, initially as a stationary or slightly mobile school and then as a highly mobile school. These polarized behaviors were observed over a fairly wide range of Δr_a levels. Similar emergent classes of aggregation were observed when vectors and velocities were allowed to increase and decrease in a continuous rather than incremental manner across zones of repulsion, orientation, and attraction (Viscido et al. 2004).

Individuals are likely to dynamically adjust their zones of attraction and alignment according to their size, physiological condition, and environmental situation. For instance, in laboratory studies, schools of Atlantic herring and Atlantic cod *Gadus morhua* reduced nearest neighbor distance with increased school velocity (Pitcher and Partridge 1979); individual packing also increased with increased school

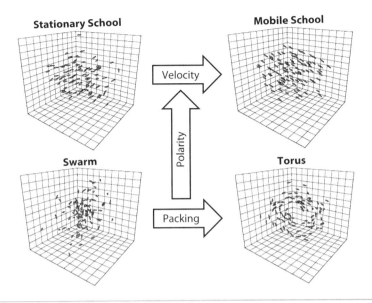

Figure 2.8. Collective schooling states emerging from individual movement propensities associated with nearest-neighbor distances (packing), velocity, and polarity (directional orientation). Schooling graphics reprinted from Couzin et al. 2002, with permission.

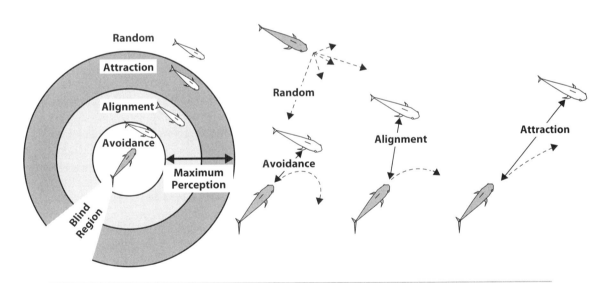

Figure 2.9. Self-organizing movement rules based on zones of avoidance, parallel orientation, attraction, and random unoriented movement. Through adjustments in the regions of repulsion, alignment, and attraction, school properties emerge. Figure concept from Inada 2001 and Couzin et al. 2002.

size for cod and saithe (Partidge et al. 1980). Using sophisticated imaging of individual starling position within flocks of up to 8,000 individuals, Ballerini et al. (2008) observed individuals moving constantly throughout the flock rather than maintaining their position.

Simulation modeling uncovered an intriguing emergent property: an unexpected sequence of aggregation states. In scenarios where Δr_0 was sequentially increased from an initial swarm state, a stable torus developed over an intermediate range of Δr_0, as expected (Couzin et al. 2002; fig. 2.10). On the other hand, when Δr_0 was decremented from large to small, the aggregation transitioned directly from a polarized school to a swarm with no intermediate torus structure. The emergent dynamic is one of path dependency in which the historical trajectory of the system (increasing or decreasing Δr_0) influenced its structure (a.k.a. hysteresis). This analysis also showed that transitions occur as thresholds as aggregations move from one dynamically stable state to another (Couzin et al. 2002; Viscido et al. 2004; chap. 7, Resilience Theory).

Information Systems

Schools and flocks persist as systems of information exchange. The position of individuals within schools and flocks influences social transmission of information about the structure of the aggregation and also the surrounding environment. Furthermore, information from many neighbors—rather than, say, only the closest individual—influences an individual's movements. In Partridge's (1981) observations on saithe schools, two subgroups emerged, classified according to individual velocities, but each group contained individuals throughout the school rather than groups of nearest neighbors. As already highlighted, experiments have shown that the transmission of school and flock information can supersede individual propensities. Individual saithe conditioned by a flashing light to race around a circular raceway did not similarly respond when placed in schools of individuals that had not been so conditioned (reported in Partridge 1980). Similarly, pigeons conditioned to home showed increased homing accuracy when released as flocks (see Consensus Decisions below).

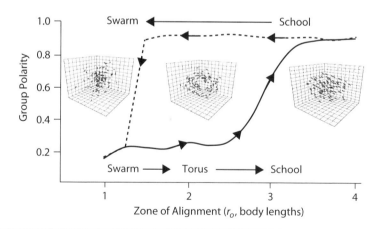

Figure 2.10. Path dependency for the influence of a movement rule (zone of alignment) on aggregation states (swarm, torus, and school; Couzin et al. 2002). Path dependency is shown by changed states that depend on whether the zone of alignment is sequentially increased or decreased. When transitioning from a school to a swarm, no torus forms (dotted line), whereas when transitioning in the opposite direction (solid line), a stable torus occurs. Redrawn from Couzin et al. 2002, with permission.

Individual leaders, those out front, have disproportionate influence on migrating aggregations. Simulation experiments using aggregation rules show that moderate increases in velocity will cause certain individuals to reorder themselves in frontal regions (Couzin et al. 2002). As they accelerate from middle and rear positions, the position of neighbors within their zones of attraction and alignment increase, resulting in increased influence by these individuals on group movements (Inada 2001). The influence of neighbors will also depend upon the overall size of the aggregation. Information can be efficiently transferred among all individuals within very small aggregations, but as group size increases, high rates of information transfer among members can result in conflicting propensities, leading to increased collision rate and loss of polarity. Conversely, decreased numbers of influential neighbors can contribute to increased school fragmentation. Viscido et al. (2005) concluded that increasing the number of influential neighbors will increase school size and density to some threshold level, after which the school structure is destabilized.

A small minority of individuals with a common directional preference can exert strong influence on school/flock polarity and structure (Conradt and Roper 2005). Such individuals may be more experienced—for instance, individuals that have been conditioned through positive reinforcement to undertake certain behaviors or others that have adopted behaviors through social transmission (see Navigation Capacities below). Leaders can represent a particular demographic subset of differing size, maturity state, or gender (Huse et al. 2002). In simulations, Couzin et al. (2005) assigned strong preference for a specific direction to a minority of individuals within aggregations while weighting this preference (by 50%) against aggregation rules similar to those described above. They observed that small minorities of discriminating individuals (<5%) in larger schools ($n \geq 100$) influenced group polarity. When preference was stronger and less influenced by interaction rules, school fragmentation was increasingly likely.

In laboratory experiments using a facsimile of a fish designed to scoot along the floor of a tank under remote magnetic guidance, this robo-fish entrained small numbers of sticklebacks *Gasterosteus aculeatus* to follow it to and from forage patches (Faria et al. 2010; Ward et al. 2012). The propensity for individuals to follow the robo-fish was positively influenced by the number of leaders and negatively related to the number of experimental fish. These responses were strongly nonlinear, indicating threshold responses to social conditions in adopting oriented movements. Guttal and Couzin (2010) showed through simulations that such "quorum" behaviors were dependent on the mix of school members, whether they were strongly oriented (independent) or strongly social (copycats).

What if leaders within aggregations show more than a single directional preference? Couzin et al. (2005) ran numerical simulations and discovered that under conditions of equal preference for two directions, the school or flock will orient toward the mean direction. On the other hand, if even a slight majority of leaders favored a single direction, then positive reinforcement arising from spacing interactions caused the entire aggregation to rapidly adopt that preference. Thus schools can rapidly respond to information gradients without discriminating between leaders. School or flock structure, velocity, and direction emerge through self-organization, the result of a small set of movement interaction rules (Conradt and Roper 2005).

Consensus Decisions

Consider a scenario where all individuals in a flock or school show a degree of directional preference, without obvious leaders. Similar to the simulations described above, Codling et al. (2007) conducted trials in which individuals showed preference for the same navigational target, but they did so with a degree of navigation error. To move beyond simulations, Gökçe and Şahin (2010) designed dozens of small mobile robots programmed to exhibit small random errors around a mean geomagnetic heading. As expected, when instructed to interact (align) with each other, larger flocks of robots showed increased accuracy in their collective heading. This statistical bet-hedging phenomenon has been termed the "many wrongs principle," where "pooling of information from many inaccurate compasses yields a single more accurate compass" (Simons 2004). Schools and flocks are expected to increase navigation accuracy, particularly where navigation targets are small, such as breeding grounds (Conradt and Roper 2005; Mueller et al. 2013). The collective compass depends strongly on the ability of individuals to align to one another and is predicted to perform poorly in turbulent and other types of environments that occlude perception of neighbors. In these instances, independent navigation is favored (Codling et al. 2007). Consensus decision making is effectively studied in homing pigeons that are conditioned to return to their roosts. Homing pigeons released away from their roost showed less variance in the direction of their return migrations when released as small flocks than when released singly (Keeton 1970; Tamm 1980).

Complex Aggregation Rules

Repulsion, attraction, and alignment represent simple interaction rules that can explain substantial diversity in the aggregation dynamics of birds and fish. Still, because individuals cannot be easily observed in situ in their schools and flocks, evidence is circumstantial and mostly derived from modeling studies. In a review of interaction rule models of animal aggregations, Schellinck and White (2011) concluded that central assumptions and predictions were rarely tested. In addition, differing sets of rules can lead to similar aggregation states, suggesting that aggregations were "multiple realizable phenomena." Further, such models are oversimplifications, unlikely to fully capture realism in the spatial structures and behaviors of aggregations. In response, modelers have endeavored to add realism to simple aggregation rules and thereby simulate more complex aggregation and migration behaviors (Schick et al. 2008).

Additional propensities were added to fundamental interaction rules for spawning aggregations of Atlantic herring, including predation (seabed avoidance) and spawning motivation (seabed attraction; Vabø and Skaret 2008). These movement propensities were linked to the progression of gonad development during the spawning season. The individual-based model (chap. 3, Alignment of Larval Dispersal with Mating Systems) included an additional 15 parameters beyond those describing simple aggregation rules. An interesting vertical aggregation pattern emerged of (1) a pelagic upper layer of reproductively spent herring and a midlevel layer of mature herring, (2) a seabed layer of spawning herring, and (3) narrow columns of newly spent herring moving up into the water column toward the surface. Similar layered structures had been observed in spawning shoals of Atlantic herring (Axelsen et al. 2000).

In an attempt to relate simple aggregation rules to directed spawning migrations, Barbaro et al. (2009) simulated the dispersal of capelin *Mallotus villosus*, a small pelagic species (<25 cm). Starting with large numbers of adults (>10^5) in prespawning locations in the northwest shelf waters of Iceland, they simulated dispersal as individuals (1) obeyed basic aggregation rules, (2) entrained into a strong clockwise cyclonic current field around Iceland, and (3) exhibited movement affinities associated with a preferred temperature range. Simulations showed that the spatial fates of aggregations were approximately similar to those of spawning capelin. But the study's conclusion that homing and spawning migrations were simply explained by aggregation rules must be balanced against the dominant role the cyclonic current played in explaining overall distribution patterns. Small capelin would expend considerable energy resisting this current (~20 cm s^{-1}). Still, aggregation rules and temperature apparently served as aggregation controls in maintaining boundaries of high-concentration patches of dispersing individuals.

Perhaps more faithful to the notion of self-organizing phenomena stemming exclusively from local interactions, Hildenbrandt et al. (2010) were remarkably successful in replicating the complex pulsating behaviors of starling flocks through only slight modification of interaction rules. Owing to aerodynamics, turning starlings lost velocity. When neighboring starlings similarly turned, variance in speed was dampened among individuals throughout the flock. This in turn brought alongside individuals in closer proximity to one another, and repulsion rules caused the entire flock to widen as it turned, creating pulsating patterns in density and width as the flock undertook turns. Simulated three-dimensional videos of turning flocks were quite similar to the high-speed amoeboid movements observed in nature (see www.oxfordjournals.org and supplementary movies for Hildenbrandt et al. 2010).

Navigation Capacities

Finding Their Way: Orientation

The repeated use of breeding and wintering habitats by birds and fishes involves homing migrations: a behavior central to complex life cycles (chap. 5, Obstinate Nature: Imprinting), one that drives much of the research on orientation mechanisms. Directed movements require an animal to plot a direction against its current location (navigation) and then while on the fly maintain that direction (orientation). Compass orientation allows movement along certain bearings and requires a directional sense. For fishes and birds, orientation is confounded by winds and currents, contributing to diverse compensating behaviors and apparent navigational errors (Newton 2008; J. W. Chapman et al. 2011). Celestial reference points (sun, skylight polarization, moon, and stars) and the earth's magnetic field provide global compasses, which fishes and birds can use on a regional basis to orient and navigate. Animals use day length and sun inclination to ascertain global position, but in many instances (e.g., moving at night or in deeper waters), other global compasses are likely more important (Åkesson and Hedenström 2007). Local piloting involves visual landmarks and sensory cues. For instance,

Pacific salmon are known to rely on odor in their final approaches to spawning streams. Precise homing must rely on bicoordinate navigation and implies a map sense, which is inherited or acquired later in life through learning (Dittman and Quinn 1996; Berthold 2001; Ramenofsky and Wingfield 2007). As highlighted in the previous section, orientation and navigation can be collectively reinforced through social transmission of information.

Orientation mechanisms are better understood in birds than in fishes. For numerous and broadly representative bird species (songbirds to cranes), laboratory and field experiments support (1) multiple orientation compasses and navigational abilities, (2) inheritance of homing behaviors, and (3) learned migration routes. Studies on marine fish orientation are dominated by those on salmon, trout, and charr species (Salmonidae). In his book *Orientation and Navigation in Vertebrates*, Rozhok (2008) references fish 24 times, 21 for salmons, trouts, and charrs. Birds and turtles were featured in the book as model systems in navigation; fishes were not. Thus the dilemma in comparing bird and fish orientation is weighting a vibrant and expanding avian subdiscipline against a relatively narrow literature on fish orientation. This was not always the case—decades ago, the work of Hasler and colleagues on the role of olfaction in homing salmon was transformative (Hasler and Scholz 1983; chap. 5, Obstinate Nature: Imprinting). Recent research on bird orientation has opened similar avenues for studying marine fishes, particularly those using the expanding capabilities of telemetry. In this section, magnetoreception is introduced as a global compass because solid experimental evidence exists for this capacity in both birds and fishes. Then the roles of inheritance, experience, and learning in orientation are examined. Finally, we look at what happens when birds and fish make navigational mistakes.

The Elusive Sixth Sense: Magnetoreception

Magnetic field lines radiate from earth's north and south poles. At highest latitudes they radiate orthogonal (perpendicular) to earth's surface, but with decreasing latitude these field lines incline, becoming parallel to the earth's surface middistance between the magnetic poles near the equator. The intensity of the geomagnetic field also varies proportionately with latitude. The angle and intensities of field lines thus result in a latitudinal compass, omnipresent in both marine and aerial ecosystems and quite stable (millennia) in comparison to celestial compasses, which continuously change (Rozhok 2008). For this reason, the geomagnetic compass represents a primary compass, more likely than others to be inherited rather than learned (Wiltschko and Wiltschko 1996; Berthold 2001). Over geological epochs (0.1–10 my), magnetic poles undergo reversals, which Lohmann et al. (2008a) speculated could lead to catastrophic changes in animal distributions. Against this view is the fact that such reversals occur gradually over thousands of years (Rozhok 2008; Wu and Dickman 2012). More importantly, animals use a geomagnetic compass together with other georeference systems in complementary, sequential, and redundant ways (Alerstam 2006). Regional geomagnetic anomalies caused by local continental or seabed geology yield substantially less intensity than the global dipole field but are within the range of animal sensory abilities and in some instances are likely to play a role in navigation (Klimley 1993; Tsukamoto et al. 2003; Lohmann et al. 2008b).

The ability to perceive and respond to magnetic fields is common to vertebrates and other animals, but to many of us this perception seems "fantastic," beyond our human experience (Berthold 2001; Åkesson and Hedenström 2007). Current models of magnetoreception can be grossly simplified as a vision-like sense and a mechanoreception-like sense. In the former, magnetic waves convert the electronic spin state within photopigment molecules (e.g., rhodopsin) located in the retina (Rozhok 2008). In bird models, gravitational field lines are focused through the pupil to specific portions of the retina, which signal their angle of inclination during horizontal flight (Rozhok 2008). The pineal gland also contains photoreceptors, and

evidence supports its role in magnetoreception in birds and amphibians. In the other model, intercellular magnetite is attached to cell membrane molecular plugs through microstrands. As the magnetite crystals oscillate and align themselves to the ambient magnetic field, the strands pull and relax their hold on these plugs, opening ion channels and causing local polarity changes across the cell's membrane (Rozhok 2008; Eder et al. 2012). Impulses of direction and amplitude of the resultant shear are transmitted by associated neurons. Concentrations of magnetite are found in the nasal region of the neurocranium in association with the acoustic-lateral line system (Walker et al. 1997), and most recently in the otolith (hearing) organs of pigeons (Wu and Dickman 2012).

For birds and fishes, a geomagnetic compass was first demonstrated in experimental arenas where the magnetic field was deliberately altered. German avian scientists uncovered a "restless" behavior immediately preceding seasonal migrations, termed Zugunruhe (Berthold 2001). In birds whose flight was prevented, Zugunruhe was manifest by repeated perching or hopping in the direction of the intended migration. By placing experimental birds in specially designed chambers, investigators measured the preferred direction of activity, recorded as use of perches or talon strikes arrayed around the periphery of the chamber (Berthold 2001). Electrical coils arranged about an experimental arena caused robins *Erithacus rubecula*, a nocturnal migrating species, to alter their directional preference in response to deflections in magnetic north (Wiltschko and Wiltschko 1972). Robins were also responsive to differing levels of magnetic field intensities. Evidence for a geomagnetic compass now exists for >20 avian species (Wiltschko and Wiltschko 1996; Berthold 2001). Recently, Wu and Dickman (2012) showed specific transduction of signals from the otolith organ of a pigeon associated with changed geomagnetic intensity and direction, a noteworthy advance in demonstrating the capacity of birds to map Cartesian coordinates on the basis of geomagnetic reception.

Geomagnetic preference studies for marine

fishes were initially conducted on juvenile ("fry") sockeye salmon *Oncorhynchus nerka*, which exhibit strongly oriented movements following their emergence from gravel nests toward nurseries that vary by breeding population (Quinn 1980). Two populations of fry use different lakes as nurseries, and their early movements differ in direction (magnetic north-northwest versus south-southeast). In a choice arena with four axial arms radiating from a central holding tank, Quinn (1980) observed that when celestial cues were obscured, fry from the two populations oriented as expected. Alteration of the magnetic field by 90° resulted in a corresponding 90° rotation in directional preference by the sockeye fry in one population (Cedar Lake) but not the other (Cultus Lake; fig. 2.11). In that population, alteration of preferred direction was slight and only occurred in the absence celestial cues. That there were population-level differences to geo-magnetic and celestial cues supported (1) inheritance of a magnetic compass in sockeye salmon and (2) simultaneous use of magnetic and celestial compasses (Quinn 1980). This pioneering study also produced more nuanced results that have been largely ignored: (1) modified orientations often deviated substantially from the expected 90°, suggesting partial or overcompensation for the changed field, and (2) results differed depending on whether individual or groups of fry were considered as the experimental unit. Still, this single study is highly cited as a primary example for the role of magnetic orientation in the oceanic migrations of salmon and other fishes. Conditioning experiments have shown response to magnetic cues by salmons, eels *Anguilla* spp., yellowfin tuna *Thunnus albacares*, and sharks (Walker et al. 1984; Formicki et al. 2004; Nishi et al. 2004; Meyer et al. 2005).

Relative to the landscapes over which terrestrial birds fly and navigate, oceans are comparatively devoid of features that a pelagic (oceanic) fish or bird might use in developing navigational maps. Lohmann et al. (2008a) proposed that such animals possess cognitive maps that rely on geomagnetic fields. A map sense would require a separate reference system specific to longitude to complement magnetic inclinations and intensities that reference latitude. Geomagnetic fields intersect continental shelves, shorelines, and islands in a manner that provides specific sets of coordinates that could be used for navigation. Consider a fish species that spawns in nearshore or estuarine waters. Resulting juveniles might imprint to the local geomagnetic signal and then depart for deeper and more distant waters. During homing migrations, the returning adult would initially detect shallow shelf waters and then, sensing the local magnetic field, reorient north or south subject to the magnetic field it imprinted upon. Such navigation depends on an east-west migration until landfall (or "shelf-fall") is made and is most efficient when continents and islands are perpendicular to latitudinal axes. In support of this view, sockeye salmon returning to spawn in the Fraser River circumnavigated Vancouver Island through northern or southern entrances depending on which entrance deviated less in magnetic intensity in comparison to the Fraser River (Putman et al. 2013; chap. 5, fig. 5.1). More precise landfalls likely entail other compass systems. For instance, maturing chum salmon *Oncorhynchus keta* migrating westward from the Bering Sea to the coastal waters of Japan exhibited cyclical surface ascents during the daytime, which Friedland et al. (2001) attributed to solar navigation. Although speculative, signals related to geomagnetic intensity and inclination could permit bicoordinate navigation (Quinn 2005; Lohmann et al. 2008a; Rozhok 2008; Putman et al. 2013).

Initial efforts to artificially manipulate geomagnetism in the field have thus far failed to confirm its role in long-distance ocean migrations. Small electromagnetic coils were attached to four wild captured chum salmon during homeward bound migrations to coastal Japan (Yano et al. 1997) and their position tracked over a 16-h period. Reversing the polarity at 11-min intervals had no detectable influence on their behavior. Mobile magnets designed to interfere with magnetoreception were attached to nine wandering albatross that

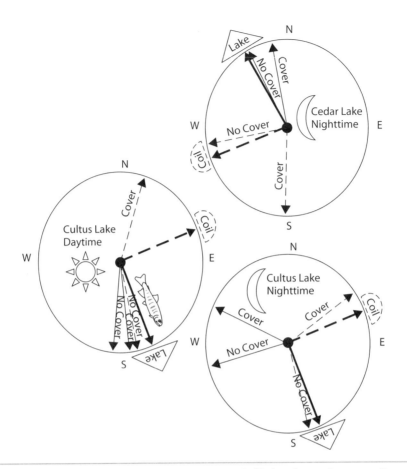

Figure 2.11. Mean vectors of orientation by young juvenile (fry) sockeye salmon as influenced by experimental manipulations of geomagnetic field and access to celestial cues. Expected orientations in choice chambers for two populations were toward lake entrances (bold continuous arrows) or—under manipulated magnetic fields—toward a heading that was shifted 90° counterclockwise (bold dashed arrows). Corresponding experimental results are shown as thinner lines with (No cover) or without (Cover) access to celestial cues. Moon crescent indicates nighttime experiments. Control data do not include treatments where the coil was present but turned off because Quinn (1980) reported that these results were biased. Graphs from Quinn's (1980) data on experiment-wise results.

were subsequently tracked during their pelagic foraging migrations for 2–12 d (Bonadonna et al. 2005). No detectable difference was found between movement pathways (range 730–8,260 km) of experimental and control birds. Both studies represent important initial experimental forays but were limited by low sample sizes, the manner in which magnetic fields were altered, and other design elements. Further trial

and error will be necessary to yield more definitive results.

Navigation Maps: Inherited, Imprinted, or Learned?

The precise nature of homing by migratory birds and fishes is well known from >100 y of tagging, but how navigation maps are acquired cannot be ascribed to a single mechanism: inheritance, imprinting, and learning all play important roles depending on taxa and population status. "Birds of two worlds" (Greenberg and Marra 2005) are those birds that migrate seasonally from temperate grounds to tropical overwintering regions. The flyways, stopover points, and breeding and wintering habitats of such birds are remarkably consistent across years and generations. Many fishes are also of two worlds, seasonally migrating from higher-latitude breeding habitats to lower-latitude feeding and wintering areas. Alternatively, many seabirds and marine fishes seasonally migrate between inshore and offshore habitats (Sackett et al. 2007; Newton 2008). How birds and fish find their way to seasonal habitats is a central question in movement ecology, where navigation itself can be tested against null hypotheses of random walk and other simple movement rule behaviors (see Lévy Flights above). Against such null models, there is considerable evidence for navigation abilities acquired through inheritance, imprinting (early experience and subsequent memory), and learning. Here we examine how navigation maps are first acquired by juveniles, reserving the topic of natal homing for chapter 5.

Direct comparison of navigation capabilities of birds and fishes is confounded by seminal differences in their life cycle. The juvenile bird is only moderately smaller (less than tenfold; Rahn et al. 1975) than the adult and has all the same basic equipment for feeding, flight, and navigation. Indeed, in some instances the fledgling is larger than its parents. The larval or juvenile fish, on the other hand, is orders of magnitude smaller than its parents (chap. 3, Fecundity and Adult Size). Further, indirect development (i.e., separate metamorphic phases) in most fishes (elasmobranchs are a noteworthy exception) means that larvae and juveniles do not possess the same physiological and behavioral wherewithal to undertake the seasonal migrations of conspecific adults. Further, comingling by small larvae and juveniles with adults would cause cannibalism in marine ecosystems where predation is strongly size de-

pendent (chap. 4, Size Spectrum Theory). Still, as we will learn in chapter 3, the small larva depends on an impressive battery of sensory and swimming capabilities in its dispersal to nursery habitats. But thereafter little is known as to how juveniles leave nurseries and adopt the migrations of their parents. In contrast, quite a bit is known about how juvenile birds initially adopt the migration circuits of previous generations.

For many birds, differing navigation abilities exist between naïve first-time migrants and experienced adults. A dominant theory is that juveniles inherit an orientation compass and can sometimes change compass bearings during their first seasonal migration, but that a navigation map must be acquired either by imprinting (early experience) or learning (Berthold 1996; Newton 2008). White storks *Ciconia ciconia* provide an excellent example. Among the largest flying birds (2–4 kg), they rely on soaring flight during their migration from northern Europe to overwintering grounds in North Africa. The gregarious storks migrate as flocks. In telemetry experiments (Chernetsov et al. 2004), juveniles were experimentally isolated from their parents in the Baltic region of Russia and released as small groups (<12 each) at sites 2,000 and 3,000 km east of the breeding area (Volga River region and western Siberia). Control juveniles were allowed to flock with adults from the same breeding ground and exhibited little variance in subsequent flight routes as they circumvented the Mediterranean Sea through the Israeli flyway toward traditional wintering grounds in North Africa. Displaced juveniles showed much more scattered routes, with an overall southerly flight that took them to the Middle East, thousands of kilometers distant from their traditional wintering ground.

A small number of field experiments on reef fish show that juveniles can learn spatial behaviors from experienced conspecifics (Dodson 1988; Brown and Laland 2003). In a classic experiment, naïve juvenile French grunts *Haemulon flavolineatum* learned and adopted diel foraging migrations if allowed to comingle with experienced adults, but they did not show these migrations in the absence of adults (Helfman and Schultz 1984). Spatial learning in marine fishes is commonly ascribed to schooling fishes where rules of aggregation favor a collective compass (see Schooling and Flocking above). Cushing (1962a), who later became a strong advocate of imprinted navigational abilities (Cushing 1995), initially proposed that a common pool of juvenile North Sea herring adopted the migration circuits of separate populations of adults. Adopting a particular migration circuit was influenced by where and when a juvenile attained a critical length that would favor its comingling with adults (Burd 1962; Cushing 1962a). This so-called adopted migration theory (McQuinn 1997a) has been exceptionally difficult to prove. Still, circumstantial support has been mustered (Corten 2001; Huse et al. 2002; Petigas et al. 2007, 2010). For Gulf of St. Lawrence herring, McQuinn (1997b) found evidence that spring-spawned juvenile herring adopted migrations of a dominant autumn adult spawning aggregation, each identified through otolith microstructure. Social transmission of migration circuits for schooling fishes has strong conceptual appeal and remains a dominant hypothesis in the acquisition of navigational capacities in marine fishes despite the lack of direct evidence. We will return to adopted migration as an explanation for natal homing and life cycle closure in chapter 5.

Acquisition of navigation capacities can occur absent social transmission. In contrast to the gregarious white stork, small white-crowned sparrows *Zonotrichia leucophrys* undertake solo winter migrations yet also acquire navigation abilities during their first year. Small numbers of adults and juveniles (15 each) were transported across the continental United States in a closed airplane compartment from the northwestern United States (breeding ground) to the eastern United States (New Jersey), outfitted with small radio beacons, released individually, and tracked by small aircraft for 3–10 d (Thorup et al. 2007). Displaced adults migrated in a southwest direction, which would have steered them toward their winter-

ing habitats located in the southwestern United States and northwestern Mexico. Juveniles migrated in a southerly direction, however, a course that would have steered them toward the wintering ground prior to displacement. Rather than possessing a learned navigational map, Thorup et al. (2007) suggested that juveniles possessed only an orientation compass. An initial migration and winter occupancy were needed for the juvenile to acquire a navigational map. Remarkable and inexplicable is how the adult sparrows initially developed a continental-wide map and sensed their distant displacement. Recall also the pelagic Laysan albatross, which possessed a capacity to navigate from distant sites some occurring outside its normal range of movement (see Tagging: Connecting the Dots above).

In a displacement study, Edrén and Gruber (2005) found that compass orientation in juvenile sharks apparently took precedence over other navigational capabilities. Elasmobranchs (sharks, rays, skates) are similar to birds in that juveniles are born fully developed, principally differing from adults in their size, immaturity, and inexperience. Still, like other fishes, juvenile elasmobranchs often occupy nursery areas that provide early trophic resources and a refuge from predation. Displaced juvenile lemon sharks *Negaprion brevirostrus* showed strong homing abilities within mangrove habitats in subtropical waters. Surprisingly, regardless of direction of displacement (4–16 km north, south, and west), juvenile sharks all initially exhibited the same eastward compass bearing. This behavior often removed the sharks still farther from their original mangrove territories. Within hours, however, the displaced sharks changed their orientation again, and the majority relocated to their home range. Edrén and Gruber (2005) suggested that this innate compass was similar to that observed in homing pigeons, which can impose orientation controls that are seemingly at odds with the immediate goal of homing. Compass systems (celestial and geomagnetic) are also implicated by the spike dives that occur during long-distance seasonal migrations (>10,000 km) by juvenile southern bluefin tuna *Thunnus maccoyii* (Willis et al. 2009).

Imprinting, acquiring a navigational map through early experience, is given circumstantial support by the ability of migrating juvenile fishes to home seasonally to specific habitats. Well known for freshwater and reef fishes (Odling-Smee and Braithwaite 2003), telemetry studies have increasingly shown repeated seasonal habitat use by juvenile marine fishes that undertake extensive coastal migrations. Juvenile striped bass *Morone saxatilis* in the Hudson River Estuary (New York, USA) persist in fairly restricted regions of the estuary (<5 km range) during summer and fall; in winter they depart for coastal waters >150 km distant and then return to those same feeding territories the following summer, ignoring other potential habitats <10 km distant (Wingate and Secor 2007). Small striped bass captured and released in coastal shelf waters exhibit much the same homing behavior to localized regions (Clark 1968; Mather et al. 2009). Another example of juvenile homing is the interannual use of the same small estuary by summer flounder *Paralichthys dentatus* following seasonal offshore winter movement (Sackett et al. 2007). Within seasons, displaced juvenile flatfish showed strong homing to narrow beach regions (<100 m) despite the potential for more distant longshore dispersal (Burrows et al. 2004).

Although numerous avian studies support an acquired navigational map through the combination of an innate compass and experience (imprinting), intriguing examples exist for fully inherited or acquired navigation maps (Berthold 2001). The Eurasian cuckoo *Cuculus canorus* is a brood parasite, relying on the care of often much smaller songbird parents. Precocious juvenile cuckoos undertake migrations to Central Africa independent of the migration behaviors of their foster parents (Seel 1977). In contrast, fully adopted migrations were observed in the goose *Anser erythropus*. In a conservation effort, goslings of this species were introduced into the nests of the more common Barnacle goose *Branta leucopsis* and were reared by that species. The goslings subsequently adopted the autumn migrations of the parent species (Alerstam 1991). Similarly, in deliberate cross-foster experiments, nestling black-backed gulls *Larus fuscus*, a resident species in Britain, adopted the longer migrations of the manipulated parental species herring gull *L. argentatus*. Also relevant are efforts by US federal scientists to artificially imprint homing behaviors on juvenile whooping cranes *Grus americana* by causing them to follow ultralight aircraft during their initial autumn migration (Olsen 2001). As whooping cranes recovered, the number of older experienced adults increased, which in turn improved flock navigation performance absent the ultralight aircraft (Mueller et al. 2013).

Circumstantial evidence for an inherited navigation map for Pacific salmon relates to the precise nature of juvenile emigration from coastal watersheds to marine habitats. Directional preference experiments similar to those on sockeye fry (see fig. 2.11) provide evidence that larger emigrating juveniles exhibit an inherited compass (Quinn and Brannon 1982). After they depart nursery lake systems, juveniles must undertake a complex temporal sequence of navigational turns to gain access to major tributaries that lead to the ocean (Burgner 1991). Further, the downstream migration coincides with a sequence of physiological changes that transform the stream juvenile (parr) into an oceanic juvenile (smolt). Indeed, the precise itinerary of migration and physiological change argues for inheritance of an entire migration syndrome (developmental program) for Pacific and other salmon species (Burgner 1991; Dingle 1996; Quinn 2005; chap. 5, Homing on a Fixed Itinerary: Fraser River Sockeye Salmon). Once in coastal ocean waters, juveniles likely orient again by innate compass and become entrained into major current systems that in turn lead them to subadult migration circuits (Pearcy 1992; Quinn 2005; Hedger et al. 2008).

Losing Their Way

For highly migratory fishes and birds, it is difficult to conceive how juveniles learn and memorize en route landmarks and gradients that allow them to develop a navigational map. One way to understand such adaptations is to consider homing behaviors when they are imprecise, circuitous, or otherwise off the mark. Here we examine what happens when birds and fish make directional "errors" and undershoot or overshoot their home journeys. Straying is an important type of navigational error, reserved for chapter 5.

As already noted, anomalous occurrences of individuals outside their traditional areas have received far greater attention among avian ecologists than among fish ecologists. Observatories, bird lists, and banding studies all emphasize that so-called vagrants (irregular visitors) do in fact occur at moderate frequencies (often >10%; Newton 2008). The causes of vagrancy include weather effects (storms, winds), density effects, and human introductions, but also navigation errors, such as overshoots, mirror-image migration, and reverse-direction migration (Newton 2008). These errors typically attend juvenile migrations, often in an east-west direction, consistent with the limitations of geomagnetic navigation (see above). Wind and other environmental and social influences can cause naïve juveniles to misjudge travel distances and overshoot wintering or breeding grounds (Newton 2008). Interestingly, favorable flight conditions can also cause overshoots, when reduced travel costs cause juveniles to mistime when to disembark.

In their homing migrations to breeding habitats in the Columbia River, Chinook salmon frequently overshoot, missing the entrance of their natal stream and in rare instances swimming hundreds of kilometers farther upstream against strong currents before falling back to reorient. In a study of >5,000 radio-tagged returning adults, Keefer et al. (2008) observed moderate instances of overshoot (10%) and temporary occupancy of nonnatal tributaries (15%). Overshoots were apparently exacerbated when natal tributaries were adjacent to dams and impoundments. Still, these behaviors were self-correcting, as permanent straying was estimated at 2.5%. Adjustments occurred through a "guess-and-go" piloting behavior: first sense

for navigation cues and then reorient. Telemetry indicates that striped bass adults return to their natal estuaries along a coastal "flyway" and repeatedly nose into nonnatal estuaries, consistent with undershooting and self-correcting guess-and-go behaviors (Grothues et al. 2009). Self-correcting behaviors are also observed during return migrations of Pacific salmon along British Columbia where islands, bays, and complex bathymetry obscure the mouths of major spawning tributaries. Pascual and Quinn (1991) proposed the "pinball" hypothesis, where impeded salmon reverse course 180°, swimming away from obstacles and then reorienting at a moderately different directional angle, take "another swim at it."

Inherent variability in initial orientation preferences is expected to cause greater scatter in pathways and habitat distributions in juveniles than in adults, but a symmetrical error highlights the difficulties of east-west orientation for juvenile birds (Newton 2008). In reverse migrations, the juvenile chooses the correct north-south bearing but incorrectly reverses the east-west orientation. This type of behavior is observed in several songbirds. Tests in directional choice chambers for blackpoll warblers *Dendroica striata* indicated that navigational errors were innate: both correct and reverse orientations were observed absent environmental cues (De Sante 1983, as referenced in Newton 2008). European songbirds that migrate in a principally east-west direction between breeding and wintering areas exhibit reverse migrations. For instance, the Pallas's warbler *Phylloscopus proregulus* breeds in Siberia and typically migrates to Southeast Asia to overwinter. Reversing that fall migration 180° west yielded a "vagrancy shadow" in the British Islands, where this species is irregularly observed during autumn months (Rabøl 1969). Long east-west gradients across Eurasia and North America could contribute to increased incidence of reverse migrants for those continents.

The incidence of reverse migrations in marine fishes is purely speculative, but a place to look would be coastal species that have pan-oceanic distributions and occur along eastern and western ocean margins. Typically, distributions of the same species along two or more ocean margins are viewed from the lens of range expansion/contraction during epochs of glaciation and other environmental change (e.g., Grant et al. 2006), but navigational errors might also play a role in explaining species ranges. An intriguing discovery was the colonization of North American Atlantic sturgeon *Acipenser oxyrinchus* hundreds of years ago into northern European seas, where it was previously recognized as a separate species, the Baltic sturgeon *Acipenser sturio*. Ludwig et al. (2002, 2008) proposed a relatively rapid pan-oceanic displacement from Canadian coastal waters to the North and Baltic Seas, consistent with a propensity for navigational errors along an east-west trajectory. Regrettably, these North American sturgeon vagrants were driven to extinction by overharvest and habitat loss several decades ago.

In navigation errors and modes of self-correction we see evidence of complex systems in which interactions and redundancies compensate for variable flight conditions (i.e., the influence of wind and currents; J. W. Chapman et al. 2011) and more rigid elements of navigation such as innate compasses. As Groot (1982) observes, "The redundant nature of orientation systems and the wide range of interactions between components during evolution have resulted in complex spectra of direction finding in Pacific salmon."

Summary

In this chapter, comparisons were made between fishes and birds in fundamental aspects of locomotion, analysis of movement pathways, models of movement and aggregation, and navigation. Birds and marine fishes show strong convergence in all these aspects, perhaps more so than other paired comparisons between vertebrate classes. Moving through fluids, albeit of divergent densities, birds and fish exhibit similar vertebrate design elements, including streamlined bodies, oscillating propulsive limbs, and fin/wing aspect ratios associated with trade-offs in velocity versus maneuverability. Controlling buoyancy and lift and exploiting local flows in their respective fluid media (e.g., winds, thermals, tides, and other ocean currents), fishes and birds are capable of moving hundreds to thousands of kilometers. For small larval fishes, viscosity in higher-density seawater imposes strong limits to swimming. In contrast, in low-density air, lift requirements curtail flapping flight performance in the largest birds and indeed grounds those birds >20 kg. Fish can fly (glide) and birds can swim, but these are special adaptations that lead to nonlinear trade-offs. The flying fish shows decreased gliding performance at small and large sizes. For the diving guillemot the exigencies of flight result in suboptimal diving appendages in comparison to diving animals that do not fly.

Advances in the analysis of movement and migration and related discoveries during the past several decades are strong justification for this book: this is an exciting time to be a student of animal migration. Approaches classified as distribution snapshots, tagging, telemetry, and life cycle tracers have uncovered previously hidden movements and life cycles. Important developments include (1) increased spatial and temporal scales for detecting aggregations through integrated radar and sonar platforms; (2) increased accessibility of long-term animal survey and tagging databases; and (3) new tracer and telemetry technologies that permit individual animals to be tracked over their entire geographic range. Thus the previously obscured pan-global movements of snooty shearwater are now known to connect subpolar feeding areas in wide pan-Pacific figure-eight flying routes. South America's most important fishery resource, blue whiting, is now known to be partitioned by breeding populations on either side of Drake Passage. These and other discoveries came through technological breakthroughs. In the future, there will be important gains in cross-technology studies that compensate for biases and exploit opportunities specific to each approach and the data it generates. Flexible statistical approaches (maximum likelihood and Bayesian estimation procedures as well as state-space models) will see greater application, in keeping with the analytical and conceptual challenges of the ever-increasing data streams on animal distributions and movements.

We also found commonality in how birds and fish follow behavioral rules related to (1) the propensity by individuals to match search patterns to scales of environmental resources (e.g., random walks and Lévy flights) when in isolation or (2) how individuals aggregate according to zones of repulsion, alignment, and attraction. In Lévy flights, individual movement pathways were summed to show a pattern that reflected back to the individual—the pursuit of resources though frequent short movements interspersed with infrequent longer-distance forays. In models of individual interaction rules, collective schooling and flocking behaviors emerged from interactions rather than individual agency. Feedbacks between zones of avoidance, alignment, and attraction amplify small interactions between neighbors into large collective changes, including stable and dynamic aggregation states and path dependency between those states. Arguably, most evidence for these self-organizing behaviors by schools and flocks comes from a rich modeling literature, which has been inadequately matched by field and laboratory testing. On the other hand, modest modifications to simple aggregation rules have emulated with remarkable precision some of the most complex and dynamic behaviors among animal aggregations (i.e., starling flocks), indicating that this literature deserves greater attention than it has received in past treatments on bird and fish migration.

Birds and fishes exhibit homing behaviors that require them to navigate (plot its direction against its current location) and orient (maintain a given direction). Dominant sensory cues that guide orientation are celestial and geomagnetic. Because stable gradients in geomagnetic field inclinations persist regardless of environmental conditions, they are deemed

a primary compass. Birds are model systems in vertebrate orientation studies, far more so than marine fishes, but limited experimental evidence suggests similar mechanisms, including (1) multiple orientation compasses and navigation abilities within taxa, (2) inheritance of homing behaviors, and (3) learned migration routes. Among birds that seasonally migrate between high and low latitudes, juveniles typically migrate according to an inherited compass, but in some instances this compass is modified or supplanted by socially transmitted behaviors. Most marine fishes start life as small larvae (see chap. 3), limiting their comparison to birds. How juvenile fishes adopt adult migration circuits is critically important to models of population structure that underlay fisheries assessment but remains largely unknown. Navigational errors in birds support the view that movements across latitudes are more precise than those along an east-west direction, consistent with a primary geomagnetic compass. Vagrants (irregular visitors) associated with misdirection or overshoots are moderately common in birds, well documented by direct observations. They are similarly likely to be common in marine fishes, despite a limited observation basis owing in part to decades of deliberately ignoring such incidences as anomalies. Self-correcting behaviors often attend navigation errors, including redundant compasses and adaptive sampling behaviors ("guess-and-go" and "pinball" tactics).

Segue

Complex group dynamics result from relatively simple interactions among individuals. Self-organizing systems are those where "simple repeated interactions between individuals produce complex adaptive systems" at the group level (Sumpter 2006). The term self-organizing is not completely accurate, as such repeated interactions can also lead to seemingly maladaptive behaviors ("self-disorganizing behaviors"; chap. 4, Schooling through Food Webs). In his review of collective animal behaviors, Sumpter (2006) lays out differing agencies that stem from collective behaviors that are supplemented in this book (chap. 1, table 1.2): (1) individual integrity, (2) positive and negative feedback, (3) response thresholds, (4) leadership and consensus, (5) redundancy, and (6) synchronization, to which are added (7) carryover and path dependency as well as (8) ecological overhead. This chapter has highlighted how schools operate as self-organizing systems as a result of the interactions of independent fish (integrity) operating largely in redundant ways, contributing to rapid synchronous behaviors owing to efficient information transfer. Schools are not fully redundant, with certain individuals exhibiting stronger influence. This leadership can cause conformational changes in an aggregation as it moves from swarm to torus to schooling stable states separated by threshold conditions. The recent history of the aggregation will influence whether the intermediate torus state occurs or not (path dependency). Homing and other navigational behaviors by the school will be increasingly precise through the combined navigation capacities of its members through a sampling effect (redundancy), which serves to reduce variance in navigational outcomes. These collective agencies were first presented in chapter 1 (table 1.2), and further examples will be given throughout the remainder of the book.

Next we turn to how marine fishes differ from birds and other taxa in their migration ecology—the remarkable diversity of mating systems and the dominant mode of starting life small as larvae. (Elasmobranchs are a noteworthy exception but will also receive due treatment.)

3   Mating Systems and Larval Dispersal

The spawning fish, on account of their much more definite cravings after certain external factors, are brought together over a much smaller area than that over which the species is ordinarily distributed.

—Schmidt (1909)

Mating Systems

Sex is a principal driver of fish migration. The same can be said for other vertebrates, of course, but a fish must capture its mate's gamete in a systemically diffusive environment that tends to separate sperm from egg. Fertilization requires periodic intimacy that in turn drives seasonal migrations and mating systems, the latter exceeding other vertebrates in number and diversity and mirroring the range of environments into which >25,000 fish species have radiated (Breder and Rosen 1966). Five examples follow.

Five Mating Systems

In dark waters 300 m deep on the Grand Banks off Newfoundland, bottom-dwelling Atlantic cod grunt to each other after a long roundabout migration. An immense aggregation arrives at the shelf break after an initial tack northward against the continental branch of the Labrador Current and then falls back along the slope branch: a wide 600-km loop around the Grand Bank (deYoung and Rose 1993). The springtime spawning aggregation concentrates spawners at 1-m spacing (Rose 1993). Males locate and circle individual females in courtship displays; their grunts signal size and readiness (Hutchings et al. 1999; Windle and Rose 2007; Fudge and Rose 2009). Then, from a 2-km-wide dome of densely packed fish, a column of aligned mating pairs ascends (Rose 1993), followed closely by satellite males (Hutchings et al. 1999; Bekkevold et al. 2002). Breeding cod erupt from the pack like chimneys, pushing as much as half-

way toward the surface, expelling clouds of gametes at the apogee of their ascent (fig. 3.1A; Rose 1993; Lawson and Rose 2000).

Mature Japanese eel *Anguilla japonicus* depart from the seagrass beds of Hamano Lagoon in temperate Japan bound for equatorial Pacific waters, toward a spawning region that lies several hundred kilometers due west of Guam. Silver eels vacate the lagoon after many years of sedentary growth, now transformed to a pelagic migratory adult. In the next 6 months, energy stores will be severely taxed during the >2,500-km journey, culminating in females and males expelling 60% and 20% of their biomass as gametes: a fatal one-way trip (Tsukamoto 2009). The spawning location, known through the collection of embryos and spawning adults (Tsukamoto et al. 2011), is situated in a vast >2,000-m deep seascape. Yet, without previous experience, adults throughout temperate Asia return to where life began. The recent capture of mature Japanese eels in this region (Chow et al. 2009; Tsukamoto et al. 2011), a coup for a generation of dedicated scientists, still only hints at the underlying migration, courtship, and spawning behaviors. On new moons, eels apparently spawn near sea mounts within the North Equatorial Current (Ishikawa et al. 2001; Tsukamoto et al. 2003), shedding gametes in flows that will convey their offspring west and northward toward shelf and freshwater habitats in China, Korea, and Japan. But what path(s) lead adult eels back from Hamano Lagoon and other parts of Asia remains a profound mystery (Tsukamoto 2009).

Big flashy striped bass (40–130 cm total

length) announce their spawning with surface splashing. After migrating hundreds of kilometers from coastal waters into the estuaries and rivers of the Chesapeake Bay, they spawn in tidal freshwater reaches—some of these <100 m across (Mansueti 1958). During spring months, ripe females hold in deeper regions of upper estuaries, making brief daily incursions into freshwater, testing spawning habitat conditions (Hocutt et al. 1990). Cued by rising temperature, spawning waves of thousands of females ovulate and release billions of eggs all at

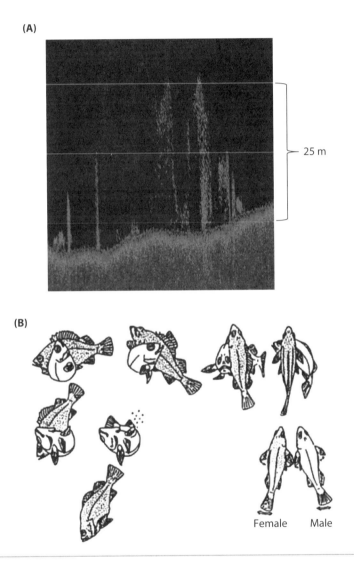

(A)

25 m

(B)

Female   Male

Figure 3.1. Mating systems for (A) Atlantic cod and (B) Japanese sea perch. Spawning columns of Atlantic cod were observed in an echogram off Cape St. Mary's, Newfoundland, Canada (Lawson and Rose 2000). The depth of the echogram ranged 50–55 m and spanned 2.5 km. Observations of mating behaviors of Japanese sea perch were made in shallow waters of Kyushu, Japan (Shinomiya and Ezaki 1991). Following courtship displays, the pair ascends, coils, and copulates. Echogram and sea perch illustration reprinted with permission.

once in waters just up-estuary to the saltfront, which retains eggs and larvae within the freshwater nursery (Rutherford and Houde 1995; Secor and Houde 1995a). Multiple males in the train of a single female churn the water and release prodigious amounts of sperm (Sullivan et al. 1997).

Each year thousands of kilometers of shallow coastal waters in British Columbia change from temperate dark blue-green to tropical aqua, the result of thousands of tons of sperm and eggs shed by Pacific herring *Clupea harengus pallasi*. Ovulated eggs stick to the surfaces of seagrass, macroalgae, and rock in waters shoal of 10 m and are fertilized by a bath of sperm (Hay et al. 2009). Few such overt exhibitions of fish reproduction exist in nature: the herring's spawn can be viewed from a distance of several kilometers. In a given year, 2% of the entire coastline of British Columbia is painted by a wash of fertilized eggs (Hay et al. 2009). Mature herring in aggregations that number thousands to millions signal to each other through pheromone and prespawning milt and then shed all their gametes in single weeklong spawning waves in sheltered waters (Hay 1985; Ware and Tanasichuk 1989). Resulting egg densities can exceed 100,000 m^{-2}. The herring carefully synchronize spawning, moving from distant offshore feeding grounds months in advance and then holding at full maturation for days to weeks until temperature and tides are right. Maturation and ovulation proceed so that eggs are deposited during neap tide to limit their desiccation on subsequent low tides (Hay 1990). This timing also favors dispersal of offspring, many of which will hatch as free embryos during the next neap tide (Hay 1990).

Japanese sea perch *Sebastes inermis* do not migrate to spawn but show restricted movements among structured habitats in shallow coves of Kyushu, Japan. Solitary male rockfish (<20 cm total length) transition from roving amidst structured habitats to defending much smaller territories of 10-70 m^2 during the winter spawning season (Shinomiya and Ezaki 1991). Courtship of individual females entering these territories includes complex circling displays when males release urine, signaling their territory and readiness to spawn (Shinomiya and Ezaki 1991). Females reciprocate with changes in body color. The consenting pair ascends snout to snout. Gaining purchase with paired fins and opercular spines, the male twists and aligns his vent, injecting sperm into the female's gonoduct through modified urogenital papillae (fig. 3.1B; Helvey 1982; Shinoyima and Ezaki 1991). Sperm is kept inactive and stored by the female within her ovaries for several weeks (Wourms 1991; Nichol and Pikitch 1994; Sogard et al. 2008); she will mate with other males and give birth to tens of thousands of small pelagic larvae, the progeny of one to several paternal lines (Hyde et al. 2008).

Mating systems as described above are stories of adaptation to environmental circumstance. But probing further into these and other mating systems, the science of recent decades has uncovered substantial diversity and a more nuanced view of the interplay between physical constraints and two important components of reproductive success: fertilization and larval dispersal. In this chapter, we first dissect what happens inside the gamete clouds left in the wake of broadcast spawners (fish that spawn in the water column)—how fertilization success is influenced by attributes of the sperm and modified by mating behaviors. We then turn to what females contribute in terms of the quality and number of their eggs and how this allotment carries over to the dispersal of eggs, embryos, and larvae. This discussion is followed by examination of the alignment between mating systems and larval dispersal as well as consideration of how selective harvest of spawners can influence this alignment.

Gamete Clouds

Peering into the gamete cloud left in the wake of spawning cod moves us into a different physical realm. At this microscopic scale, viscosity and microturbulence hinder progress by the beating spermatazoa (Riffell and Zimmer 2007; Cosson et al. 2008a,b). In the initial 30 s, the cod spermatazoan will attain its highest velocities (>50 μm s^{-1}; fig 3.2B; Butts et al. 2010). Although a given spermatazoan can persist for more than 60 min (Atlantic cod spermatazoa have unusually long lifetimes in comparison to other marine fishes; Trippel and Morgan 1994; Stockley et al. 1997), velocity quickly erodes after the first several minutes (Tuset et al. 2008). The maximum distance that a spermatazoan might travel ranges from 2 mm for sea bass *Dicentrarchus labrax* to 10 mm for bluefin tuna *Thunnus thynnus* to 14 mm in Atlantic cod (Cosson et al. 2008a,b). Regions where Atlantic cod, striped bass, and Pacific herring spawn experience strong density gradients and turbulence, resulting in velocity shears that will often surpass the propulsive capacities of the spermatazoa (Rothschild and Osborn 1988). Thus fertilization success tends to be negatively associated with flow (Petersen et al. 2001).

The cod spermatazoan orbits an egg that has a circumference 40 times greater than its own length and a volume that exceeds its own by 10^9. Turbulent mixing will quickly place the egg out of range for an individual spermatazoa (Rothschild and Osborn 1988; Levitan 1993). Spermatazoa motility and ranges suggest that sperm limitation should be common in fishes, yet fertilization success by external spawners typically exceeds 85% (Trippel and Neilson 1992; Yund 2000). Clearly, synchronous mating systems that place spermatazoa in close proximity to eggs are critical, but another component is equally crucial to fertilization success: sperm excess.

Copious Sperm

The frequent and conspicuous mating behaviors of bluehead wrasse *Thalassoma bifasciatum*, a small Caribbean reef fish, have provided a unique opportunity to observe gamete release and fertilization success in situ. The wrasse spawn in pairs or in single female–multiple male (polyandrous) aggregations but do so consistently at the same shallow water locations, and in association with upward surges in swimming behaviors that are readily observed

during daytime (Warner 1995; Petersen et al. 2001). The same male will spawn many times in a given day, providing numerous observations. Further, human divers deploying large collapsible bags can fully capture the gametes of each spawning event (Shapiro et al. 1994). Enumeration of wrasse eggs and sperm yields an explanation for high fertilization rates—sperm to egg ratios in gamete clouds were ~10^5:1 (Shapiro et al. 1994). Thus the distances between individual eggs in the newly shed cloud are overcome by sperm density (Levitan 1993, 1996; Butts et al. 2009). In wrasse, cod, and other species reared in captivity, fertilization success is sensitive to spermatazoa concentrations (Trippel and Neilson 1992; Marconato and Shapiro 1996; Petersen et al. 2001; Butts et al. 2009). Increased egg size and densities will also contribute to fertilization success through increased contact surface area and encounter rate (Levitan 1993; Yund 2000).

Gatherings of reproductively active fishes promote fertilization, particularly evident in large predatory reef fishes that suddenly depart reef territories and migrate days or weeks to join others in dense fits of spawning activity. Gamete clouds may become so dense that they attract predation by large planktivores such as whale sharks and manta rays *Manta alfredi* (Heyman et al. 2001; Anderson et al. 2011). The Caribbean Nassau grouper *Epinephelus striatus* represents a prime example. Solitary and territorial throughout most of the year, they migrate tens of kilometers to gathering points (Domeier and Colin 1997). Cued by the full moon during winter months, they spawn much like northern cod, with spawners rushing up into the water column from dense demersal aggregations numbering >1,000 (Colin 1992). But this mating system differs from cod, which remain schooling throughout the year. For Nassau grouper and other large predatory coral fishes, successful foraging depends on solitary territorial behaviors (Domeier and Colin 1997).

Sperm is cheap. Owing to their small size of sperm, individual males can convey huge numbers of sperm at modest metabolic cost.

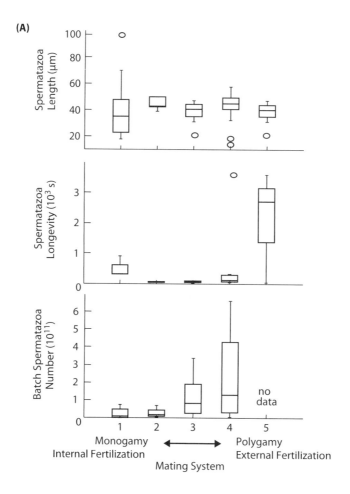

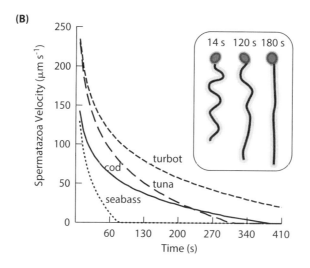

For instance, the mature testes of a 100-kg Atlantic bluefin tuna comprise 1% to 5% of their mass, yet at 4×10^{10} mL^{-1} semen density (Cosson et al. 2008a,b; Heinisch et al. 2008), an individual male can potentially shed $>10^{12}$ spermatazoa over a single season's spawning period. For British Columbia Pacific herring, Hourston and Rosenthal (1976) sampled the water column amidst a single large spawning event (the aggregation occupying >50 ha). At measured sperm density and attached egg densities, the authors estimated 24 sperm encounters per egg.

In comparing demersal (e.g., Pacific herring) to pelagic (e.g., Atlantic cod, eels, groupers) spawning behaviors, we might expect that fertilization success would be higher in the former. In pelagic systems where both spermatazoa and eggs are subject to rapid dilution, higher concentrations of spermatazoa are likely required. Indeed, tropical reef fishes typically release 10^3–10^6 sperm per egg, depending upon mating system and species (Petersen et al. 1992; Marconato and Shapiro 1996). Why, then, spawn pelagic eggs? This topic is reserved until later in this chapter. For now, we accept the premise that fertilization success is typically high (>85%) because males produce sperm in excess. Very high sperm releases, ~10^{11} spermatazoa, occur not only in pelagic spawners but also across a wide range of mating system types (fig. 3.2A), including demersal spawning fishes such as salmons and sturgeons (Stockley et al. 1997), indicating that sperm occurs in excess for reasons other than providing sufficient densities to achieve fertilization.

Sperm Competition

"Broadcast," "promiscuous," or "free" spawning are misleading descriptions because they imply that gamete clouds contain a homogenous representation of the members in a given mating system (Taborsky 1998). Delving further into the gamete cloud, we discover important diversity—a range of sperm performance and paternity that relates to how individual males allocate sperm to each spawning event. Thus cod, eels, and striped bass might be classified blithely as broadcast spawners, but each of these species' mating systems involves multiple courtship roles for males. Genetic markers that track parentage confirm that paternity and fertilization success varies substantially among males, dependent upon their size, condition, mating behaviors, and amount of shed sperm (Rakitin et al. 1999; Bekkevold et al. 2002; Hoysak et al. 2004). Individuals within the same species and population trade off one reproductive trait against another as a means to capitalize on fertilization opportunities (Gage et al. 1995; Taborsky 1998). This phenomenon, termed *sperm competition*, is as widely distributed in fishes as it is in other vertebrate taxa (Parker and Pizzari 2010).

Two modes of male mating behaviors are common in fishes (Taborsky 1998). One mode seeks to monopolize fertilization opportunities by limiting access to spawning females.

Facing page

Figure 3.2. (A) Spermatozoa length, longevity, and number among types of mating systems and (B) exponential decay in velocity after time of spermiation. In (A), a mating system continuum is indicated on the bottom plot. Box-and-whisker plots show median, first, and third quartiles; whiskers are ±1.5 times the interquartile range, and circles are outliers. In (B), decay rates in velocity are estimated from data in Cosson et al. (2008a). Phases of cod spermatozoa motility are shown in the inset graph modified from Cosson et al. (2008b). Data in (A) from Stockley et al. 1997 as well as additional sources: seven sturgeon species and one paddlefish from Psenicka et al. 2008, bluefin tuna from Cosson et al. 2008b, and Atlantic herring from Evans and Geffen 1998.

Examples include species in which males are selected by females, establish and defend territories associated with spawning or nursery habitats, internally fertilize eggs, or otherwise secure monogamous mates. The second mode exploits fertilization opportunities that exist in otherwise monogamous mating systems. Examples include small satellite males, sneakers, and streakers, and males that exploit mating opportunities on a transient basis (pirates; Fitzpatrick et al. 2007). Studies on bluehead wrasse, which exhibit both paired and polyandrous spawning, have allowed critical comparisons of the trade-off engendered by monopolizing and exploitive tactics. Larger males that held territories were preferentially selected by females and, for each spawning event allocated sperm in proportion to female size, shedding 10^3–10^4 spermatozoa per egg (Shapiro et al. 1994). In contrast, within polyandrous aggregations, individual males released six times this amount. Across a diversity of fishes the two mating systems are reflected in spermatazoa attributes—greater speed, longer flagella, and shorter life spans tend to occur for monogamous than satellite types to better capitalize on egg proximity (fig. 3.2A;Parker and Begon 1993; Stockley et al. 1997; Thünken et al. 2007).

Alternative reproductive and migratory traits often go hand in hand, epitomized by species of Pacific salmon. The monopolizing anadromous male ("hooknose") attains a larger size at maturity owing to oceanic forage migrations that greatly enhance growth rates; in comparison to resident (nonanadromous) sneaker males, they are typically tenfold or greater in weight. Gross (1985, 1991) has suggested that adoption of either tactic is frequency dependent in accordance with expected reproductive success (chap. 6, Developmental Plasticity). For instance, an excess number of a given reproductive mode will reduce its reproductive success relative to the alternate mode. Thus disproportionate removals of migratory salmon through ocean harvests are expected to disrupt this mating system, with negative consequences to population stability and persistence.

Fertilization and Migration Ecology

For the dominant mode of external fertilizing marine fishes, spawning initiates a lifetime (ontogenetic) sequence of dispersal and migration that begins with the fertilized egg. Thus spawning location and conditions and fertilization success have important carryover effects on the early migration ecology of offspring. Spawning in the water column causes offspring to be transported by currents that aid or curtail their delivery to, or retention within, suitable habitats for growth and protection. Temporal aspects that favor larval transport can also be selected, such as in reef fish, where spawning is cued to lunar phases and tides that may promote their exodus from predator-laden reefs (Johannes 1978) or otherwise influence their transport. Marine fishes that spawn on structured seabed habitats gain some enhanced protection for young, but most of these too will be influenced by local physical forces that cause displacement or retention of hatched larvae.

A second widespread mating system among marine fishes relies on internal fertilization or viviparity (bearing live young; Taborsky 1998; Fitzpatrick et al. 2007). Together with species exhibiting parental care (which can attend either internal or external fertilization), these mating systems are often, but not always, associated with limited spatial ranges that aid in securing mating opportunities and specific rearing habitats (e.g., black rockfish). Still, important exceptions occur. For instance, some sharks will undertake long migrations between foraging, mating, and pupping habitats (McFarlane and King 2003; Jorgensen et al. 2012a).

Provisioning Eggs

Throughout a teleost's adult life, oocytes proliferate. This process differs from other vertebrates—elasmobranchs, birds, and mammals—which all carry a fixed store of eggs from early development forward (Tyler and Sumpter 1996; Lombardi 1998). The result is high plasticity in fecundity occurring both among and within fish species. Some species produce litters of one, while others are capable of shedding billions of eggs over a lifetime. Still, each ovulated egg is a semiclosed system, carrying materials, energy, oxygen, and maternal instructions that will provision and guide development of the embryo and larva (Heming and Buddington 1988). How to provision eggs represents a fundamental trade-off for the spawning female: whether to allocate her store of materials and energy toward relatively few large eggs or many small eggs (Smith and Fretwell 1974). Among vertebrate groups, fish represent this trade-off in extremis, ranging from a litter of several 1.5-m-long white shark *Carcharodon carcharias* neonates (Uchida et al. 1996) to clutches of hundreds of millions of <1-mm-diameter eggs in ocean sunfish *Mola mola*. Further, within the same species or population, ovulated eggs can vary several-fold in mass, influenced by season, maternal effects, and population source (Zastrow et al. 1989; Fleming and Gross 1990; Marteinsdottir and Begg 2002; McIntyre and Hutchings 2003). In chapter 4 (Gleaning Production), we will consider how this trade-off affects early growth and survival. Here we examine how fecundity and egg size relate to adult migration and larval dispersal.

Females allocate energy reserves to reproduction principally as vitellogenin, which is stored in the liver and transported to the ovaries as precursor ingredients for yolk and oil constituents of the ripe egg (Tyler and Sumpter 1996). As a fraction of the female's weight, this transfer of energy stores can be substantial. Atlantic cod, striped bass, Pacific herring, and black sea bass will spawn between 15% and 30% of their weight in egg mass (Hay and Brett 1988; Secor et al. 1992; Tyler and Sumpter 1996; McIntyre and Hutchings 2003; Plaza et al. 2004), which is still considerably less than Japanese eel (60%; Tsukamoto et al. 2011). Oocyte development in these species is synchronous; that is, a single cohort of vitellogenic oocytes develops coinciding with each spawning event. The duration of vitellogenic provisioning varies substantially among synchronous spawners, ranging from days to months (Tyler and Sumpter

1996; Bobko and Berkeley 2004). Asynchronous spawners experience continuous oocyte development and include species such as anchovies and sardines that spawn frequently throughout a spawning season. Energy allocation to gonads can be remarkably high in such species. The bay anchovy *Anchoa mitchilli*, which spawns every several days, will spawn three- to four-fold its weight in eggs over a spawning season, provisioning them on the fly through high consumption rates (Luo and Musick 1991).

Discrete cohorts of different-sized oocytes are observed in the ovaries of synchronous spawners, each representing a future spawning event. For striped bass and other species that spawn once per year, cohorts of immature oocytes advance at an annual time step (Hempel 1980). This interval could cause an imbalance: an excessively high or low number of oocytes destined to mature in a future year against the female's capacity to provision that year's allotment. In fact, synchronous spawners make real-time adjustments on how many mature eggs will be ovulated. The mechanism responsible for regulating egg number is atresia, the death and resorption of immature eggs within the ovary's follicles (Hay and Brett 1988). In the extreme, atresia can cause "skipped" spawning owing to cessation of gonad development and nonparticipation in mating or a spawning migration (Rideout et al. 2005). On the other hand, accelerated oocyte development can recruit additional eggs to a given spawning event. For instance, Tyler et al. (1994) removed one ovary from rainbow trout *Oncorhyncus mykiss* and found that the remaining ovary compensated with accelerated oocyte recruitment within a 4- to 5-month period. Clearly, fish exhibit a remarkable degree of adaptation in provisioning their eggs, responsive to their energetic circumstance.

In addition to adjustments to egg number, viviparous fishes can modify offspring number through changes in fertilization rates. Black rockfish and other *Sebastes* store sperm, but in many instances not all eggs are fertilized. Bobko and Berkeley (2004) observed that ~20% of the eggs of mated *S. melanops* were not fertilized. In viviparous fishes, both unfertilized eggs and embryos can serve as a supplemental source of nutrients for sibling embryos. In this same species, Boehlert and Yoklavich (1984) estimated that metabolism during the monthlong gestation period would exhaust endogenous yolk reserves of embryos, indicating that a supplemental nutrient source was used. The vitelline envelope that surrounds the developing embryo is particularly thin in *Sebastes*, which could facilitate uptake of nutrients available from catabolized eggs (Boehlert and Yoklavich 1984; Yamada and Kusakari 1991). Sand tiger *Carcharias taurus* and other sharks exhibit intrauterine cannibalism between sibling embryos, an extreme form of energy provisioning among offspring. Also common among sharks is direct provisioning of maternal nutrients through intrauterine secretions or through a placenta (Hamlett et al. 1993), which occurs in ~20% of sharks. Similar to sharks, there are examples of parental provisioning through secretions (e.g., within the brood pouch of male seahorse *Hippocampus* spp.) or through placenta-like organs in teleosts, but these reproductive modes are rare (Wourms 1981).

Among teleost species, pelagic eggs are typically smaller and show lower variance in size than those scattered upon, or attached to, bottom substrates and surfaces (Duarte and Alcaraz 1989; Fuiman 2002), although in a review of reef fishes, Thresher (1984) detected no difference in mean size. The larger the oocyte (i.e., yolk content), the less effective buoyancy becomes (see below), which could explain in part the larger size and wider variation in demersal-spawned eggs. Outside of the viviparous sharks, egg diameter in fishes ranges from 0.3 mm (the small viviparous shiner perch *Cymatogaster aggregata*, which contains no yolk) to 8 cm (the coelacanth *Latimeria chalumnae*; Tyler and Sumpter 1996).

Fecundity and Adult Size

In contrast to terrestrial vertebrates, the transport of reproductive stores by fishes as they migrate is constrained by density rather than

gravity. The density of water exceeds that of air by nearly 1,000-fold, resulting for selection against large blunt bodies that increase turbulent resistance (chap. 2, Locomotion in Fishes). Although swimming morphologies are diverse, the quintessential hydrodynamic form of least resistance is that of cruising tunas and mackerels (Scombridae), which is fusiform with an anterior skew in volume and tapering caudal region (Lindsey 1978). Other functional morphologies favor maneuvering (high lateral profiles) and acceleration (wider posterior trunk and tail). Regardless of the class of swimming style (chap. 2, Locomotion in Fishes), large gonads will fully occupy the visceral cavity, potentially distending the abdomen, compromising form, and resulting in less efficient swimming. In Scombridae, heavier musculature further limits the space that gonads can occupy (Magnuson 1978). During final stages of maturation, abdomens of other species like striped bass and *Sebastes* spp. become flaccid and can attain a very round aspect. Still, because the visceral cavity typically comprises <25% of a fish's volume (estimated from common dressed weight ratios), selection for efficient swimming forms is expected to set design constraints on fecundity and offspring size (Wooton 1984). Thus, in aquatic vertebrates, maximum reproductive investment should be limited by the animal's volume (Schmidt-Nielsen 1984).

This physical constraint on fecundity and offspring size was examined for a set of 288 species, representing 39 taxonomic orders, including both teleosts and elasmobranchs (fig. 3.3A). Under the expectation of an egg-size fecundity trade-off, we would predict that larger females would carry more eggs, conditioned on egg size. This prediction holds: the larger the fish and visceral cavity, the greater the number of offspring, conditioned on offspring size (fig. 3.3B). The strength of the relationship ($r = 0.93$) is similar to ones produced by Wooton (1984) and Duarte and Alcaraz (1989), who plotted reproductive effort (total egg mass per fish) against fish length and explained a similar amount of variance (Wooton, for Canadian freshwater fishes:

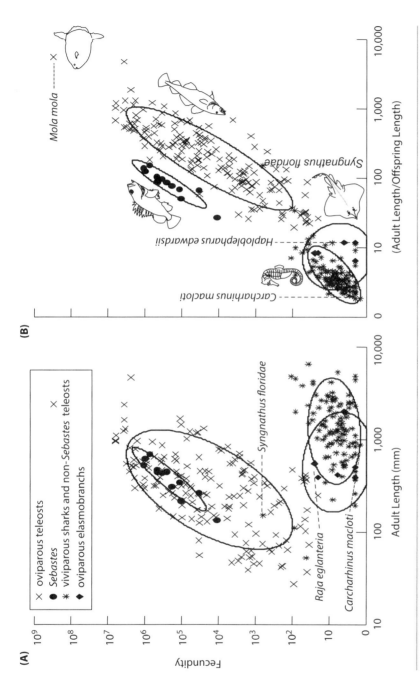

Figure 3.3. Adult size constraints on fecundity. (A) Batch fecundity is scaled to adult length for 288 species of teleosts and elasmobranchs. (B) Scales batch fecundity to adult length conditioned on offspring size. Groups, circumscribed by 67% confidence ellipses, are designated for four reproductive modes: (1) oviparous teleosts, (2) *Sebastes* spp., (3) viviparous sharks and viviparous teleosts other than *Sebastes*, and (4) oviparous sharks. End-member species in the data set include *Mola mola* (ocean sunfish), *Syngnathus floridae* (dusky pipefish), *Raja eglanteria* (clearnose skate), *Carcharhinus macloti* (hardnose shark), and *Haploblepharus edwardsii* (puffadder shyshark). Offspring size refers to hydrated oocyte diameter for teleosts and length of neonate for elasmobranchs. Where multiple adult sizes were given for a given species, a median length was used. Data are from an unpublished analysis by Kraus and Secor (with permission from R. Kraus, US Geological Survey) and several principal data sets, including Rounsefell 1957, Compagno 1984, Duarte and Alcaraz 1989, Love et al. 1990, and Shuozeng 1995.

$r = 0.87$; $n = 162$; Duarte and Alcaraz, for marine and freshwater fishes: $r = 0.94$; $n = 152$). Duarte and Alcaraz (1989) concluded that fish of similar sizes are constrained energetically to allocate similar reproductive effort. Through a different lens, we observe that similar-sized fish are physically constrained to have similar sized gonads. Thus the fish's swimming form limits reproductive effort—a generalization that applies to (1) both teleosts and elasmobranchs and (2) a wide range of mating systems (fig. 3.3B). Beyond this general pattern, noteworthy deviations are apparent. Both oviparous and viviparous elasmobranchs show a flatter response, indicating that offspring number is less sensitive to female size. Also, note that the *Sebastes* group shows a significantly elevated fecundity at adjusted female size compared to other teleosts. These are sedentary fish, and ripe females in this genus are known for their distended abdomens.

Incubation: Hatching as an Ecological Event

For most fishes, incubation is a period of vulnerability. Egg constituents affect the embryo's spatial fate: for instance, oil droplets impart buoyancy; a sticky chorion adheres to substrates. Mating systems also affect the embryo's fate. Examples include liberating eggs into favorable incubation conditions, internal rearing, and parental protection and care of offspring (Thresher 1984; Heming and Buddington 1988). But the encapsulated embryo can do little or nothing to control its own ecological fate (one noteworthy exception is intrauterine cannibalism by some sharks reviewed above). Indeed, among teleosts, development and hatching rates follow a fairly rigid schedule determined by initial egg size and temperature. Meta-analysis of different data sets of freshwater and marine species (Pauly and Pullin 1988; Duarte and Alcaraz 1989; Pepin 1991; Fuiman 2002) showed that incubation times were well predicted by initial egg size and temperature (albeit Pauly and Pullin 1988 failed to detect an egg size effect). Across species, temperature controls rates of cell proliferation, differenti-

ation, and yolk conversion in a strong linear manner.

Species meta-analyses such as those presented above (fig. 3.3) are informative from the standpoint of understanding phylogenetic constraints. For instance, we might interpret that fecundity is constrained by volumetric limits, or that embryo development rate is controlled by temperature and tissue substrate. When we look at such relationships within a species, however, different relationships can emerge: females can increase both egg size and fecundity with size (Zastrow et al. 1989; Fisher et al. 2007); final stage adjustments can be made to fecundity and egg size (Hay and Brett 1988); and, as physiological thresholds are approached, nonlinear temperature effects on stage duration and yolk conversion are expected (Hempel 1980; Heming and Buddington 1988). The difference in relationships observed *across* versus *within* species is of primary concern in drawing inferences—in the former instance we are concerned with phylogenetic constraints, and ecological ones in the latter. Confusion on this scaling issue can lead to poorly specified or overlooked concepts (Pepin and Miller 1993).

For oviparous fishes, hatching terminates an important period of embryo vulnerability because at this point the free embryo or larva initiates some control over its spatial fate. Within the vitelline envelope, the embryo has undergone a program of cell proliferation, differentiation, and growth, resulting in salutatory development of skeletal, muscular, neural, vascular, respiratory, excretory, digestive, and sensory systems. Hatching is not a distinct event within this ontogenetic progression (Balon 1984), but from an ecological viewpoint the time and location of hatching strongly influence early dispersal and migration.

Viviparous species liberate advanced-stage larvae or juveniles, with clear advantage to the offspring. By internally incubating the offspring, the female not only provides protection against predation and environmental stressors, but also has contributed to the offspring's spatial fate. For instance, internal incubation by *Sebastes* spp. in the California Current decreases the period of embryo dispersal within this strong current system, favoring their regional retention in shelf waters that serve as nursery habitat (Parrish et al. 1981; Bjorkstedt et al. 2002). Other viviparous fishes such as sharks convey embryos from spawning to nursery (pupping) grounds (e.g., Campana et al. 2010).

Embryo and Larval Dispersal

Embryo Dispersal

In the wake of pelagic spawners, ovulated eggs drift up and down according to their specific gravity. The egg's principal constituents are proteinaceous yolk material, water, and a hardened chorion, which provides protection for the developing embryo (Kjørsvik et al. 1990; Tyler and Sumpter 1996). Under these loads, the egg (or embryo) will sink if the specific gravity of the yolk and chorion is not offset by hydration, lipid inclusions, or vertical turbulence. Hydration often occurs in the final stages of egg maturation and in some species causes substantial increases in egg volume (Hempel 1980). Oil inclusions, particularly as wax esters, are also common in pelagic spawning fishes (but perhaps no more so than in demersal spawning fishes; Heming and Buddington 1988). Following fertilization, water influx within the space between the chorion and embryo (the perivitteline space) affects specific gravity. This occurs as part of the cortical reaction in which the egg's vitelline envelope becomes impervious, prevents the entry of multiple sperm, and separates from the external chorion, which is semipermeable. Osmolytes, sandwiched between these two layers, cause water influx (Hempel 1980), which alters the specific gravity of the egg and results in a larger surface area that slows the sinking rate and interacts with turbulent flow.

Egg buoyancy mechanisms are well exemplified by striped bass, which spawn relatively large pelagic eggs in freshwater portions of estuaries. Although their mature oocyte is only slightly larger than most marine spawners (~1.2 mm diameter), following fertilization, the perivitelline space expands to a volume greater than eightfold that of the oocyte (2–3 mm diameter). The ovulated egg also contains a large oil globule of wax esters, comprising 50% of the oocyte's mass as dry weight (Eldridge et al. 1991) and providing positive buoyancy. Despite the large oil globule, the specific gravity of the fertilized egg (1.002 g cm^{-3}) remains slightly greater than ambient waters (<1.0001 g cm^{-3}; Bergey et al. 2003). Physical conditions in the spawning habitat retain eggs and embryos in the water column. Strong changes in physical density occur at the interface between freshwater and brackish water, which results in vertical circulation and turbulent mixing along the saltfront (North and Houde 2001).

Thresher (1984) notes that "at hatching, larvae from pelagic eggs are little more than floating balls of yolk with a sliver of 'protofish' attached." In terms of its bulk dry weight, the free embryo is considerably smaller than the yolk from which it developed because of past metabolism during incubation (Heming and Buddington 1988). With limited organ system development, the integument is the principal site for respiration, osmoregulation, and excretion (Varsamos et al. 2005). For many species, the pelagic free embryo (a.k.a. yolk sac larvae) remains relatively inactive as it completes its metamorphosis into a larval form, although embryos of some species are capable of startle reactions that evade predators, buoyancy regulation, and some degree of sustained swimming. We therefore consider dispersal of the free embryo and larva jointly in the next section.

Larval Dispersal

Dispersal of the free embryo or larva is impelled by physical transport according to the offspring's specific gravity, swimming behavior, and ability to acquire energy and oxygen. In recent decades the dominant view that fish larvae were passive agents swept by ocean currents (e.g., Harden Jones 1968) has been revised by two important discoveries. First, larval attributes—specific gravity, vertical position, and swimming abilities—influence their dispersal fates (Norcross and Shaw 1984; Reiss et al. 2000; Leis 2006). Second, the currents and hydrography in which fish spawn can cause retention of embryos and larvae in seemingly open systems (Sinclair 1988). We consider these in turn, first by addressing larval attributes of two example species, Atlantic cod and Japanese eel, with particular attention to ontogenetic changes in specific gravity and swimming performance. We then consider Leis's (2006) question on whether pelagic fish larvae constitute drifting plankton or active nekton. Finally, for several case study species, we align categories of larval dispersal with mating systems and oceanographic settings (continental shelf, open ocean, upwelling current, and estuary) to examine how physical ocean forcing influences larval migration ecology.

The Atlantic Cod Larva

For Atlantic cod, incubation is slow in the cold waters off Newfoundland's coast, delaying hatching of the free embryo until 1 month after fertilization (Pepin and Helbig 1997). Enzymes from hatching glands located on the embryo's head have eroded the enveloping chorion, and through whole-body contortions the embryo frees itself. Shed of its chorion, the embryo is positively buoyant in seawater (Yin and Blaxter 1987). At ~5 mm in length, its jaws are fused with the epidermis; only nonfunctional rudiments of gills, alimentary tract, paired fins, and the swim bladder exist (fig. 3.4A). Prominent features unique to this stage include a large yolk sac from which the embryo will continue to derive energy, oxygen, and materials for development as well as a finfold, a large transparent keel of epidermis that surrounds nearly the entire body (larval cod descriptions here and elsewhere are from von Herbing et al. 1996). Lacking mouth or anus, the embryo's connection to the environment is through its integument and a small vent (the gill pit) at the junction of the head and yolk sac. The embryo is transparent

except for a few pigment spots. It can detect light, sound, physical shear, acceleration, and its osmotic status—but does so via sensory systems that are only partially developed.

Over the first 5–7 d of life (at 5°C), the free embryo has several unique adaptations to sustain itself while it undergoes the most rapid phase of development of its entire life. Chloride cells line the gill pit and together with others distributed across the integument (Varsamos et al. 2005) allow the embryo to start to regulate its internal chemistry. The finfold substantially increases the lateral profile and surface area of the embryo and early larva, transferring greater resistance and thrust to whole-body flexions. The enlarged finfold also increases the surface area over which respiration occurs as the larva transitions from yolk-derived oxygen to gill respiration, the latter still several weeks away. The yolk remains the single store of energy for development, maintenance, and locomotion; over the first week of life, half of it will be consumed.

Because of the free embryo's small size, it resides in a viscous medium (low Reynolds numbers; chap. 2, Locomotion in Fishes), and locomotion is characterized by pulses of tail flexion that are interspersed by periods of rest. Maximum speeds are just under one body length per second (BL s^{-1}). Without gill ventilation, locomotion relies on anaerobic metabolism. For long periods of time (20%–50% of its life span) the embryo is inactive. The duration of pulsed swimming events are short (<10 s), suggesting limited ability to evade a persistent predator or swim into a favorable environment. Still, by rapid displacement of several body lengths, an embryo can move out of many predators' visual ranges (Webb and Corolla 1981; Shepherd et al. 2000). At 3–4 d after hatch, the mouth and gill slits open, which permits inspiration of water and small protozoa and phytoplankton, triggering assimilation within the intestine (von Herbing et al. 1996, 2001). The embryo begins to feed and displays orientation toward small planktonic prey.

The period of transition between the em-

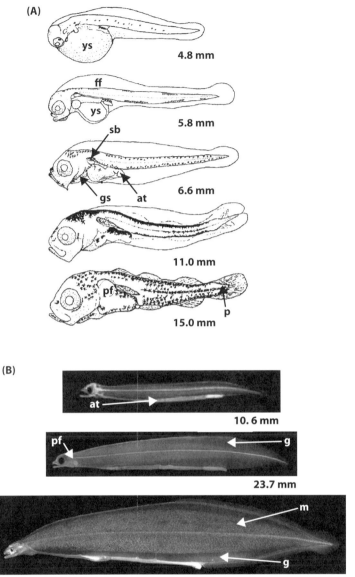

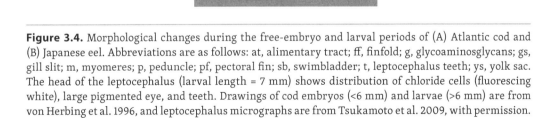

Figure 3.4. Morphological changes during the free-embryo and larval periods of (A) Atlantic cod and (B) Japanese eel. Abbreviations are as follows: at, alimentary tract; ff, finfold; g, glycoaminosglycans; gs, gill slit; m, myomeres; p, peduncle; pf, pectoral fin; sb, swimbladder; t, leptocephalus teeth; ys, yolk sac. The head of the leptocephalus (larval length = 7 mm) shows distribution of chloride cells (fluorescing white), large pigmented eye, and teeth. Drawings of cod embryos (<6 mm) and larvae (>6 mm) are from von Herbing et al. 1996, and leptocephalus micrographs are from Tsukamoto et al. 2009, with permission.

bryo and larva, lasting ~1 week for cod, is variously defined in the literature by the initiation of feeding or exhaustion of the maternal yolk. The mouth becomes increasingly functional with an articulating lower jaw, permitting active capture and manipulation of larger prey. The gut and intestines grow and differentiate, the liver develops, and digestion begins. Gill arches, respiratory epithelium, and filaments appear, and circulation of blood cells begins. The swimbladder, a gas-filled organ contributing to buoyancy, becomes partially functional, as do paired pectoral fins. By the end of this period, the yolk is gone, much of it converted to the larva's increasing length, which is now 35% longer than at hatching.

Swimming during the transition period remains intermittent, but the larva now exhibits a beat-and-glide pattern of swimming (von Herbing and Boutilier 1996). Although still working against laminar flow, increased size and a more efficient swimming form allow the larva to separate propulsion (beat) from attack (glide) behaviors in its pursuit of prey (von Herbing et al. 2001). Gliding also permits efficient use of inertial forces in covering greater distances. Higher uptake of oxygen across the integument supports increased (aerobic) metabolism and locomotion. An important constraint, however, is the stagnation of water surrounding the larva. This is particularly critical in a viscous system where the larva's immediate boundary layer can become depleted in oxygen and concentrated in its own waste. For instance, Weihs (1980) calculated a 40% loss of surrounding oxygen supply to a stationary northern anchovy *Engraulis mordax* embryo within a 2-min period. So, in addition to feeding and escape, intermittent swimming by embryos and early stage larvae also serve an important function in respiration and excretion by dissipating the self-polluted boundary layer, increasing convection and oxygen exchange across the integument's surface area, and causing the larva to enter a new parcel of oxygenated water (Weihs 1980; von Herbing and Boutilier 1996).

As the larva feeds, it becomes negatively buoyant and sinks in the water column. Vertical velocity by a passive larva (w) can be calculated using Stoke's terminal velocity formulation,

$$w = 2gr^2((p_s - p_l) \cdot (9v)^{-1}),\qquad 3.1$$

where g is acceleration due to gravity (9.81 m s^{-2}), r is the radius of the larva were it a sphere (a 5.75-mm larva has an equivalent radius of 1.66 mm; Sclafani et al. 1993), p is the density of the seawater (s) and larva (l), and v is viscosity (1.61 10^{-3} kg m^{-1} s^{-1}). Thus the larva moves up and down within the water column according to its relative density, counteracted by viscosity, which remains a relevant force for first feeding larvae. Feeding larvae will build denser tissues (e.g., muscle) and increase their specific gravity, whereas starving larvae will retain the buoyant characteristics of the embryo (Sclafani et al. 1993; Anderson and deYoung 1995). Under a range of specific gravity estimates for cod larvae reared in Newfoundland seawater, equation (3.1) yields a sinking rate of 3.7 and 5.6 mm s^{-1}. As a passive particle, the larva would sink hundreds of meters each day. To overcome this sinking rate, Sclafani et al. (1993) suggested that first-feeding larvae might sustain swimming speeds of 10 mm s^{-1} for 12 h each day, a rate more than tenfold higher than laboratory rates observed by von Herbing and Boutilier (1996). Mesoscale circulation and turbulence are other factors important in determining the larva's specific gravity and position in the water column, to which we will return later in the chapter. In contrast to a growing larva, the starving larva is positively buoyant. Unless the larva expends its ever-diminishing resources on downward locomotion, the larva will be exposed to the surface, which incurs survival risk owing to poor feeding conditions and increased predation (Neilson et al. 1986; Sclafani et al. 1993).

During the next 2 months, the larva will develop most functional systems that it will retain throughout its lifetime. Developmental landmarks (sensu von Herbing and Boutilier 1996) include a fully functional swimbladder; continued growth and elaboration of the alimentary system; differentiation of cranial and brachial skeletal architecture, leading to a buccal-opercular pump for gill ventilation and suction feeding; proliferation and vascularization of gill filaments; differentiation of blood cells; loss of the finfold; and formation of dorsal and ventral fins, caudal fin and peduncle, and paired pelvic fins. The rate at which each of these functional landmarks is attained depends principally upon the larva's feeding success and temperature. Feeding success yields a positive feedback where increased foraging causes higher rates of growth and development, which in turn leads to improved feeding success (von Herbing and Boutilier 1996). But because the larva now depends on aerobic metabolism, how it allocates a limited supply of oxygen to foraging, growth, energy stores, and swimming will have significant survival consequences. Temperature controls metabolic reactions, many of which will affect allocation decisions (Batty et al. 1993). Temperature also affects water density and viscosity, which will interact with larval size in controlling swimming performance. For instance, the 5-mm fish larva at 25°C will experience 60% of the viscous forces encountered by a similar-sized larva at 5°C (Denny 1993).

By the end of this period, the larva will have increased in length by approximately twofold and in mass by a hundredfold (Fiksen and Folkvord 1999), affecting its inertial properties. At this point, inertial forces dominate and lead to a much greater range in velocity, acceleration, and maneuvering by the fish. The late-stage larva can deploy its mass and swimming form (axial and paired fins, trunk and tail conformation, buoyancy control) in directed movements. Its style of sustained swimming is termed *sub-carangiform* after the order Carangiformes (e.g., amberjacks), which leverage tail flexion against a large head and trunk mass centered forward (chap. 2, Locomotion in Fishes).

The Japanese Eel Leptocephalus

Japanese eels likely spawn at 200- to 300-m depths, but their eggs are slightly buoyant, drifting upward into the warmer surface layer waters (50-100 m) of the North Equatorial

Pacific Ocean (Tsukamoto et al. 2011). At temperatures ranging 25°C-28°C, the free embryo emerges looking entirely different than the larvae of cod and other teleosts. Indeed, the larva merits its own moniker—leptocephalus—on account of its transparent leaflike form: the principal phylogenetic trait of the superorder Elopamorpha (~650 species), which includes species ranging from gulper *Eurypharynx* sp and moray *Muraena* sp eels to bonefish *Albula* sp and tarpon *Megalops* sp (Miller and Tsukamoto 2004; Miller 2009). The leptocephalus is completely transparent, a thin layer of epidermis surrounding a thin layer of muscle, which encases a large store of polysaccharides. This class of carbohydrates, known as glycosaminoglycans (GAGs), is shared with coelenterates (the mesoglea of "jellyfish") and represents the principal constituent in the growing leptocephalus. This constitution stands in sharp contrast to protein and lipids that are the typical bulk materials of teleost fish larvae, and in comparison to these materials, GAGs are metabolically inert and relatively cheap to produce (Pfeiler 1986). Overlaying a gelatinous sheath onto a vertebrate body plan contributes to extreme larval durations (0.25-3 y) and dispersal distances (>1,000 km) by Japanese and other temperate eel species (*A. anguilla, A. rostrata*) in comparison to larvae of other teleosts (Miller 2009) or indeed invertebrates.

The larva's somewhat bizarre morphology—a relatively small head, a jaw with teeth projecting at seemingly awkward angles, large surface area, thin epidermis, and gelatinous composition (fig. 3.4B)—caused early speculation that energy was obtained by absorption through the skin rather than feeding (e.g., Pfeiler 1986). Careful histological analysis, artificial rearing studies, and field trophic analysis confirm that leptocephali do indeed feed (Mochioka et al. 1993; Otake et al. 1993; Tanaka et al. 2001; Chow et al. 2010). Further, they occupy an unusual feeding niche, consuming the discarded houses of larvaceans (small planktonic protovertebrates that construct mucilaginous "houses" to entrap small organic particulates and microbes). These houses are major constituents of marine snow (detrital particulate matter), which is ubiquitous in tropical ocean surface layers. The outwardly projecting teeth serve to snag the houses and other particulate matter (Miller 2009). Recent findings demonstrate that leptocephali can also pursue and consume small zooplankton (Okamura et al. 2002). But because ocean basins are littered by marine snow, researchers presume it is the principal source of energy that supports the larva's months- to yearlong dispersal through tropical and temperate latitudes to juvenile habitats in Japan and elsewhere.

In warm tropical surface waters, development is fairly rapid (Okamura et al. 2007). Within the first several days after hatching, the head projects forward, the mouth opens, and the eye becomes pigmented (fig. 3.4B). By ~10 d after hatch, the yolk is consumed; the alimentary track has grown and differentiated and, together with functioning teeth and jaws, supports initial feeding at ~10 d after hatch (Otake 1996; Miller 2009). At first feeding, the leptocephalus has doubled its length (~7 mm) and exhibits a remarkable array of sensory capabilities. The eyes are large and heavily pigmented, a large optic lobe is visible in the brain, nasal organs are conspicuous (Miller 2009); the larva responds to local water displacements through mechanoreceptors. Among these receptors are precursors of the lateral line system, but also specialized organs that are unique to the leptocephalus and appearing first in the free embryo, which allow it to respond to local fluid displacement, say, those caused by a predator (Okamura et al. 2002).

Over the many months of dispersal, the leptocephalus grows from 3.5 mm at hatching to ~50 mm at metamorphosis, where upon it will transform into the juvenile "glass eel" stage. Such growth represents an ~1,000-fold gain in mass. Outside of the superorder Elopamorpha (within which leptocephali can attain lengths as large as 300 mm), the only fishes that rival this gain in size are several flatfish species. The Dover sole *Microstomus zachirus*, for instance, can attain 60 mm in length during its 1- to 3-y larval duration (Toole et al. 1993). Given the long

larval duration and the large size that the leptocephalus attains, mortality must be low. Miller and Tsukamoto (2004) speculate that the lack of pigmentation plays an important role in reducing predation. The long and laterally compressed leptocephalus epitomizes "anguilliform" style locomotion, in which large-wave undulations propagate down the entire body (chap. 2, Locomotion in Fishes). In contrast to other modes of swimming, anguilliform swimming permits both forward and backward locomotion, providing the leptocephalus another means of escape.

Leptocephali lack a swim bladder, instead relying upon a buoyancy system that is probably common and underappreciated in teleost larvae: chloride cells distributed on the integument. Chloride cells are the principal means by which fishes regulate internal osmolyte (salt) concentrations. In larvae with high surface area to volume ratios, integument chloride cells can be particularly effective in controlling specific gravity through changes in water flux. Chloride cells distributed on the integument of larvae have been documented for a diverse set of teleosts (Varasamos et al. 2005). Tsukamoto et al.'s (2009) recent discovery of high densities of integument chloride cells in the larvae of eels, presumed to be ancestral taxa, suggests that integument chloride cells may be general to teleosts.

Because the leptocephalus contains so much water (GAGs are 90%–95% water; Pfeiler 1986), integument chloride cells are particularly effective in regulating buoyancy. By 10 d after hatch, they occur from head to tail (fig. 3.4B). The specific gravity of leptocephali, although varying with stage and size, is close to that of the seawater in which they occur (Tsukamoto et al. 2009). Thus they are well classified as gelatinous zooplankton. Over a larva's long life in the plankton, regulation of buoyancy through water balance will contribute to its feeding and transport by retaining it in the upper mixed layer (Miller 2009). For other fish larvae, buoyancy via water transport through the epidermis is likely to be important only early in life, when surface area to volume ratios and water content are relatively high, and gills and swim bladders are undeveloped (Varasamos et al. 2005). For these larvae, the convective flows generated by mouth respiration (the bucco-opercular pump) make chloride cell placement on gill filaments much more effective in regulating water balance than cells placed on the integument. Further, owing to the large difference in specific gravity between water and air, the swim bladder represents a powerful means to regulate buoyancy. Because leptocephali attain such extreme larval sizes, it is tempting to invoke paedomorphosis (arrested development of a larval trait; Gould 1977) in its retention of a mechanism for buoyancy and osmotic regulation that most other teleosts utilize only during the early larval period (Hiroi et al. 1997; Varasamos et al. 2005).

Plankton or Nekton?

The central issue underlying the broad question posed by Leis (2007)—"are fish larvae plankton (drifters) or nekton (swimmers)?"—is how much control fish larvae have over their spatial fates (Bradbury et al. 2003). Leis and others have shown that larvae of tropical fishes are strong swimmers (~10 BL s^{-1}; Leis et al. 2007), exhibiting directed movements over kilometers and often culminating in their arrival to nursery reef habitats. But, as we have seen for cod and eels, larvae can also control the rate at which they sink or rise through changes in their specific gravity and the action of a swimbladder. Leaving aside the question of directed swimming behavior until later, let us turn to buoyancy as a means by which larvae can adjust their ecological fates related to feeding and transport.

Larval Transport

Because water is 830 times denser than air, gravity acts with much less force and is more easily overcome in aquatic ecosystems. By matching specific gravity with the surrounding water, fish can maintain their position in the water column. Similarly, by causing specific gravity to depart from ambient conditions, the fish moves vertically. For a fish larva, reg-

ulation of buoyancy is controlled by its ability to control water balance, the development of its swim bladder, and its body composition (nutritional status; see review by Govoni and Forward 2008). Viscous forces at small larval sizes will also retard vertical movements due to buoyancy. For instance, the expected sinking rate of spermatozoa is close to nil. For pelagic larvae, even incomplete controls on buoyancy are predicted to influence its transport.

Aquatic ecosystems are structured by density: layers of heavier water underlying lighter ones with density determined by depth, temperature, salinity, and seafloor topography. These are not static layers but ones influenced by the dynamic nature of wind, tides, and the earth's rotation (Bakun 1996). The action of these combined forces results in systems of circulation, currents, gravitational flow, fronts, eddies, and turbulence that vary between mesoscale (1–500 km) and macroscale (>500 km). These forces cause differential movement among the density-structured layers of the water column known as vertical shear. Thus a larva will experience wind-driven transport to a greater extent if it resides in surface waters rather than deeper ones. Further, owing to the earth's rotation (the Coriolis effect), changes are expected in both the amplitude and direction of wind-forced transport as we move from surface to deeper waters on account of Ekman transport. By altering their position in the water column, particularly over extended periods of time, larvae from the same spawning event can experience differing dispersal fates that might separate them by tens to hundreds of kilometers (Cowen et al. 2000; Bradbury and Snelgrove 2001; Petersen et al. 2010), far outstripping what could be accomplished by swimming alone. Similarly, by moving dynamically among parcels of water, a larva can arrest its transport, resulting in regional retention (Paris et al. 2007).

A large advance in understanding larval dispersal came through numerical modeling, which allows larvae to be inserted into mesoscale circulation features and their individual fates tracked. Newly hatched larvae can be released as passive particles of certain size and specific gravity, and can then be allowed to grow, survive, or die. Alternatively, they can be given specific behaviors, such as diel vertical behaviors, that cause them to interact dynamically with parcels of water that vary in specific gravity and flow. In numerical models, larvae are drawn from virtual populations and given attributes (e.g., size, specific gravity, vertical behavior) according to underlying probability distributions. They are then tracked individually over days, weeks, and months at time steps typically ≤1 d. Each time step of the model advances the larva according to how that interval's velocity fields interact with attributes of the larva. Additionally, these models incorporate latent effects: what has happened in the recent past will influence near-term fates (Bartsch et al. 1989; Paris et al. 2007). The spatial fates are then compiled collectively for cohorts of thousands of individuals. Application of numerical models will be reviewed later in this chapter, but it has become clear that these models are now central to understanding in situ larval dispersal. A century of observing larvae through ichthyoplankton studies has provided invaluable insight, but these investigations cannot simultaneously follow parcels of water and individual larvae. Vessel-based sampling of larvae and velocity fields entails averaging over temporal and spatial scales that unfortunately obscure variance in the biophysical interactions that control larval transport (Taggartt and Frank 1990; Sentchev and Korotenko 2007; Irisson et al. 2010). Numerical modeling permits an observed spatial pattern—e.g., arrival of larvae at a habitat distant from a known spawning region—to be explored with scenarios of physical forcing and larval behavior, which would not be possible through field study alone (Miller 2007; Sentchev and Korotenko 2007). Thus research on larval dispersal during recent decades has increasingly coupled field observations with predictions made from numerical modeling.

What behaviors within a larva's sensory, buoyancy, and swimming arsenal might affect its dispersal? Numerical modeling can explore possible larval movement cues. For instance, Sentchev and Korotenko (2007) evaluated alternative vertical movement cues for the European flounder *Platichthys flesus*, simulating the release of thousands of early stage larvae into the English Channel. Modeled fates of those larvae over 18 d showed that movements cued to tides (changes in sea height or tidal current velocity) resulted in spatial distributions of larvae that better emulated observed distributions than vertical movements cued to daily light cycles. Laboratory experiments have shown that larvae of this species occur at greater depths during ebb than flood tides (Burke et al. 1995), consistent with the predictions of the numerical modeling experiment.

That marine fish and benthic invertebrate larvae often coincide in the plankton invites comparison between these broad categories of taxa. In their review of planktonic larvae of teleosts and invertebrates (~100 species each), Bradbury and Snelgrove (2001) found that invertebrate and vertebrate larvae interact similarly with ocean current systems, particularly through their vertical distributions. They state that the "advantage of this [common] strategy is that vertical currents tend to be weaker than horizontal currents and a relative weak swimmer has a greater chance of regulating its transport through vertical movement than through horizontal swimming." Their review also casts some important distinctions. Most benthic invertebrates are sessile or sedentary as adults. While they can choose the timing of larval releases, they cannot choose the place. As reviewed previously, most marine fishes migrate and select specific times and places for spawning. Further controlling dispersal fate, fish larvae tend to be stronger and more efficient swimmers with more advanced sensory systems (Kingsford et al. 2002). Invertebrate and teleost larval durations overlap broadly and are similarly responsive to temperature (Bradbury and Snelgrove 2001), but invertebrates can extend their planktonic durations by arresting growth and development prior to settlement to benthic habitats (a.k.a. the precompetent period). Some tropical fish species (e.g., bluehead wrasse; Victor 1986) do this as well, but the phenomenon is apparently rarer among teleosts. The more directed behaviors of teleost adults and larvae would seem to justify past discipline emphases on larval dispersal as part of a population's life cycle in marine fishes, whereas larval dispersal is viewed principally as a vector of colonization in benthic invertebrates. Still, this dichotomy is exaggerated because larval dispersal contributes to colonization and metapopulation dynamics in marine fishes as well (chap. 5, Between Closed and Open Populations: Connectivity).

The Active Larva

Swimming performance by fish larvae must overcome two physical thresholds: viscosity and ambient currents (Leis 2010). As the larva grows, its increased length together with elaboration of its swimming form (increased musculature, streamlining, swimbladder, and fin development) will release the larva from viscous forces, where it must deform microparcels of water as it slips forward to an inertial environment where it can more efficiently utilize its momentum. Temperature also plays an important role; colder water is of greater density and viscosity, slows growth and development, and retards muscle contraction rate (Fuiman and Batty 1997; von Herbing 2002). Marine larvae generally transition between viscous and inertial forces between 5 and 11 mm depending on ambient temperature and taxa, but some tropical species become inertial larvae much sooner owing to warmer temperatures and an advanced body form (Leis 2010).

Swimming performance is measured as "critical speed." The larva is forced to swim at ever-increasing speeds until exhaustion, and critical speed is calculated as the weighted average of the speeds over the experimental interval. There is no direct ecological interpretation of critical speed, but the index is useful in species comparisons of swimming performance. In an effort to provide some ecological context, Leis (2010) reported human diver observations of larval speeds that were ~50% (range 29%–91%) of laboratory-derived critical speeds for

a set of 10 tropical and temperate species. In a review of species from 15 families, Leis (2010) reported that critical speeds measured as BL s^{-1} were substantially less at larval sizes 5–7 mm than at larger sizes, consistent with the presumed influence of viscosity on performance. Still, several tropical families (e.g., Ephippidae, Carangidae, Pomacentridae, Apogonidae) showed no apparent decline in swimming performance at the smallest larval lengths.

Swimming performance by the inertial (active) larva matches that of later juvenile and even adult stages. Leis (2010) reported in situ rates ranging between 3 and 18 BL s^{-1} for tropical and temperate species. Under the assumption that the swimming rates reported by Leis are sustainable, simple calculations would have a 10-mm larva capable of swimming 4.3 km d^{-1} at 5 BL s^{-1} and 13.0 km d^{-1} at 15 BL s^{-1}. Consistent with these calculations, endurance experiments that cause larvae to swim continuously at a sustained speed showed 10-mm larvae of a tropical species (e.g., Ambon damselfish *Pomacentrus amboinensis*) capable of swimming 10–18 km nonstop. These are quite relevant distances in terms of the larva's dispersal fates, but they are still far from realistic because we have not yet considered ambient currents and the larva's ability to navigate.

The active larva is capable of overcoming ambient currents, but this capability by no means implies a head-on attack. Rather, as Leis (2010) points out, even a small departure from the encountered flow fields, if consistently applied, can result in substantial net displacement and in this way constitutes a directed movement (for review of animal orientation in flow, see J. W. Chapman et al. 2011). The active larva is not too different than the juvenile or adult in the battery of sensory organs deployed, nor their acuity. As reviewed already, vision, hearing, and mechanoreception begin early in ontogeny (Blaxter 1986; von Herbing et al. 1996; Kingsford et al. 2002; Miller 2009); olfaction occurs later in larval development (Kingsford et al. 2002), and there is some evidence that larvae can perceive pressure and thereby regulate their depth (Govoni and Forward 2008; Huebert 2008). In support of this evidence, Leis (2006) showed that larvae of Japanese croaker *Argyrosomus japonicus* reared to different sizes in the lab and released into the wild descended to depths that decreased with length at release—driven apparently by an ontogenetic schedule. Similar to swimming performance, larval size is a key determinant in the range of vision, hearing, mechanoreception, olfaction, and other senses (Kingsford et al. 2002). Because we have defined the active larva with respect to ambient flow conditions, it is tempting to speculate that orientation is mostly driven by how the larva responds to flow and pressure (swimming at angles to flow and through tidal vertical movements), although light (e.g., diel vertical movements), sound (e.g., noise associated with a flow-buffeted reef), and smell (e.g., estuarine flume) have also been shown to influence larval orientation.

The increased control a larva asserts over its spatial fate is manifest as increased patchiness with increased stage of development and size. Patchiness (P) can be indexed as

$$P = 1 + \left(\sigma^2 - \bar{X}\right)\bar{X}^{-2}, \qquad 3.2$$

where $\bar{X}$ is the overall mean concentration and σ^2 is variance. The index scales linearly with variance in concentration (Stabeno et al. 1996), and when $P = 1$, larvae are homogenously distributed (McGurk 1987). Mating systems cause eggs and free embryos to be initially distributed as concentrated patches. Free embryos and early stage larvae are largely passive such that transport, diffusion, and turbulence erode these patches into diffused, more uniform distributions that vary with weather and hydrography (McGurk 1987). Increased swimming speed and orientation by active larvae should cause reemergence of patchiness according to dominant behaviors. Bradbury et al. (2003) detected this predicted U-shaped pattern in *P* (high-low-high patchiness) for four species, including Atlantic cod, in the semienclosed Placentia Bay (Newfoundland). Early stage larvae were more homogenously distributed than embryos, while larger larvae were displaced up-

bay, resulting in an increased *P*. Early schooling behaviors might also contribute to increased patchiness by active larvae (Bradbury et al. 2003). An element of uncertainty, however, is the role of mortality in the distribution of larvae. As an example, bay anchovy eggs are concentrated during summer months in several regions of the Chesapeake Bay. Young larvae under the action of estuarine currents become more homogenously distributed, but more advanced larvae (>10 mm) are predominantly captured in up-estuary regions. The species exhibits the classic U-shape in patchiness, but is this due to directed dispersal or differential mortality (i.e., increased survival of up-estuary larvae)? Using otolith strontium as a tracer of up-estuary dispersal (chap. 2, Life Cycle Tracers), Kimura et al. (2000) concluded that dispersal was the more likely cause, estimating net up-estuary displacement rates of ~5 BL s[-1] for larvae originating in the lower part of the Chesapeake Bay.

Starting Life on the Seabed

Like benthic invertebrates, the pelagic larvae of many marine fishes result from eggs that are incubated in demersal rather than pelagic habitats, leading to the following question: Does the dispersal fate vary according to incubation location? With pelagic spawning, the embryo is "launched" into the water column and often within a circulation feature that will have a strong influence on the larva's fate. The pelagic embryo and early larva are also vulnerable to unfavorable transport and predation (Bradbury et al. 2003). The demersal egg and embryo may have been scattered or attached to a substrate or may have been brooded (guarded against predation and ventilated). Following incubation, the offspring typically emerges as a larger, further developed, and better swimming larva in comparison to larvae originating from pelagic spawning (Thresher 1984, 1988; Webb 1988; Fuiman 2002). These more robust larvae are more likely to overcome ambient flow conditions or to undertake directed vertical movements (McGurk 1989; Bradbury et al. 2003). Thus the pelagic spawned embryo is given a spatial head start, whereas the demersal embryo is given a developmental head start (Leis 2010). McGurk (1989) speculated that by forgoing passive dispersal in the pelagic environment, demersal spawners could increase the likelihood of certain transport outcomes.

Alignment of Larval Dispersal with Mating Systems

In the following examples, mating systems result in concentrations of embryos and larvae in varied oceanographic settings that influence subsequent dispersal. These are but a handful of examples that reflect upon those mating systems reviewed previously and represent a range of larval dispersal behaviors and ecosystems for which numerical models have been applied. More comprehensive reviews of larval dispersal include those by Heath (1992), Shanks (1995), Bakun (1996), and Cowen and Sponaugle (2009).

Shelf Spawning: Atlantic Cod

The interactions of coastal currents and ocean boundary currents cause sea level heights to vary tens of centimeters over the continental shelves of the Canadian Maritimes, where adult cod spawn. Owing to the high density of seawater and the earth's rotation, changes in sea surface topography cause strong flows within water column layers of similar specific gravity (a.k.a. geostrophic currents; Bakun 1996). On the Scotian Shelf, Reiss et al. (2000) used a numerical modeling approach to match predictions of larval occurrence in contiguous water masses defined by specific gravity against field observations of larval distributions. In a remarkable manner, larval concentrations and sizes were structured by specific gravity (fig. 3.5). Higher abundances of smaller larvae occurred in warmer saltier waters in comparison to larger larvae, but regardless of size the cohort's larvae all occurred along the same density contour (isopycnal; fig. 3.5A). We can thus conceive of a cohort of larvae regulating their buoyancy to maintain themselves in a specific parcel of water for extended periods. In a nu-

merical experiment, Reiss et al. (2000) released virtual larvae from known spawning bank habitats, "instructing" larvae to merely follow streamlines of equal water density. Vertical profiles of water density over the shelf were used to estimate sea level height and to interpolate geostrophic flow within the surface layer (≤ 40 m). Model results generally corresponded with observed concentrations of larvae over shelf banks (fig. 3.5B). The study was elegant in demonstrating that a simple mechanism might largely explain early larval distributions in a complex shelf habitat (one comprising numerous basins, banks, and channels) without including, for instance, effects from tidal currents, fronts, wind, measured coastal current speeds, and larval swimming and buoyancy. Still, other shelf systems are bounded by stronger coastal currents and gyres, for which other mechanisms warrant careful consideration.

The Labrador Current is three to five times stronger (0.3–0.5 m s^{-1}) than the coastal currents on the Scotian Shelf, yet cod spawn within the current's main inshore branch in northern regions of the Newfoundland Shelf. Although the Newfoundland Shelf contains several banks, retention of bank-spawned offspring is unlikely under prevailing southeasterly flow conditions. Indeed, a principal issue is whether larvae originating from spawning anywhere on the Newfoundland Shelf can be retained on the same shelf. To address this question, Pepin and Helbig (1997) deployed dozens of drifting drogues at the northwest extent of the shelf and tracked their progression through satellite telemetry during fall months (cod spawn spring to fall). From this data they empirically modeled surface flow fields across the shelf, with periodic wind and weather events simulated as model variance terms. Utilizing ichthyoplankton data, future distributions of larvae were projected (fig. 3.6). Pepin and Helbig (1997) predicted that a substantial fraction of larvae was transported off the shelf. A subsequent numerical tracking study confirmed little shelf retention by passive particles released into the main offshore Labrador Channel (Pepin et al. 2011). Important was the variance structure under-

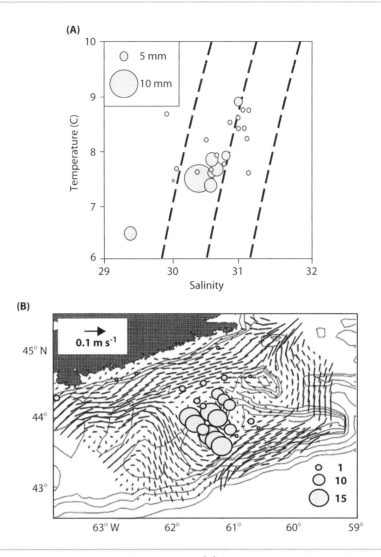

Figure 3.5. Distribution of Atlantic cod larvae among (A) density-structured water layers and (B) across the Scotian Shelf (Canada). Cod larvae are distributed by size across masses of water characterized by differing density (temperature × salinity). Note that larvae of divergent sizes occur within the same isopycnal. In (B), densities of larval Atlantic cod (number per 100 m³, scaled by circle size) on the Scotian Shelf overlie flow fields predicted from sea surface topography. Flow vectors are indicted by arrow direction and length. Larvae are associated with bank structures where low flows occur. Graphics from Reiss et al. 2000, with permission.

lying predicted dispersal, which resulted in cross-shelf transport of larvae due to simulated episodic wind events. In forward projections, a large number of simulated larvae became transported into inshore bays, a somewhat surprising occurrence in that the coastal zone was not explicitly included as part of the circula-

tion model or ichthyoplankton sampling. These results suggested to the authors that, within the coastal zone, minimal currents or local recirculation systems could cause high concentrations of larvae (fig. 3.6). A subsequent coupled ichthyoplankton-grapefruit drifter study showed cyclonic flow and larval retention in a large Newfoundland embayment (Bradbury et al. 2000), consistent with Pepin and Helbig's prediction.

An interesting alignment between mating systems and larval dispersal and survival of Newfoundland cod has been proposed by deYoung and Rose (1993). During cold years, spawning aggregations shift to the south, resulting in increased transport away from the shelf and decreased survival of larvae (fig. 3.6). In such years, local inshore egg and larval production may be of increased importance to persistence of Newfoundland cod.

As larvae are transported across shelf habitats, they are also influenced by diffusion as a result of the action of water particles. As a patch of embryos or larvae is transported along a surface current, the shape of that patch broadens into an ellipse, the long axis of which represents both diffusion and transport, the shorter axis principally determined by diffusion. Diffusion alone can result in net displacements of ecological consequence (Heath 1992).

For haddock *Melanogrammus aeglefinus* off Nova Scotia, for instance, Campana et al. (1989) observed abundance contours of progressively older larvae that conformed to the mutual influence of advection and diffusion, with some shelf-spawned larvae drifting toward coastal zone habitats (fig. 3.7). For Biscay Bay common sole *Solea solea*, Koutsikopoulos et al. (1991) concluded that diffusion was the principal mechanism by which larvae dispersed from shelf spawning habitats to inshore nurseries.

Ocean Spawning: Japanese Eel

Japanese eel spawn within the world's largest marine ecosystem, the North Pacific Gyre, a system of major ocean currents that drive the gyre clockwise by the action of temperate westerly and tropical easterly trade winds. Successful transport of *A. japonicus* leptocephali during their 3,000-km journey to juvenile habitats in Japan and elsewhere depends on two segments of the gyre: (1) the Northeast Equatorial Current (NEC), which forms the southern bound-

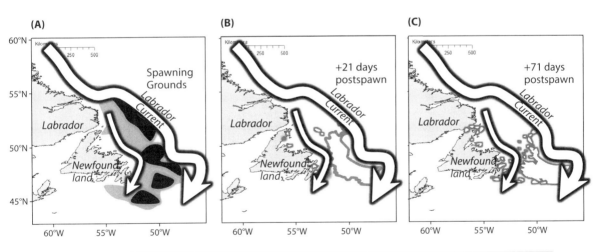

Figure 3.6. Dispersal of larval Atlantic cod in shelf waters off Newfoundland. (A) The northern stock of cod comprises subpopulations spawning in shelf waters between the main and smaller inshore branch of the Labrador Current (open arrows). Generalized adult spawner concentrations for the early 1990s are shown in black; those in the mid-1980s are shown in gray. (B and C) Larval distributions are shown as white ellipses following transport over 21 and 71 d after spawning (based on initial distributions of Stage 1 eggs) to downstream shelf and slope waters. Compiled from Taggart et al. 1994 and Pepin and Helbig 1997 and redrawn with permission.

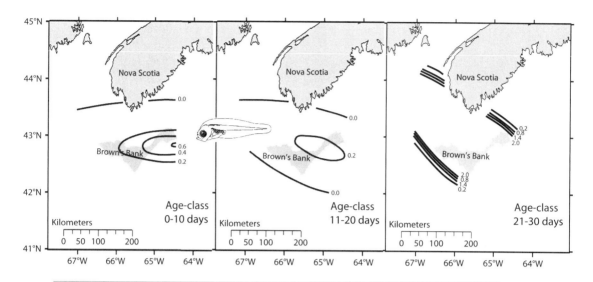

Figure 3.7. Dispersal of larval haddock following spawning near Brown's Bank, Nova Scotia, Canada. Isopleths of larval concentration (higher number, higher density) are shown by age-class, determined by otolith microstructure (n = 881). For age-class 11-20, a cubic response model was also fit (not shown), suggesting some degree of larval concentration and retention in addition to diffusion shown here. From Campana et al. 1989, with permission. Larval haddock (15-mm) illustration reprinted from Bigelow and Schroeder 1953.

ary of the gyre and conveys larvae westward from their spawning location west of Guam and the Mariana Islands, and (2) the Kuroshio Current, a northerly current that carries larvae from the NEC to subtropical and temperate latitudes (Kimura et al. 1994). But transiting between these currents is not straightforward: the NEC branches into two streams, the Kuroshio and the Mindano, and the latter heads toward Southeast Asia, where few Japanese eels have been recorded (Kimura and Tsukamoto 2006). Increased transport to the Kuroshio is expected when larvae occur in the northern portions of the NEC prior to the current's bifurcation (fig. 3.8). Drifters deployed within the spawning region at either the surface or at depth (150 m) followed very different tracks (Kimura et al. 1994). The surface drifter showed little net displacement over its 75-d deployment, but the middepth deployment showed rapid westerly displacement at a rate of 20-32 cm s^{-1}, a rate that corresponded to the rate at which older leptocephali were found in regions

progressively west of the spawning location (Kimura et al. 1994). Importantly, however, the middepth buoy drifted into the Mindano Current. Any regulation of the vertical position by the leptocephalus will clearly result in large changes in its spatial fate.

Kimura et al. (1994) postulated that through vertical migration leptocephali not only utilize the strong middepth flows of the NEC, but also take advantage of surface layer flows (<70 m) that are deflected northward. Flows move northward as a result of the force applied by trade winds: their westward force is deflected to the north in the surface layer owing to earth's rotation (Ekman transport). The leptocephalus, by migrating between middepth and surface layers of the NEC each day, can accomplish both (1) relatively rapid westward transport dependent on its time at middepth and (2) northerly displacement within the NEC and bifurcation zone dependent on its time in the surface layer. Diel vertical migrations have yet to be confirmed for Japanese eel leptocephali, but with increasing size the proportion of larvae in the surface layer during nighttime tends to increase (Otake et al. 1998). Diel vertical movements have been documented for the leptocephali of other temperate *Anguilla* spp. (McCleave and Kleckner 1982).

Adding complexity to the trip, the position of the NEC is dynamic year to year and strongly linked to El Niño events, which cause a southward displacement of the NEC bifurcation (Kimura et al. 2001; Kimura and Tsukamoto 2006). Because evidence indicates that spawning location remains fixed at a specific location, such displacements are predicted to have strong influence on the entrainment of larvae into the Kuroshio Current. To evaluate the spatial fates of leptocephali associated with El Niño, a numerical experiment was undertaken utilizing an ocean circulation model that simulated three-dimensional flow fields throughout the northwest Pacific Ocean (Kim et al. 2007). Locations of the bifurcation zone were determined for El Niño and non–El Niño years. Co-

horts of simulated larvae (n = 1,000) were released at fixed spawning sites (Tsukamoto et al. 2003) and at spawning sites that shifted north or south according to the displacement of the NEC. Larvae were given a vertical movement behavior (residing at 150-m depths by day and 50-m depths at night) and tracked over a 1-y period assuming 100% survival (most leptocephali of this species metamorphose to a demersal juvenile form by 5 months; Arai et al. 1997). Individual larvae were then scored on the basis of whether they were entrained into the Kuroshio or Mindano Currents. All numerical experiments showed highly dispersive larvae with a wide range of fates over the fairly long period of dispersal (fig. 3.8). Mesoscale eddies resulted in diffusion to regions outside of the principal ocean currents. Only a minority of leptocephali entrained into the Kuroshio Current (18%–36%). Allowing displacement of the spawning location with climate increased transport into the Kuroshio by 10%–16% (Kim et al. 2007). Some evidence suggests that adults may cue

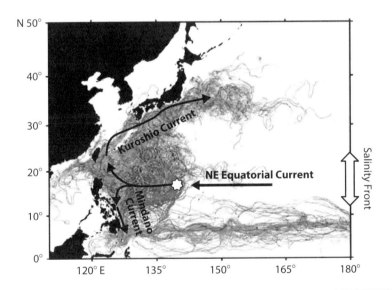

Figure 3.8. Projections of individual Japanese eel leptocephalus dispersal paths from a central spawning site (star), initiated during the May–December spawning season. Individual dispersals are shown for a non–El Niño year (1995). Major currents are indicated together with the salinity front, which is displaced southward during El Niño years. From Kim et al. 2007, with permission.

their spawning to occur near the salinity front that bounds the southern face of the NEC (Tsukamoto 2009). Given the dispersal distances, durations, and unpredictable spatial fates of leptocephali, Japanese eels presents a particularly vexing alignment between their mating system (relatively fixed site, distant ocean spawning) and larval dispersal.

Upwelling Spawning: Rockfish

The southerly winds along the coast of California cause deflection of surface layer waters to the west arising from the earth's rotation. Strong upwelling by bottom layer waters results from the displaced surface water, which injects nutrients into surface layer waters of the California Current. The Ekman transport of water offshore is most intense at Cape Mendocino, just north of San Francisco, a region where dozens of rockfish species (*Sebastes* spp.) spawn. This spawning activity seems puzzling because high offshore transport of surface water might rapidly disperse larvae into inhospitable oceanic waters with low productivity over regions with no bottom juvenile habitats (Parrish et al. 1981; Bakun 1996). On the other hand, Yoklavich et al. (1996) speculated that offshore dispersal might favor larval survival by reducing their exposure to inshore predators.

The mating systems of *Sebastes* spp. in this region affect larval dispersal in several ways: (1) by incubating larvae internally, females advance the developmental stage of the larvae at birth (e.g., release of an active larva versus an encapsulated embryo) and thereby reduce its passive planktonic period; (2) strong vertical shear exists within the top 100 m, so that by adjusting depth at which larvae are released, the female can influence larval dispersal; and (3) spawning seasonally, rockfish can exploit the winter months when southerly winds relax, disrupting upwelling and offshore transport. A numerical modeling experiment in this same region pointed to depth and season of larval release as principal determinants of larval dispersal (Petersen et al. 2010). Assuming that larval *Sebastes* do not migrate vertically, neutrally buoyant larvae were released into the water column at five different depths (0–70 m) and tracked for 35 d during periods of upwelling and nonupwelling (fig. 3.9). The simulated larvae were transported within their assigned depth layer through application of an ocean circulation model. During the upwelling season, shelf retention by larvae was favored when they were released and maintained at levels below 20-m depth. At shallower depths, larvae were transported >150 km across the shelf over the 35-d experiment. During nonupwelling months, little net dispersal occurred across the shelf, although substantial northward drift (>150 km) occurred along the coast of California. The model is necessarily simplistic in assumptions pertaining to persistent depth distributions by cohorts of larvae but provides valuable insight into the alignment of mating systems (depth and season of spawning), seasonal ocean circulation, and larval dispersal.

Estuarine Spawning: Striped Bass

Estuarine saltfronts are mesoscale features that demarcate large density differences between freshwater and salt water. This is an important retention feature for striped bass and other fishes that spawn in tidal freshwater. In a field experiment on larval dispersal and retention (Secor et al. 1995; Secor and Houde 1998), groups of millions of 5- to 10-d-old striped bass larvae were marked by staining their otoliths and then released into two Chesapeake Bay estuaries (the Patuxent and Nanticoke Estuaries). For groups released in freshwater above the saltfront, dispersal within the first 5 d was rapid, soon mirroring the distribution of coinciding wild larvae (fig. 3.10). Dispersal up- and down-estuary occurred for both groups of larvae, dependent upon rain events and related displacement of the saltfront. Representatives of these groups survived into the juvenile period (Secor and Houde 1998). In stark contrast, groups of larvae released downstream of the saltfront were never again observed. For most estuarine systems, embryos and larval striped bass <10 mm are rarely encountered in salinities >1 (Uphoff 1989; Grant and Olney 1991; Rutherford and Houde 1995; Secor and Houde 1995a;

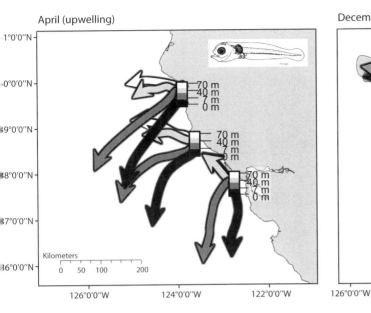

Figure 3.9. Projections of individual *Sebastes* larval dispersal paths during upwelling (April) and nonup-welling (December) seasons over the first 35 d of transport in the California Current System. Cohorts of larvae were assumed to stay retained within layers of water at 0-, 7-, 40-, and 70-m depth. In April, trajectories associated with each release depth are uniquely shaded. In December, depth of release did not influence the direction of trajectories, which are shaded by region of release rather than depth. Graphics from Petersen et al. 2010, with permission. Illustration of *Sebastes goodei* larvae from Wikimedia Commons.

Bennett et al. 2002). Still, important exceptions do occur: in higher-energy systems such as the upper Chesapeake Bay (receiving flow from the largest US Atlantic watershed) and tidal bore estuaries in Canada, larvae frequently occur at higher salinities (1–5; Robichaud-leBlanc et al. 1996; Rulifson and Tull 1999; North and Houde 2001, 2006).

At the saltfront, seaward flowing freshwater overlies the denser bottom waters, which flow from the ocean. The result is a strong convergent zone, which entraps passive particles depending on the strength of the vertical shear and the specific gravity of the particles. Often associated with the saltfront is the estuarine turbidity maximum: a broad region (>5 km) of high suspended solids and plankton, which can enhance larval production (Laprise and Dodson 1989; North and Houde 2001; Winkler et al. 2003; Shoji and Tanaka 2007). Passive

and active retention by embryos and larvae at the saltfront can only occur if adult striped bass spawn in freshwater. North et al. (2005) simulated spawning into an idealized estuary based on a three-dimensional circulation model, which accounted for wind, flow, salinity structure (vertical shear), and small-scale turbulence. The probability that embryos were retained by convergent flows was highest when spawning occurred 0–6 km up-estuary of the saltfront. High down-estuary winds and flow caused transport of embryos to regions below the saltfront, particularly for low and high specific gravity eggs. The numerical model simulations provided perspective on documented spawning behaviors. Striped bass typically spawn in regions 0–10 km up-estuary to the saltfront during periods of declining flow; these behaviors would favor embryo retention according to the modeled estuarine circulation

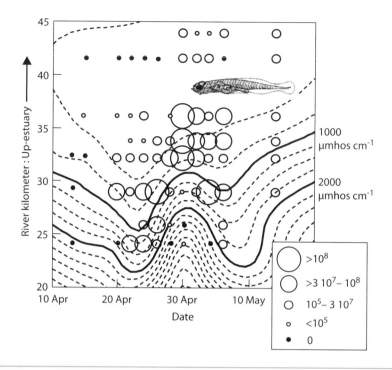

Figure 3.10. Distribution of larval striped bass in relation to the saltfront during spring. Salinity front is designated by the contour, 2,000 μmhos cm^{-1} conductivity (salinity ≈ 1). Larval abundance by river segment is indicated by circle area. Two large precipitation events (those of April 22–23 and May 9) caused downriver excursion of the saltfront together with a shift in the distribution of larvae. Unpublished data (Nanticoke River Estuary, Chesapeake Bay, 1997) by the author. Larval striped bass (6.3-mm) drawing from Mansueti 1958.

features (Grant and Olney 1991; Rutherford and Houde 1995; North and Houde 2006).

As the yolk sac and oil globule are consumed, the striped bass larva gains weight and now must rely on active measures to persist in regions up-estuary to the saltfront. In the upper Chesapeake Bay, North and Houde (2001) observed that striped bass larvae were distributed in the upper and lower segments of the water column during flood and ebb tides, respectively, a pattern consistent with retention over a tidal cycle. They speculated that striped bass larvae manifested retention not through responses to density or flow but by tracking zooplankton prey, which were passively retained at the saltfront. For striped bass larvae of the Sacramento–San Joaquin system in California, Bennett et

al. (2002) detected distribution patterns of larvae consistent with tidal vertical movements during dry years. During a moderately wet year they detected a reverse diel migration (surface orientation by larvae during the daytime) and proposed that this vertical behavior resulted in lateral horizontal retention in flank shallow nursery areas above the saltfront.

Coastal Spawners: Estuarine Ingress

Estuarine larvae of some species do not originate in tidal freshwater regions but are spawned in coastal waters and subsequently disperse to regions up-estuary to the saltfront. The Japanese sea bass *Lateolabrax japonicus* spawns in nearshore coastal waters, and larvae utilize estuarine residual currents, vertical behaviors, and swimming to reach tidal fresh-water regions prior to juvenile metamorphosis. This is a remarkable journey in the Chikugo Estuary (near Nagasaki), where tides are in excess of 6 m (Matsumiya et al. 1982; Shoji and Tanaka 2007). Similar patterns of estuarine ingress by active larvae have been described for flatfish species (fig. 3.11; Rjinsdorp et al. 1985; Tanaka et al. 1989; Champalbert and Koutsikopoulos 1995). A commonly invoked mechanism for ingress into estuaries and inshore coastal systems is tidal transport (Norcross and Shaw 1984), in which the larva moves up into the water column during flood tide and retreats to the bottom during ebb tide, thus minimizing its exposure to flows that would carry it outside the estuary (fig. 3.11). For larvae, this behavior must be inferred through discrete depth sampling, which is often of insufficient spatial and temporal resolution to provide straightforward interpretations (Neilson and Perry 1990; Miller 2007). Further, the underlying cues that lead to diel or tidal vertical movements have been poorly studied, apparently differ between those taxa studied, and often change during ontogeny

(Olla and Davis 1990; Burke et al. 1995; Forward et al. 1996; Hurst et al. 2009). Thus tidal transport should be considered as just one possibility amidst a repertoire of behaviors and circulation systems that active larvae can exploit to transport themselves up-estuary (Kingsford et al. 2002; Bolle et al. 2009).

Selective Harvest during Spawning Migrations

The directed spawning migrations by many marine fishes provide opportunities for targeted fisheries on adult schools and aggregations. Large bottom trawls have historically targeted the dome-shaped spawning aggregations of Atlantic cod, located through hydroacoustics. During a period of extreme overfishing in the Chesapeake Bay, gauntlets of gill nets removed up to 90% of male striped bass during their spring spawning runs. Traditional weir fisheries in the St. Lawrence River harvested up to 40% of the emigrating American eels beginning their journey to the Sargasso Sea. The return migrations of Atlantic bluefin tuna to

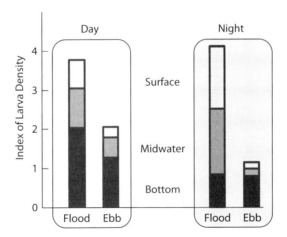

Figure 3.11. Tidal stream transport evident in the vertical distribution of plaice larvae in a North Sea estuary between tidal stages and day and night (Rijsndorp et al. 1985). Depth of collection shown as black (bottom), gray (midwater), and white (surface). Distributions of pooled larvae from samples taken during February–May 1981 and 1983. Redrawn from Rijnsdorp et al. 1985, with permission.

spawning regions in the Mediterranean follow well-known pathways; these fish have been efficiently harvested with traditional trap nets and modern purse seines. For some species, it is the eggs themselves that are targeted. The attached demersal eggs of Pacific herring (usually on seaweed fronds) support a lucrative market in Japan. Caviar harvests have resulted in extirpated or depressed sturgeon abundances throughout the world (Limburg and Waldman 2009).

To evaluate the impact of these fisheries on exploited stocks, we need to consider vulnerability, removal rates, impacts on behavior, and the potential for artificial selection (Dunlop et al. 2009). Fish that move along predictable routes and show highly aggregated spawning behaviors will be more vulnerable to population-level effects than those that show more variable migrations and dispersed or protracted spawning behaviors. Thus diadromous fishes such as striped bass, sturgeons, salmons, shads, and *Anguilla* spp. that rely on river corridors for spawning migrations are exceptionally vulnerable to overexploitation (Waldman 2013). Alternatively, fishes that reproduce over large regions and over protracted periods (e.g., Atlantic bluefish and menhaden *Brevoortia tyrannus*) will be more resilient to fisheries that principally target spawners.

With sufficient political will and information, managers can sometimes utilize patterns of vulnerability to effectively survey and control harvests during spawning migrations. It is feasible, for instance, to survey entire runs of Pacific salmon and to designate sustainable harvest levels (a.k.a. escapement-based regulations). This is a precautionary approach in that the harvest can be directly matched with the stock's reproductive potential regardless of climate, oceanographic, and other influences on stock productivity (Walters and Parma 1996). Historical management of Caspian sturgeons represents an important case study. Simultaneous to development of large hydroelectric projects in the Volga River, the Soviet Union implemented strict controls on sturgeon har-

vests based upon escapement of spawners. Traditional trap fisheries that intercepted spawners of Russian *Acipenser gueldenstaediti*, stellate *A. stellatus*, and beluga *Huso huso* sturgeons operated as both survey and management sites in support of export caviar fisheries, which were important to the country's economy. Managed removal of spawners (20%–30% y^{-1}) was likely higher than sustainable rates (<10% y^{-1}; Boreman 1997), but sturgeons skip spawning (see Provisioning Eggs above; Secor 2008), and during most spawning seasons, adults resided in the Caspian Sea where harvest was prohibited. Escapement-based regulations coupled with an exploitation refuge in the Caspian Sea led to highly productive caviar fisheries following dam construction at Volgograd and elsewhere (Secor et al. 2000). (Soviet hatchery programs produced detectable but comparatively small effects on stock sustainability.) With the dissolution of the Soviet Union in 1991, the prohibition on Caspian Sea harvests became unenforceable, the escapement-based management system broke down, and poaching proliferated. Numbers of adults plummeted in the Volga River, Ural River, and elsewhere from the overharvest of large juveniles and adults from the Caspian Sea. The Convention on International Trade in Endangered Species of Wild Fauna and Flora and Caspian Sea nations now work to rebuild sturgeon stocks, but efforts to survey and regulate fisheries in the Caspian Sea are substantially less effective than the tightly controlled escapement-based fisheries that preceded them.

Often it is not the overall harvest of spawners that influences sustainability of a fishery but rather selective removal or conservation of specific types of fish. Recovery of the Chesapeake Bay striped bass stock presents an example of how migration behaviors of different stock components can influence the effectiveness of fishing regulations. In the 1970s, targeted exploitation during spawning season by both commercial and recreational sectors resulted in extreme rates of harvest, approaching 90% for small males, which were particularly vulnerable to gill nets and angling (Secor 2000a). Large females tend to be coastal migrants and primarily vulnerable only during their spawning runs. Historical regulations in the Chesapeake Bay prohibited the harvest of these large females in part because markets and anglers preferred smaller pan-sized striped bass (Richards and Rago 1999). Although males and small females were decimated during the 1970s by overfishing, the large female striped bass persisted in the stock for >20 y. Following restrictive regulations on harvest levels in the early 1980s, the stock recovered rapidly owing to reproduction by relatively few, but old and fecund ocean-going, females (chap. 7, A Tale of Two Species: Northern Atlantic Cod and Chesapeake Striped Bass).

As described previously (see Mating Systems above), Atlantic cod form dense prespawning and spawning aggregations. Hydroacoustic visualization of a bottom trawl passing through a large aggregation showed a newly formed void in the shoal approximately five times the width of the trawl (Morgan et al. 1997). In a telemetry study of a much smaller aggregation disturbed by gill net deployments in the Gulf of Maine, Dean et al. (2012) observed complete abandonment of the spawning ground. Evaluating the densities of spawning aggregations on the Grand Banks during a decade of high exploitation and population collapse in the 1980s, Taggart et al. (1994) and Hutchings (1996) observed that the number of high-density outer shelf aggregations remained relatively constant, while lower-density midshelf aggregations declined in number with time. Membership in dense aggregations apparently conveyed energetic or reproductive benefits, which caused renewal of these aggregations each year despite fisheries that targeted them. So, although the overall abundance of Grand Banks declined as a result of overfishing, this decline was not detected because surveys and fishery assessments were biased high owing to the high-density aggregations. As dense aggregations coalesced, fisheries more readily exploited them, causing exploitation efficiency to rise despite the col-

lapse in the population (Hutchings 1996; Rose and Kulka 1999). At smaller sizes, aggregations may be particularly vulnerable to disruption of spawning behaviors due to fishing gear deployments (Dean et al. 2012). The long-term consequences of spawning site abandonment remain unknown.

Harvest patterns can influence reproductive traits and behaviors (Law 2000). Fishing typically targets larger juvenile and adults, which results in higher overall mortality on this segment of the population. Associated with long-term harvest is earlier maturation and increased fecundity (Rjinsdorp 1993), life history responses that in part compensate for increased adult mortality. These traits trade off against larger size at maturation, longer reproductive life span, and in some instances larger offspring in unexploited populations (Reznick et al. 1990; Rjinsdorp 1993; Hutchings 2005). As one example, a stock of the reef-associated red porgy *Pagrus pagrus* decreased its female size of maturation by 33% over a two-decade period of high exploitation rates (Harris and McGovern 1997). Red porgy are protogynous, where adults are initially female and then depending on size, age, and social environment develop into males, which make up only a minority of the adult population. Because recreational fisheries targeted large individuals, sex ratios became skewed toward females, potentially affecting fertilization success owing to the limited availability of mature males (Johannes 1978). Whether such changes are temporary or long lasting depends upon underlying genetics. Laboratory and field manipulations have shown that the traits are heritable to varying degrees, and selection regimes imposed by fisheries (e.g., size-specific removals) can alter long-term expression of reproductive traits regardless of environmental conditions (Heino and Godø 2002). A single laboratory study on Atlantic silversides *Menidia menidia* (a small semelparous estuarine species) found that shifts in reproductive traits were not permanent after selection was relaxed; however, more generations were required to reverse the shift than to initially induce it (Conover et al. 2009). In summary, fishing selectivity patterns are predicted to lead to smaller and younger spawners and higher individual fecundity, which can persist many generations after exploitation pressures are relaxed. A shift in the age and size composition of spawning adults has numerous implications for spawning behaviors and reproductive success, which will be reviewed further in chapter 4 (The Storage Effect).

Selective harvest can also cause short- and long-term changes to where and when fish spawn. Intensive harvest of spawning aggregations of Atlantic herring was followed by long periods during which historical spawning banks remained fallow (Corten 2001; Petitgas et al. 2010). A period of high exploitation and population collapse by Newfoundland northern stock cod coincided with a distributional shift from midsized aggregations in midshelf regions in the mid-1980s to large aggregations in outer shelf regions in the early 1990s (fig. 3.6; Taggart et al. 1994). High exploitation in estuaries and rivers can lead to extirpation of entire spawning runs of anadromous species (Limburg and Waldman 2009). Salmon hatcheries represent a classic example of how selective harvest can alter the timing of spawning runs. Hatchery managers, anxious to achieve a production quota for eggs and juveniles each year, target early spawners. Over generations this practice has caused spawning adults of Atlantic and Pacific salmon species to return several weeks earlier (Quinn et al. 2002). One can easily conceive that fishing patterns on later or earlier portions of a spawning run would impose a similar selection regime. Further, salmon harvest and hatchery practices can alter the mating systems of resident satellite and migratory territorial males (Gross 1991). Harvesting can remove specific behaviors from populations, such as more exploratory or "bold" behaviors that result in faster growth but increased vulnerability to capture (Biro and Post 2008). The consequences of the loss of specific behaviors associated with migration and reproduction depend upon whether they are transmitted through genetic or cultural inheritance (Dodson 1988; Warner 1990; Cury 1994; McQuinn 1997a). Regardless of their inheritance, spatial diversity in spawning habitats represents a critical element in the stability and resilience of populations and metapopulations. Transmission of spatial behaviors and their diversity within and across generations will be central topics in chapter 5 (Recalcitrant Nature: Adopted Migration and Tradition) and chapter 7 (Diversity in Spatial Structure: The Portfolio Effect).

Summary

Variation and modalities in migration and dispersal are of ecological consequence. Although mating systems, egg provisioning, and larval dispersal can be generalized for specific species or ecosystems, a recurring subtext is the prevalence of minority behaviors and diverse outcomes. The Newfoundland Atlantic cod stock spawn primarily in deep shelf waters with larvae swept by the strong Labrador Current, but adults also spawn in much shallower embayments of Newfoundland, where larvae are retained. Shelf temperatures and fisheries that influence where most spawning occurs drive how these modalities play out. Spawning cod aggregations are made up of monopolizing and satellite males, each type altering their chance of fertilization through release of spermatazoa that vary in number and motility. How the female provisions each egg and how far up in the water column eggs are released will influence embryo dispersal, the early growth and survival of the larva, and the rate at which it becomes an active swimmer.

As exemplified by Japanese eels and Pacific rockfishes, modes within mating systems and larval dispersal fates are influenced by climate. The spawning location of Japanese eels, so distant from juvenile habitats, shifts north and south with El Niño; small changes in the match between spawning locations, larval vertical behaviors, and the dominant North Pacific Gyre result in a stunning variety of dispersal fates across tropical and temperate East Asian waters. Fates of larval rockfish, whether they are retained in inshore waters of the California Current or not, depend upon where they are launched into the water column and whether offspring are spawned during upwelling or nonupwelling seasons. The strength and seasonality of upwelling also depend on El Niño, which can lead to climate oscillations and ecological regime shift (chap. 7, Resilience to Fishing and Climate Change: Uncharted Territory).

Egg provisioning, although constrained by the female's size and related swimming mechanics, shows modes of adapting to local circumstance. Pacific herring, *Sebastes*, and other fishes will adjust oocyte numbers against their ability to provision each batch of ripe eggs. With greater size and age, females often increase both egg number and offspring size, but offspring size can be highly variable between and within populations, as observed for striped bass. Variable provisioning affects embryo and early larval buoyancy and dispersal through allocation to oil inclusions and yolk. Through internal rearing, viviparous *Sebastes* and elasmobranchs give their offspring a developmental head start in subsequent dispersal and migration behaviors. Similarly, fish that spawn demersal eggs, which tend to be larger and have longer incubation times, may confer dispersal advantages to their young. But these behaviors come with associated trade-offs exemplified by the limited movement of fecund female *Sebastes*, relatively low fecundity for elasmobranchs, and egg predation (in the absence of parental care) for demersal spawners.

Fish larvae are not passive. Between and within species, a diverse repertoire of environmental cues and behaviors drives their dispersal. Vertical shear can impel long-distance dispersals by embryos and larvae, which make small changes in vertical position through adjustments to specific gravity, resulting in much greater dispersal than horizontal swimming alone. Further, fish larvae, in comparison to similar-sized invertebrate plankton, are effective swimmers. The active larva likely uses multiple vertical positioning cues and swimming behaviors as it disperses. Indeed, for striped bass larvae, vertical position cues apparently vary between wet and dry years within the same California estuarine ecosystem.

Fishing can influence the alignment of mating systems with early dispersal by directly disturbing spawning behavior and influencing reproductive traits within fished stocks. Exploitation will tend to reduce certain groups within a population—those whose patterns of migration and aggregation cause them to be particularly vulnerable as well as those that are selectively targeted by the fishery. Such exploitation can lead to disruption of spawning behaviors in the short term, and over longer periods can cause artificial selection for traits that affect reproductive schedules and population dynamics. Fisheries that cause stocks to shift toward younger adults will affect provisioning to offspring and spawning behaviors. Selectivity by fisheries (and hatcheries) can shift when and where spawning occurs, which in turn will influence larval dispersal behaviors by changing the suite of transport settings that larvae encounter. It is worth noting, however, that escapement-based harvest limits, a traditional management measure that often targets spawners, can lead to sustainable harvest rates in the face of changing levels of population productivity affected by climate and other factors. Still, in this and other management practices, unbalanced harvest toward certain population segments will likely diminish population stability, resilience, and persistence (chap. 7, Collective Agencies and Biodiversity).

Segue

For many marine fishes, mating systems explosively launch numerous larvae into an open and variable marine ecosystem. Examples include the countless dispersal trajectories that Japanese eel larvae experience. Newfoundland cod larvae jet away from natal shelf habitats, swept by the strong Labrador Current. Similarly, *Sebastes* larvae spawned off central California during seasons of strong upwelling can be rapidly transported to offshore regions where few juveniles occur. Although larval swimming and vertical behaviors will contribute to some larvae finding their way to nursery habitats, what about the others? In chapter 5, we will address mechanisms that lead fish back to natal habitats and the consequences of lost larvae. But first we ask in chapter 4 why so many marine fishes produce small offspring, and how small and widely dispersed fish larvae contribute to complex and diverse life cycles. These issues are addressed in the context of how the life cycles of teleosts and elasmobranchs are adapted to glean production across multiple trophic levels in size-structured marine food webs.

4 Complex Life Cycles and Marine Food Webs

Migrating through Size Spectra

In the ocean, for planktonic organisms such as fish larvae, "food heaven" almost invariably equates to "predation hell."
—Bakun (2006)

Marine Food Webs: Big Things Eating Little Things

As they migrate, fish traverse food webs. Over its lifetime the teleost larva is destined to increase its mass by three to six orders of magnitude, with most growth occurring during the first year of life. Early on, ranks of larvae wither under extreme mortality, yet they exploit the most productive part of the marine food web—that composed of small plankton. To meet their ever-changing demands for prey, larvae, juveniles, and adults migrate to new habitats and food webs. Prey resources typically vary on a seasonal basis, dependent on transient plankton blooms and underlying oceanography and climate. The set of life stage- and season-specific patterns of habitat use and the migration pathways that connect them define an overall complex life cycle (Harden Jones 1968; Baker 1978).

Marine food webs are size structured. Over its life cycle an individual fish balances the high consumption rates needed for maintenance, growth, and reproduction against risks associated with predation, starvation, disease, and environmental stresses. Early in life, high mortality and poor resource matchups are common (Hjort 1914; Cushing 1990; but see also Leggett and DeBlois 1994). The periodic life history mode (late maturity, high fecundity, and low early survival) is pervasive among teleost fishes. Storing gametes in relatively large and invulnerable adults offsets early losses. Spawning behaviors also hedge against losses, where embryos and larvae are placed into favorable conditions through directed adult migrations. In contrast, elasmobranchs exhibit a different life history mode (late maturity, low fecundity, and relatively high early survival), but here too life cycles must be considered against a trophic seascape where "big things eat little things."

In this chapter, complex life cycles are placed into frameworks of (1) how marine food webs are structured, (2) how fish balance growth and survival trade-offs in complex life cycles, and (3) how populations of marine fishes persist given high variability in early survival. The question of why fish start life small, although not appropriate to all fishes, leads to broad inferences on how life cycles of fishes are adapted to unique attributes of marine food webs. The patchy distribution of resources bears on this question and allows us to begin to explore the diversity of life cycles among species and populations. We start this chapter by contrasting marine food webs with terrestrial ones, emphasizing production and biomass structure among trophic levels. We then move on to size spectrum theory, complex life cycles, ontogenetic habitat shifts, and exploitation of transient and patchy prey resources. Fish schooling is reviewed as a principal collective agency in the way migration responds to changes in food webs. Conservation of demographic attributes that favor stability and resilience is considered in the context of modes of life cycle traits and spatial behaviors that covary with age structure and longevity.

Marine and Terrestrial Food Webs:
How Different?

The free embryo of the Atlantic cod initially consumes single-celled organisms, including phytoplankton. Later, with development of a functional jaw and increased pursuit, the young larva begins consuming small-sized zooplankton, such as early stage copepods and other crustaceans. As its length and jaw's gape increases, ever-larger zooplankton are pursued and consumed. And so it continues into juvenile and adult stages: the cod consuming larger invertebrates and small fishes, exploiting ever-increasing prey size as it disperses and migrates among habitats during its life cycle. Importantly, the cod's gape typically limits it to prey ≤1% of its mass. Now consider the cod larva's terrestrial analogue by size: an ant (1–5 mm) consuming carbon from its primary habitat and food source, a 10-m oak tree. Over multiple generations, the ant's colony will be sustained by this single source of carbon and energy, which on an individual basis is ~10^9 times the mass of the ant consumer. This comparison of the trophic dependencies by two organisms, each starting life at a similar size, epitomizes how and why marine and terrestrial food webs are so fundamentally different (Cohen 1994).

Terrestrial food webs contain a substantially higher density and stock of energy and carbon than do marine food webs, despite the latter's much greater area and volume (table 4.1). Assuming that the great majority of oceanic biomass occurs within the upper 500 m, the earth's oceans support a much more diluted carbon stock, 1.1×10^7 g km^{-3}, in comparison to 7.5×10^{10} g km^{-3} estimated for terrestrial ecosystems (estimates here and elsewhere were taken directly or derived from Cohen 1994). The distribution of carbon also deviates: in terrestrial systems, carbon is chiefly stored in relatively large metazoan plants such as trees, whereas in marine systems, it occurs more transiently in short-lived phytoplankton (fig. 4.1A). (Macroalgae, such as kelp, and coral reefs might be conceived as tree analogues and comprise a significant amount of marine carbon stocks,

~30%, but these have twentyfold shorter turnover times than terrestrial vegetation; Valiela 1995). Further, photosynthesis is less efficient in marine systems where nutrients sink away from sunlight and are rapidly lost to the euphotic (sunlit) portions of the water column. Single-celled algae that occur here have life spans of hours to weeks. In comparison, carbon sequestered in terrestrial plants turns over at durations of months to centuries, translating to global estimates of carbon turnover of ~1 month in marine systems and 11 y in terrestrial ones.

The difference in turnover rates by primary producers confers very different shapes to marine and terrestrial food webs. The quintessential triangular shape of terrestrial systems, exhibiting the decreasing transfer of carbon stocks to higher trophic levels, does not similarly apply to marine food webs. More columnar shapes are common, where transient carbon is more efficiently welled up into stocks of consumers. Thus the base of the terrestrial food web is defined by storage of carbon, whereas in marine food webs it is determined by high turnover rates and production of carbon (fig. 4.1A). The challenge, then, for a consumer in a marine system is to capture dilute sources of energy that are mostly produced by small-sized plankton. Cohen (1994) highlighted adaptations that concentrate small marine prey such as by entangling structures or filter feeding. Fish exhibit yet another strategy, termed *life history omnivory* (Pimm and Rice 1987), where a fish transits through increasing trophic levels over its life cycle. Importantly, starting life small as plankton leads to efficient conversion of ecosystem production into larval and juvenile biomass. Life history omnivory also entails complex life cycles because changing trophic demands can rarely be met by staying in place.

Evidence for the link of fish to phytoplankton production comes from strong positive correlations between ecosystem primary production and fish production or harvests (Nixon 1988; Iverson 1990). Marine ecosystems differ by orders of magnitude in primary production, and so does fish production (fig. 4.1B; Houde and Rutherford 1993). Conversion efficiency of

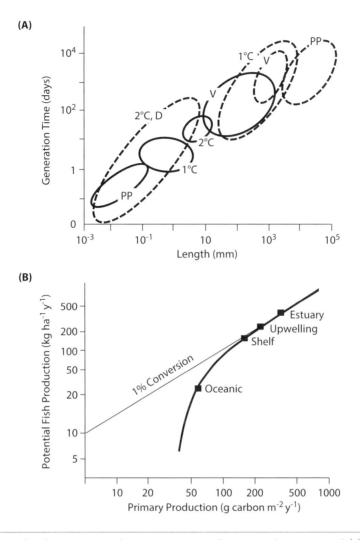

Figure 4.1. (A) Food web comparisons between marine and terrestrial systems, and (B) conversion of primary production to higher trophic levels in marine ecosystems. Dashed and solid ellipses are for terrestrial and marine food webs. Groups included primary producers (PP), invertebrate primary and secondary consumers (1°C, 2°C), vertebrates (V), and detritivores (D). (A) redrawn from data and graphics in Steele 1991, and (B) from Houde and Rutherford 1993, with permission.

primary production to fish production (mean trophic position ~2.5) varies as well, from ~1% (estuarine, upwelling, and shelf ecosystems) to 0.4% (oceanic ecosystem). Houde and Rutherford (1993) predicted global fish production rates of 71, 356, 116, and 817 10^{12} g carbon y^{-1} for estuarine, upwelling, shelf, and oceanic ecosystems. Although such estimates are interesting from a comparative framework, they represent gross simplifications. By assigning a single trophic position estimate (2.5), we ignore how fish change trophic positions throughout life. To better engage the relationships between primary production, fish production, and complex life cycles, we turn to size spectrum theory.

Table 4.1. Global comparison of marine and terrestrial food web attributes

Attribute	Marine	Terrestrial
Earth's surface (%)	71	29
Mean depth of life zone (m)	500	50
Volume of life zone (10^8 km^3)	1.8	0.0075
Living biomass (10^{15} g carbon)	2	560
Living biomass density (10^7 g carbon km^{-3})	1.1	7500
Net primary production (10^{15} g y^{-1})	20–44	48
Residence time of carbon in biomass (y)	0.08	11.2
Animal biomass (% total)	10	0.01
Predator animal body mass (g)	4,700	140
Number of trophic links	high	low

Note: Estimates are from Cohen (1994). Mean depth of life zone and related statistics differ from Cohen's stipulated depth of 1.37 km.

Size Spectrum Theory

Imagine filtering the ocean with a mesh that collected biota from phytoplankton to whales—how would numbers vary across this tremendous range in mass (~10^{12} µg)? The ability to sample a large part of this size fraction (range in mass = 10^7 µg) became feasible through automated particle counters. Using a Coulter counter, Sheldon and Parsons (1967) summed volumes of particles corresponding to nanno-plankton, microplankton, and macroplankton and observed that total volumes for each of these groups were remarkably similar. Smaller plankton is more numerous than larger plankton, but when summed within size fractions, each size-fraction volume was equivalent. As we have already reviewed, turnover rates are faster for smaller-sized plankton than larger consumers (fig. 4.1A). Therefore numerous small prey support less abundant big predators. When weights are summed within log$_{10}$-scale mass increments (which correspond coarsely with trophic levels), biomass levels across producers and grazers are similar. Subsequent inclusion of fishes (larger consumers) showed that they too conformed to the concept of an equivalent size spectrum (Boudreau and Dickie 1992; Sprules and Goyke 1994; Jung and Houde 2005).

The size spectrum theory posits that energy flows from small to large sizes through predation, in a series of predator-prey steps that begins with productive and abundant plankton (Kerr and Dickie 2001). As an emergent ecosystem attribute, fitted size spectra can supply information on overall ecosystem production, size-specific predation rates, trophic conversion efficiencies, predator to prey size ratios, and dominant trophic guilds (Rice and Gislason 1996). Our interest here, however, lies in how fish traverse size spectra over their life cycle—particularly (1) how fish during their first year of life glean production from transient blooms of plankton but avoid predation; (2) how fish life cycles play out within a size spectrum; and (3) what happens when fish migrate between ecosystems characterized by differing size spectra.

Gleaning Production

In size spectra, biomass is plotted against classes of organism weight (alternatively, energetic units can be used). In following a fish's life cycle through the size spectrum, it is convenient to consider a cohort of similar-aged fish and then audit biomass as members of that cohort grow and die in response to changing fields of prey and predators. The demographic fate of a cohort (N_t) is typically estimated from a beginning abundance (N_0) by accounting for its instantaneous mortality rate (M) over time (t).

$$N_t = N_0\, e^{-Mt} \qquad\qquad 4.1$$

Thus for any period t we can estimate a rate of loss assuming M is constant over that period. To track the cohort on the size spectrum, time periods are converted to weight stanzas; that is, what is the survival between successive weights (e.g., W_t vs. W_0)? The answer will depend on the rate at which the cohort moves between these weights: the cohort's instantaneous somatic growth rate (G).

$$W_t = W_0\, e^{Gt} \qquad\qquad 4.2$$

We can readily see that weight-specific survival (l) will depend on integrating mortality rate over the period required by the fish to grow between the two weights. This rate is termed the *physiological rate of survival* because it relates to the physiological state of an individual (i.e., weight) rather than age (Istock 1967; Beyer 1989):

$$l_{(W_t, W_0)} = e^{\,(-M/G)(W_t - W_0)} \qquad\qquad 4.3$$

or

$$l_{(W_t, W_0)} = w_t / w_0^{\,(-M/G)}. \qquad\qquad 4.4$$

Here M and G are assumed to be constant and independent to each other over the specified weight interval. The biomass of the cohort ($B_{(Wt)} = \sum N_t W_t$) over a size stanza is estimated as

$$B_{Wt} = B_{W0}\, \left(w_t / w_0^{\,-M/G} \right). \qquad\qquad 4.5$$

Weight-specific growth and mortality rates vary in important ways over a fish's life history, which we will now consider.

The size spectrum is typically portrayed as an annual snapshot, but if we depict a larval cohort, we should consider how it interacts with plankton resources and predators that vary over weeks and months. In addition, recall that phytoplankton stocks tend to be transient. In temperate and other marine ecosystems, most annual primary production occurs as seasonal plankton blooms, triggered by changed solar radiation and mixing (e.g., seasonal changes in upwelling, precipitation, surface warming, or cooling; see Match-Mismatch below). Heath (1995) observed seasonal dynamics in the size spectrum, deploying a laser optical plankton counter in a Scotland fjord. He detected a seasonal displacement of peak biomass, initially associated with a phytoplankton bloom in early April and later propagating into a wave of zooplankton grazers, which occurred ~100 d later. Heath projected this propagation wave using a cosine function and other terms to generate an annual cycle in primary production that transfers to series of larger consumers in a dampened manner (fig. 4.2A).

To explore how larval fish might exploit seasonal biomass waves, Pope et al. (1994) simulated an annual phytoplankton bloom through a cyclical probability function (similar to fig. 4.2A) and tracked biomass across a size spectrum (10^{-7}–10^5 g). Pope and colleagues then "spawned" cohorts of cod larvae at monthly intervals before, during, and after the peak phytoplankton bloom. On the basis of physiological mortality rates of larval cohorts, they predicted that optimal spawning times occurred just ahead of the phytoplankton bloom because of a fundamental trade-off: (1) acquire sufficient smaller zooplankton prey to support growth versus (2) avoid mortality due to subsequent waves of larger predators. Pope et al. (1994) coined this finding "surf riding the size spectrum," whereby "the obvious strategy is to 'surf' up the size distribution, riding the prey wave and staying ahead of the predator wave as long as possible." They called attention to equation (4.3): when growth exceeds mortality over a large range of sizes, the amount of ecosystem production that a cohort can glean

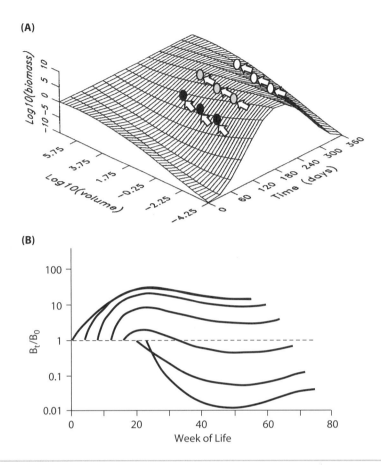

Figure 4.2. Production of larval fish cohorts as they progress through a seasonally pulsed size spectrum. (A) Primary production peaks on day 140 and propagates to larger-sized consumers ($\log_{10}$ volume) as the season progresses. Overlaid are ovals representing the production of larval cohorts that are spawned prior (black), during (gray), or after (white) the peak phytoplankton bloom. The trajectory of the early and intermediate spawned cohorts place them in a favorable size spectrum with respect to a greater abundance of smaller prey and lesser abundance of larger predators. The late cohort experiences low-prey and high-predator fields in this size-structured depiction. (B) Each line represents a larval cohort starting life at a time prior to or lagged behind peak primary production, which occurred at 4 weeks. B_t/B_0 represents biomass gain by the cohort. (A) and (B) are modified from modeled size spectra from Heath 1996 and Pope et al. 1994, respectively, with permission.

will increase the larger the stanza. A smaller initial embryo size can substantially increase resources available to a growing cohort, but this must be balanced by seasonal changes in predation levels. In the cohort-specific biomass curves, predation dominates as the larger consumer wave overtakes the larval cohort during weeks 40-60 of their simulation (fig. 4.2B). This trough (predation) can be avoided by en-

tering the system later at a larger size. Pope et al. (1994) proposed two alternate "strategies": (1) start small, glean peaks in ecosystem production, and offset high mortality through high fecundity and selection of favorable spawning times (e.g., pelagic spawning teleosts) and (2) minimize mortality by starting life big (e.g., viviparous elasmobranchs).

Migrating among Size Spectra: Complex Life Cycles

We next explore how biomass trajectories of co-horts evolve not just seasonally but over an entire life history. Mapping an entire life history onto a size spectrum requires data on stage-specific mortality and growth rates (Hartvig et al. 2011), which are available for the intensively studied Chesapeake Bay striped bass (Setzler et al. 1980; Houde 1996). For cohorts originating in two hypothetical spawning tributaries A and B, biomass was tracked according to equation (4.5), from the embryo stage (10^{-3} g) to the death of the oldest adult at ~4 10^4 g (fig. 4.3A). For both tributaries, the cohort loses substantial biomass during its early life. Indeed, during the embryo period, mass is lost owing to both high mortality and weight loss of individual embryos, which by definition cannot recoup losses from consumption. For tributary A, the mass spawned by the population is not fully recovered until the adult period, but its return to this level indicates that the cohort has replaced itself over a single generation. The cohort for tributary B does not fare as well—declines in biomass persist for a longer period, and as a result the mass of spawned eggs is never recovered through later growth and survival. Following each life cycle to its terminus, a sharp decline occurs in the size spectrum where maturation shifts energy allocation away from growth to reproduction. Basal metabolism demands an ever-increasing fraction of the fish's consumed energy, and somatic growth declines (Pauly 1981; Andersen et al. 2008). The adult now moves slowly through the size spectrum, utilizing surplus energy for reproduction, maintenance, and in some instances migration (see The Storage Effect below). This stasis stands in contrast to early phases of life that are characterized by rapid biomass turnover through growth and predation.

The spectral life history emphasizes a multiphasic life cycle, and indeed many explanations for the evolution of complex life cycles focus on distinct metamorphic stages (a.k.a. indirect development; Wray 1995). Istock (1967) argued that natural selection should operate independently on each stage, such that complex life cycles would be favored when long-term fitness components were equivalent between stages. For instance, a hypothetical ancestor with only an adult life history stage might divide its life cycle as a means to exploit a new set of resources. For most teleosts, resource expansion seems a likely contributor to the development of complex life cycles, where larval and juvenile growth phases exploit an extremely wide and diverse spectrum of prey within marine food webs. Still, given the widespread nature of complex life cycles in fishes, it seems implausible to ascribe a single major cause to their evolution.

Most changes in the cohort-specific size spectra occur during the first year of life. For tributary B (fig. 4.3A), we observe that after the loss of larval biomass the tributary supports higher rates of juvenile production than tributary A (weight stanza ~10^0–10^2 g). One plausible explanation is density-dependent growth, which begins to exert influence during the juvenile period. Regardless of cause, a "discriminating" fish from tributary A would transit to tributary B during the early juvenile period (fig. 4.3B). Here the relative rank of alternative habitats changes with size or ontogenetic stage. The rank order of alternate habitats depends on cohort production and can be assessed by the ratio M/G; see equation (4.4). The habitat with a lower M/G ratio should be favored, which some authors have associated with reproductive success (Werner and Gilliam 1984). The ranking between tributaries changes for juveniles >5 g. Coincident with this changed ranking, cruising striped bass (5 BL s^{-1}) can swim ~1 km h^{-1}, giving the early juvenile a range of tens of kilometers over which to explore, evaluate, and occupy new habitats.

Complex life cycles are often a starting point in explanations for the evolution of migration. For instance, in fauna from bees to humans, Baker (1978) presents an encyclopedia of circumstantial evidence in support of the view that migration is driven by lifetime reproductive success (see also Istock 1967; Werner and Gilliam 1984; chap. 1, Migration: A Trait, Syn-

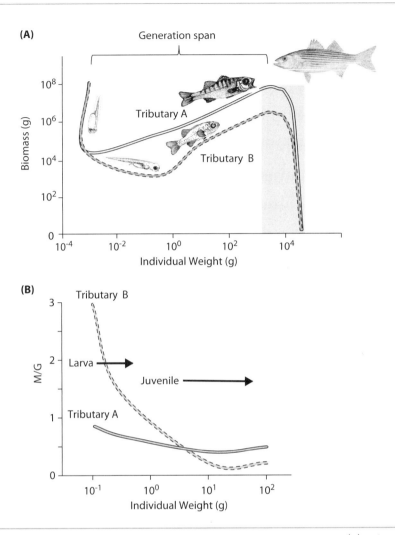

Figure 4.3. Depictions of two striped bass cohorts' life histories as size spectra. (A) Cohort biomass is shown from spawning to death in two hypothetical tributaries. The adult period is indicated by a gray box. (B) The ratio M/G (instantaneous growth over mortality) is shown for the juvenile period. After 5 g weight, M/G changes rank between the two tributaries. Data for two tributaries are generalized from data in Secor and Houde 1998 assuming populations of 50,000 females. Illustrations of yolk sac larva from Mansueti 1961; larvae and transforming juvenile from Pearson 1941; and juvenile from Doroshov 1970, with permission.

drome, or Complex System?). Darwinian fitness is attributed to traits and behaviors that favor use of habitats contributing to reproductive success either directly (e.g., fertilization success in reef spawning fish) or indirectly (e.g., lagoon growth habitats for juvenile Japanese eels) after migration and other costs associated with movement to new habitats are appro-

priately discounted (Baker 1978). Related tests have emphasized trade-offs among habitats associated with forage rewards and predation risks assessed against reproductive success. A classic example was pond experiments on bluegill *Lepomis macrochirus*, a freshwater lake fish that early in its life resides in protected marsh waters (Werner et al. 1983a,b). In adjacent open water habitat, prey resources became increasingly available, and predation risk decreased with bluegill size, causing larger juveniles to move from littoral marsh to open water habitats. The size-related shift to open water habitat was matched with a change in diet and growth and higher physiological survival. The literature provides experimental evidence that in some instances fish can make immediate choices in selecting habitats that will favor their growth and survival prospects (Gilliam and Fraser 1987; Dahlgren and Eggleston 2000; Juanes 2007).

Ontogenetic Habitat Shifts

Changes in habitat that depend on fish size or stage of development are referenced as ontogenetic habitat shifts. Particularly relevant examples occur for amphibians, where aquatic tadpoles metamorphose into more terrestrial adults and immediately encounter a refuge from aquatic predators (Werner 1986). For bluegill juveniles, metamorphosis does not accompany a habitat shift; in this case, increased size alone provides a predation refuge (Werner et al. 1983a,b). For marine fishes, settlement—a shift from a more pelagic dispersive stage to a more sedentary demersal stage—is a common type of ontogenetic habitat shift. Larvae and young juveniles move from pelagic to demersal food webs with attendant changes to their foraging success and predation risk (Dahlgren and Eggleston 2000). Indeed, the large change in M/G values for striped bass juveniles (fig. 4.3B) corresponds to a habitat transition between pelagic habitats to littoral ones. Ontogenetic habitat transitions, whether accompanied by metamorphosis or not, require changed behaviors and often represent a "dangerous" period of increased exposure to predators, reduced growth, and physiological stress (Leis and Carson-Ewart 2001; Juanes 2007). Such transitions also occur for sharks where shallow nurseries offer juveniles refuge from adults, but they may be forage limited (Heithaus 2007). We next consider three examples of ontogenetic habitat shifts: those for Atlantic cod, European plaice, and Atlantic bluefish.

Newly transformed juvenile cod (~12 mm) persist in the water column, still removed from the demersal habitats where larger juveniles and adults reside. On Georges Bank, pelagic juveniles occur over a wide swath of the bank after riding its margins as larvae in tidally driven currents (fig. 4.4; Lough 2010). They shoal for protection against predation, undertake diel vertical migrations, and feed on adult copepods, mysids, and amphipods (Auditore et al. 1994; Hüssy et al. 1997). Following its transformation from the larval stage, the pelagic juvenile now has a full complement of fins and fin rays but retains a terminal mouth and a fusiform body shape conducive for feeding and swimming in its pelagic habitat. In contrast, the demersal juvenile develops a subterminal mouth, barbells, a slightly compressed body shape, and cryptic coloration adapted to a structured demersal habitat (Auditore et al. 1994). The cohort of pelagic juveniles does not settle to demersal habitats all at once, but does so over many months and over a wide size range (25–80 mm). On Georges Bank, nursery habitats occur primarily over a relatively small section of gravel-laden seabed (fig. 4.4; Lough 2010). In demersal habitats, juveniles hold territories that promote feeding on larger demersal prey, such as amphipods, decapods, small fishes, and polychaetes (Auditore et al. 1994). Still, the transition conveys risk, with losses exceeding 50% during the first 2 months (Tupper and Boutilier 1995; Lough 2010). Preference for structured habitat, diurnal shoaling behaviors, and avoidance of nearshore habitats are all behaviors that reduce predation (Lough et al. 1989; Juanes 2007). Among sand, seagrass, cobble, and reef bottom habitat types, for instance, juveniles selecting cobble and reef suffered less postsettlement mortality than those settling in sand or seagrass (Tupper and Boutilier 1995). Earlier settlers grow faster and more effectively hold territories, but postsettlement changes in habitat distributions can occur as a result of changes in local density (Juanes 2007). With increased settlement in preferred habitats, density-dependent predation, including cannibalism, can reduce or reorder habitat suitability (Sundby et al. 1989; Fromentin et al. 2001; Lough 2010).

Atlantic bluefish experience an ontogenetic habitat transition that conveys larvae and young juveniles 300–800 km from spring spawning regions in shelf waters off the southeastern United States to inshore and coastal juvenile feeding habitats in the mid-Atlantic United States. The Gulf Stream Current transports surface-oriented larvae to higher latitudes, and thereafter young juveniles disperse into inshore estuarine and coastal waters by swimming or by becoming entrained into warm-core ring eddies (fig. 4.5; Hare and Cowen 1996). The long-distance habitat transition by larvae and young juveniles conveys risk of unsuccessful transport (Hare and Cowen 1993), but it also causes them to enter high-latitude habitats at times favorable to undertake a piscivorous lifestyle relatively early in their ontogeny (Juanes and Conover 1995). The juveniles remain pelagic but already (during the late larval period) have developed canines,

a laterally compressed carangiform swimming form, and silvery pigmentation—all favoring pursuit predation (Hare and Cowen 1994; Scharf et al. 2009). In the mid-Atlantic United States, spring spawning by most coastal fishes occurs later than in the southeast US Atlantic. By the time juvenile bluefish arrive, its dominant prey, juvenile Atlantic silverside, has grown into a vulnerable size range for the foraging bluefish. Rapid growth by juvenile bluefish allows them to maintain a favorable size advantage over the growing silverside prey, effectively surfing the size spectrum. Later spawned cohorts of bluefish are well matched to a second dominant prey source, larval and juvenile bay anchovy, which originate principally from eggs spawned during summer months (Juanes and Conover 1995).

Plaice show a large metamorphic change in morphology and behavior that is associated, but not precisely coincident, with settlement to a demersal habitat. In the English Channel and southern approaches to the North Sea,

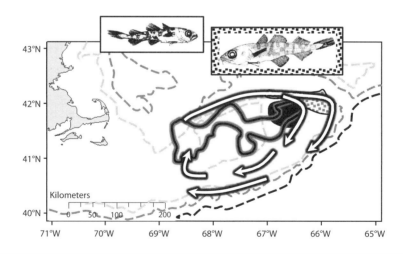

Figure 4.4. Ontogenetic habitat shift for Atlantic cod on Georges Bank. Demersal juveniles are concentrated in gravel-laden seabed habitats on northeastern Georges Bank (stippled region) and originate from pelagic juveniles that occur over a much broader region of the bank (open region). A clockwise gyre conveys eggs and larvae from the eastern spawning ground (black region) around the bank's slope. As larvae transform, they move toward shallower bank regions. Pelagic and demersal juvenile stages are shown in inset boxes. Regions of pelagic and demersal juvenile incidence are generalized from Lough 2010 for collections made in 1986-1987, used here with permission. Illustrations of pelagic and demersal juvenile cod by Elisabeth A. Broughton appearing in Auditore et al. 1994.

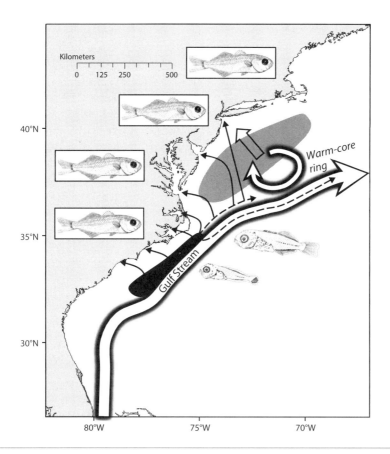

Figure 4.5. Ontogenetic habitat shift for Atlantic bluefish. Juveniles destined for US South Atlantic and mid-Atlantic estuaries originate from spring (black area) and summer (gray area) spawning areas. Larvae originating from spring spawning entrain into the Gulf Stream (dashed arrows). Cross-shelf dispersal occurs either through direct swimming (solid arrows), or through retention in warm-core rings and cross-shelf transport. Spring- and summer-spawned bluefish arrive in estuaries at a size that can exploit young-of-the-year forage fishes. Larval illustrations from Norcross et al. 1974, and juvenile illustration from Able and Fahay 1998, with permission.

adults spawn during winter months, releasing their gametes into the water column (Gibson 2005). Over the next 30–70 d, prevailing currents drive larvae into nearshore regions of the Netherlands and German Bight (fig. 4.6; Bolle et al. 2009). During this time the bilateral larva begins its transformation, becoming asymmetric by 10 mm with transition of its left eye to a middorsal position. But the transforming larva continues to feed and disperse in the plankton (Keefe and Able 1993). Late spawned cohorts in May encounter dense jellyfish blooms, which can substantially crop their ranks (van der Veer 1985). Cross-shelf transport to the shallow nurseries of the Wadden Sea may be facilitated by tidal stream transport (chap. 3, Larval Transport; Rijnsdorp et al. 1985) or through estuarine residual currents, which are landward in bottom waters (fig. 4.6; Bolle et al. 2009). Upon arrival in nursery habitats of the Wadden Sea, juveniles (>15 mm) complete their metamorphosis in large numbers (0.1–0.5 juveniles m^{-2}). In these sometimes very shallow habitats (<1 m), they are exposed to a gauntlet of predators—shrimp, crabs, juvenile cod, and cormorants—that crop the settled juveniles according to their size and density (Sogard 1997; van der Veer et al. 1997). Within a week after settling, newly settled plaice can experience >50% losses due to the dominant shrimp predator *Crangon crangon*, with estimated daily predation (2%–15% d^{-1}; van der Veer and Bergman 1987). A positive feedback between plaice density and predation can exert strong regulation on juvenile abundance, which would otherwise be controlled by transport and mortality during

the larval period (van der Veer et al. 1997; Bolle et al. 2009). Following settlement, juvenile plaice serially outgrow size-selective predation by shrimp (vulnerable sizes <30 mm), crabs (vulnerable sizes <50 mm) and juvenile fish predation (vulnerable sizes <45 mm; van der Veer and Bergman 1987; Ellis and Gibson 1995). Thereafter, survivors derive enhanced foraging benefits within the shallow nursery habitats of the Wadden Sea (Ellis and Gibson 1995).

The risk of habitat transitions for young settling juveniles depends upon the size and abun-

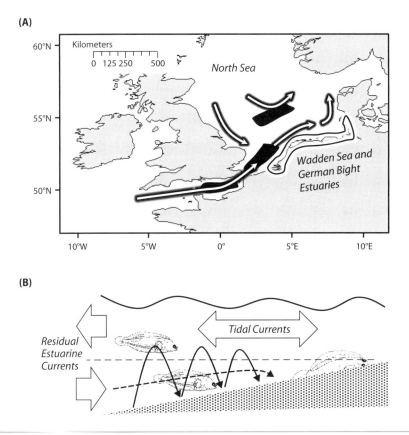

Figure 4.6. Ontogenetic habitat shift for European plaice. (A) Juveniles destined for the Wadden Sea and other southern North Sea nursery habitats originate from spawning (black areas) in the Southern Bight of the North Sea and recruit to Wadden Sea nurseries. Arrows indicate major surface currents. (B) Cross-shelf transport occurs by two means: (1) tidal transport where ingressing young swim up in the water column on flood tides (solid arrows), and (2) dispersal in bottom waters via estuarine residual currents, which are landward in partially mixed estuaries (thick dashed line). The thinner dashed line represents the pychnocline, the boundary between fresher surface waters and saltier bottom waters. Plaice illustration from Russell 1976.

dance of predators in the new habitat, which will vary year to year. For example, a cohort of 45-mm juvenile spot *Leiostomus xanthurus* settling in a shallow estuarine habitat will fare much better if it arrives earlier, when flatfish predators are not sufficiently large to be effective predators (Rice et al. 1997). The cohort's growth can continue to keep them ahead of the cohort of potential predators, until they reach a size where they are no longer susceptible.

Effect of Migration on Size Spectra

From an ecosystems perspective, how might migration and habitat shifts affect size spectra? Increased dominance of species showing size-dependent migrations would deflect the size spectrum, perhaps in a manner similar to that expected from fishing effects (fig. 4.7). Eggs and larval fishes are rarely a dominant component among zooplankton, but the arrival of juveniles into nursery habitats can affect the associated ecosystem size stanza. High abundances for this size stanza can result in a local dome in the size spectrum associated with ingress, growth, and predation dynamics (Jung and Houde 2005). Alternatively, the entire slope of the size spectrum might be deflected by changes to this size stanza. Highly productive planktivorous juveniles and adults often occur as vast schools in upwelling and shelf systems, influencing food webs in transient ways through their migrations. This group of fishes (anchovies, sardines, herrings, etc.) also plays a critical role as prey to larger piscivores. When oceanographic changes favor populations of planktivorous fish, explosive growth and changed migration patterns can occur simultaneously (see Schooling through Food Webs below). Injection of vast schools of planktivorous fishes in new ecosystem and their dominance over predators will cause large disruptions in food webs and ecosystem size spectra, particularly on a seasonal basis (Bakun et al. 2010).

For investigators applying size spectra to evaluate ecosystem attributes, the above examples should caution that migration can be a central factor biasing seasonal and annual depictions of size spectra that might otherwise have been ascribed to fishing and other ecosystem effects (fig. 4.7). Currently, a central assumption in size spectra and other food web models is stationarity, but improved methods of measuring and modeling spatial behaviors (e.g., Heath 1996) will permit their more explicit inclusion in ecosystem models.

From the fish's point of view, dispersal and migration will cause a cohort to move through multiple size spectra over their life cycle, with consequences to growth, survival, and reproduction. Still, the structuring of marine food webs by body size is but one factor contributing to complex life cycles. We now turn to another fundamental attribute of marine ecosystems—transience, or their spatially patchy and temporally periodic nature.

Marine Ecosystem Patchiness, Transience, and Periodicity

Rather than the imaginary filter that captured the ocean's size spectrum introduced in the previous section, let us now deploy a plankton net designed to capture the zooplankton prey of fish larvae in the size fraction 50–1,000 μm. Concentrations will depend upon when and where we sample varying from 10^2 to 10^3 m^{-3} in less productive marine ecosystems, such as the Sargasso Sea, to 10^5 to 10^6 m^{-3} in seasonally productive estuaries and coastal and upwelling systems; typical temperate concentrations range from 10^3 to 10^5 m^{-3} (Cushing 1995; Watanabe et al. 1998; Houde 2009). Are these ranges sufficient to support feeding and growth by fish larvae? Larvae of many species are highly susceptible to starvation early in their development (Hjort 1914; Lasker 1975; Miller et al. 1988), and larval rearing studies indicate that higher concentrations are needed, on the order of 10^5–10^6 m^{-3}. Laboratory rearing artifacts contribute to this discrepancy, but the larger probable cause is the patchiness of zooplankton and factors that increase prey encounter rates (Houde 2009). These are in fact obscured by our plankton net deployments. As the plankton net moves through the water column, it encounters voids

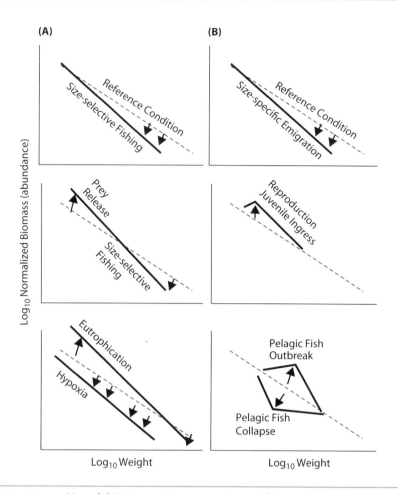

Figure 4.7. Depictions of how (A) fishing and water quality and (B) migration patterns influence size spectra. Biomass spectra are normalized to abundance (biomass/weight class). A starting reference condition is given by the dashed line. Fishing can change the slope through (1) removal of larger individuals or (2) indirect food web effects (diminished predation by harvested piscivores). Eutrophication can increase the yield of an ecosystem, particularly within the pelagic food web; however, with overnutrification, systemic hypoxia can result, causing diminished ecosystem yield. Migration of dominant species violates the assumption of stationarity in size spectrum theory and can affect large portions of the size spectrum through (1) size-specific migration; (2) large seasonal inputs of eggs, larvae, and juveniles; and (3) outbreaks and collapses in plantivorous fishes.

in zooplankton as well as dense patches, but the investigator can only estimate a single average density. Also, plankton net deployments may not be fully representative of the times and places where successful feeding occurs.

As emphasized in the previous section, complex life cycles of marine teleosts involve planktivory by larvae for most species. Al-

though speculative, Bone and Marshall (1982) proposed that the principal life history difference between teleosts and elasmobranchs—indirect versus direct metamorphic development—might be a consequence of coincident radiation of early teleosts and zooplankton (particularly copepods) during the Triassic period (200–250 Ma); elasmobranchs predated this radiation and did not later develop a free larval stage owing to phylogenetic constraints (Bone and Marshall 1982; Freedman and Noakes 2002).

How fish sample zooplankton as a patchy and transient resource entails life cycle considerations: the trade-off between embryo size and number, and when and where spawning occurs. It is not just a matter of simply starting small and gleaning biomass from a productive stock of small prey: sufficient densities of prey are patchily distributed in space and periodically distributed in time. Frontal features, for instance, can concentrate plankton 10- to 1,000-fold in comparison to regions that are immediately adjacent only hundreds of meters distant (Cushing 1995). Vertical stratification over tens of meters can cause smaller-scale patchiness (chap. 3, Larval Transport). More resolved still, turbulence can affect prey densities and encounter rates at a scale of millimeters. But on a seasonal basis, dense plankton resources can occur over broad regions. In temperate and boreal systems, spring algal and zooplankton blooms are fairly predictable and can occupy regions of 10^{-1}–10^4 km^2 (Cushing 1995). First we must consider the issue of patchiness at a mesoscale that is relevant to a marine spawning fish: that associated with frontal systems and blooms (kilometer scale). Then we move on to address the seasonal periodicity of plankton resources. Although the emphasis is on zooplankton prey, physical resources (e.g., temperature, salinity, dissolved oxygen, structured habitat) are often patchily or periodically distributed in marine ecosystems and also engender important life cycle consequences associated with early survival.

Finding Patches

Pelagic larvae quickly disperse according to diffusion and prevailing adjective forces, or by their own swimming. By diffusion alone, larvae will disperse from centers of spawning at rates of kilometers per day (Heath 1989; Reed et al. 1989; Secor and Houde 1995b), radiating into waters that vary in physical density and zooplankton concentration (chap. 3, Plankton

or Nekton?). Swimming will also substantially increase the region over which the cohort of larvae is spread and can encounter zooplankton aggregations. Thus the pelagic spawning fish sends forth its thousands to millions of offspring, which disperse and encounter patches of concentrated prey that may or may not support successful feeding, growth, and survival.

Starting life small or big will affect the efficiency with which a brood of offspring samples the environment for favorable patches (Winemiller and Rose 1993). A female is physically limited to allocate its store of reproductive tissue between offspring size and number (chap. 3, Fecundity and Adult Size). Larger offspring will be more active as larvae and search an increased volume of water, increasing the probability of encounter rates with patches, but their smaller initial number means that fewer dispersal trajectories will be realized, and loss rates will have a disproportionate effect on absolute numbers of surviving larvae. Conversely, smaller larvae have limited swimming and pursuit capabilities, which cause them to be far more susceptible to starvation and predation deaths on an individual basis. Winemiller and Rose (1993) used an individual-based modeling approach that audited feeding and growth of individual larvae, and compiled cohort mortality rates under scenarios of varying prey density and larval size (5- vs. 10-mm initial size). Importantly, prey density was not simply a concentration but simulated as levels of patchiness occurring either (1) on a daily basis (a larva encounters randomly drawn concentrations each day) or (2) for longer periods where larvae that encountered dense prey patches were subsequently retained in the patch. Overdispersion (patchiness) of prey encounters per day was simulated through adjustments of a negative binomial distribution, while overall background mean zooplankton concentration was similar among scenarios. Larger larvae were favored (total number of survivors) under low patchiness scenarios regardless of zooplankton concentration (concentrations ranging fourfold). When high patchiness was combined with high concentrations of zooplankton within patches, however, broods of small larvae were favored. This was particularly true when larvae were retained in prey-rich patches. Retention within prey patches is a reasonable hypothesis given a larva's swimming abilities, particularly as it outgrows a viscous flow regime. At 20% "patch-finding" success, survivorship by broods of small larvae exceeded by tenfold that of broods composed of large larvae. Thus even small levels of encounter with concentrated patches can offset starting life small.

That small and numerous propagules favor dispersal and sampling of heterogeneous environments is well known for marine invertebrates, algae, and plants (Palmer and Strathmann 1981). In some instances, dispersal is related to destination rather than what happens along the way. The lecithotrophic (yolk-bearing) larvae of marine invertebrates or the seeds and spores of plants disperse to new habitats where they will settle or germinate. These stages do not feed and grow as they disperse. On the other hand, larvae of marine fishes (and many invertebrates; Strathmann 1990) simultaneously disperse long distances while foraging, trading high initial brood numbers against rewards of gleaning production from plankton-rich patches, which typify their pelagic environment. Of course, we can look to the sheer variance in offspring sizes among marine fishes (chap. 3, fig. 3.3) to realize that this is a sweeping generalization: certainly other agents operate to shape maternal investments in individual offspring (chap. 3, Provisioning Eggs).

Match-Mismatch

Because of their vast volume and the high thermal capacity of water, marine ecosystems are more modulated than most terrestrial and freshwater systems in their response to short-term environmental forcing. Daily and monthly time series of sea surface temperature, for instance, are more autocorrelated and predictable than lake or river temperatures (Kaushal et al. 2010). Marine ecosystems show inertia, which is only overcome by sustained environmental forcing or strong disturbances (e.g., storms). Sustained forcing is often seasonal in

nature, associated with changes in solar irradiance, thermal stratification, precipitation, or delivery of nutrients. These seasonal patterns, although predictable to some degree, can themselves be altered by multiyear climate and oceanographic forcing, such as El Niño and atmospheric warming (see chap. 7, Resilience to Fishing and Climate Change: Uncharted Territory). Thus scientists have long sought to relate the complex life cycles and migrations of marine fishes to the periodic nature of marine ecosystems among seasons and years (Hjort 1914; Cushing 1995; Hodgson and Quinn 2002; Bakun 2006; Lehodey et al. 2006).

Most but not all large marine ecosystems and their food webs exhibit seasonal cycles. In temperate systems, spring and summer peaks in zooplankton closely follow surface layer phytoplankton blooms, brought about by water column mixing, delivery of bottom nutrients to the upper water column, and expansion of the photic zone (Davis 1987). In estuaries (including shelf outflow regions of particularly large rivers), rivers are the principal source of nutrients, and seasonal changes in precipitation rates and snow melt together with cycles of irradiance and temperature drive spring and summer plankton blooms (Malone et al. 1988). Seasonal cycles in prevailing winds in upwelling systems result in periods of higher plankton production, as discussed in chapter 3 (Bakun 1996). Ice melt and increased daylight are associated with spring blooms in polar seas, although the cycle is dampened in the subarctic Pacific, where grazing pressure by zooplankton curtails microalgal production (Frost 1987; Miller et al. 1991). Low seasonal variability in phytoplankton production also occurs in tropical oceanic gyres and other tropical waters, thought to be due to low and relatively constant nutrient levels available to these systems and high rates of grazing (Valiela 1995).

Over generations of life cycles, seasonal periods of high zooplankton prey availability are thought to select spawning behaviors timed to inject larvae into the size spectrum during a favorable period (Cushing 1969). For any particular year, however, weather and climate will modify the timing of plankton blooms, which can result in the mistimed occurrence of first feeding larvae. The match-mismatch hypothesis (Cushing 1975) postulated that relatively invariant spawning periods, albeit protracted over many weeks, occurred in or out of sync with plankton production. Spawning migrations and mating systems are aligned with natural production cycles, cued variously by environmental surrogates (e.g., day length, temperature). The match-mismatch hypothesis specifies an optimal behavior—adults should spawn on dates that most favor survival of their offspring—those dates that match the long-term peak in zooplankton abundance. But evidence indicates that mismatches are common (Leggett and DeBlois 1994; Kristiansen et al. 2011). Likely more important than the match-up in peaks is the variance in spawning times, which should be a reflection of the potential mismatch between resources and offspring. Similar to the phenomenon of lots of small larvae sampling plankton patchiness, unpredictability in the timing of resources should select for a range of dates over which offspring are spawned. Further, diversity in spawning times and locations will increase variance in larval dispersal fates. Such responses to environmental variability is termed *bet hedging*, representing a principal means by which life cycles offset risk, and in this way rarely manifest a single optimal response (Roff 1978).

Two recent analyses emphasize durations rather than peaks in matchups between cod larvae and periods of zooplankton production. Beaugrand et al. (2003) detected compelling broad seasonal associations between calanoid zooplankton abundance and juvenile North Sea cod production over a 40-y period. Kristiansen et al. (2011) simulated zooplankton production throughout the North Atlantic from observed Chlorophyll *a* and temperatures, and released virtual first-feeding cod larvae into principal spawning regions (Lofoten Islands, Iceland, North Sea, and Georges Bank). Cohorts were released year round, given foraging and growth behaviors, subjected to predation, and tracked through an individual-based model (chap. 3,

Plankton or Nekton?). Warm versus cold years resulted in protracted and curtailed periods of zooplankton production. From model results, Kristiansen et al. concluded that while strong matchups in peak periods of larval hatching and zooplankton production were beneficial, they were not necessary for elevated larval and juvenile production. More important was the total zooplankton production over the entire period of spawning and larval production (fig. 4.8).

Chesapeake Bay striped bass provide another example for bet hedging associated with spawning behaviors and larval resource availability. Spawning waves are cued by temperature over a 2-month period, which overlaps broadly with spring phytoplankton and zooplankton blooms (Martino and Houde 2010), but more critical to early survival are changes in temperature itself, which can have direct effects on growth and mortality. Spawning too early or too late can cause embryos to encounter sub- or superlethal temperatures (Secor and Houde 1995a); temperature also has indirect effects through its influence on plankton dynamics (Campfield and Houde 2010). Secor and Houde (1995a) suggested that the protracted spawning season for striped bass represented stabilizing selection for a range of hatch dates, hedging against variance in spring thermal conditions (fig. 4.8). Further, they observed a second source of bet hedging, a minority behavior in which a small proportion of the spawning population did not follow typical spawning cues and spawned during periods of low and stable temperatures. For the year of study, this behavior resulted in disproportionately high survivorship of resultant larvae. In a similar vein, Kristiansen et al.'s (2011) analysis of cod larvae indicated that, in certain warm years, autumn spawning (a minority behavior) should result in significant larval and juvenile production.

The seasonal cycles common to most marine ecosystems are nested within longer-term cycles of change, often driven by shifts in the position and intensity of atmospheric pressure cells that occur over multiyear durations. Thus the variance-dampening behaviors that result

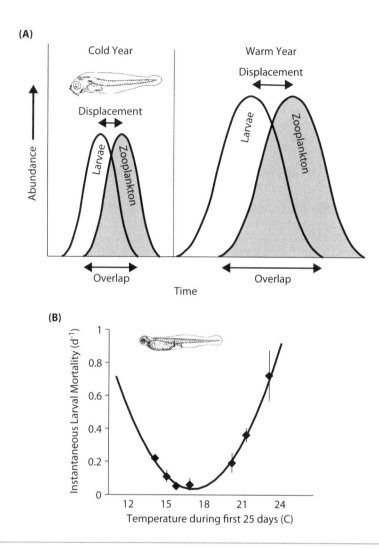

Figure 4.8. Consequences of temporal overlap between the periods of spawning and larval resources for (A) Atlantic cod and (B) striped bass. For Atlantic cod, cold and warm years result in differing degrees of peak displacement and resource overlap. Although peak displacement is higher in the warm year, the amount of zooplankton resources available to larval cod is much greater; thus overlap is a more important factor in the matchup between spawning and resources. For striped bass larvae, weekly cohorts (shown as diamonds) spawned during early or late spring (April–May) experience high mortality as a result of sub- or superoptimal temperatures. Therefore a range of spawning times improves the probability of overlap with favorable temperatures. Modified from Kristiansen et al. 2011 and Secor and Houde 1995a, with permission. Cod illustration from Bigelow and Schroeder 1953; striped bass illustration from Mansueti 1961.

from protracted spawning periods and the release of high numbers of dispersive larvae can fail in any given season. In such instances, repeated spawning across seasons or years or producing offspring with variable maturation rates can hedge against depressed production or catastrophic loss in a given season or year. In the next section we consider how life cycles are scheduled according to maturation, fecundity, early survival, and frequency of spawning.

Life Cycle Schedules

A dominant framework for understanding the evolution of complex life cycles is the "optimization approach" (Roff 1992), exemplified above for juvenile bluegill as the trade-off between foraging success and predation risk associated with a movement between habitats. Werner and Gilliam (1984) optimized a local measure of fitness, M/G, to evaluate when juvenile movement was favored. The optimization approach assumes that a single set of life history or behavioral traits will lead to an optimum fitness level: fitness measured by various representations of reproductive rate (i.e., intrinsic, finite rate of population increase; lifetime reproductive success; or reproductive values at specific stages or ages). Optimization has been particularly valuable as a deductive framework in investigating life cycle schedules. As a classic example, Charnov and Schaffer (1973) redressed a puzzling theoretical finding (Cole 1954)—that reproductive rates between semelparous (those spawning once) and iteroparous (those spawning multiple times) species were nearly equivalent—by improved model specification (reviewed in Roff 1992). Specifying juvenile survival and equating reproductive rates (finite rate of increase, $\lambda_a = \lambda_p$) between the two life history types gave

$$m_a s = m_p s + S,$$

which reduces to

$$m_a = m_p + S/s,$$

where m is offspring number, s is juvenile survival, S is adult survival, and a and p designate semelparous and iteroparous life histories. Without specifying juvenile survival, it follows that $S/s = 1$, which indicates nearly equivalent reproductive rates between life history types. But in fishes and other taxa with indirect development or biphasic life histories, adult survival will substantially exceed juvenile survival, leading to selection of an iteroparous life history.

The optimization approach can thus be used to test theory and improve specification of hypotheses, but there are often problems with its application to real populations. For instance, the model fails to predict semelparity as it occurs in Japanese eel or Pacific salmons. A leptocephalus or salmon fry clearly has much lower survival prospects than the adult, which should lead to an iteroparous life history. One argument is that the theory remains incomplete, needing additional specification. For eels and salmon, for example, an improved model could include age structure and environmental variability. Another perspective is that the optimization approach itself is flawed. Optimization typically entails an environment that is stationary and assumes that selection on traits occurs in a manner that maximizes reproductive rate. Istock (1967) assumed equilibrium conditions in his modeling of fitness components associated with complex life cycles and found that indirect development was "inherently unstable in evolutionary time," which seems unlikely given its prevalence in teleosts. An additional concern in optimization models is that life history traits are often considered as entirely plastic responses regardless of phylogenetic, genetic, physiological, and ecological constraints (Reznick 1985).

An epitome of careless application of optimization models, unfortunately rife within fish ecology and fisheries science, is the classification of species as r- or K-selected. The classification follows the logistic growth model, central to harvest theory in fisheries science and widely applied to other ecological inqueries,

$$\frac{dN}{dt} = rN\left(1 - \frac{N}{K}\right),$$

where N is population size, r is intrinsic population growth rate, and K is population size at carrying capacity. The model implies a trade-off: populations that commonly encounter resource limitation should maximize K, and those that exploit unsaturated ecosystems should maximize r. As reviewed by Roff (1992), Pianka (1970) initially applied K maximization to species living in tropical environments where competition was assumed to prevail, contrasting these to temperate-latitude r-selected species that were controlled principally by density-independent abiotic conditions. Pianka further proposed that r and K trade-offs were evident in covarying life history traits. Small offspring size, high fecundity, rapid development, early maturation, small body size, and semelparity were r-selected; slower growth, delayed maturation, large body size, iteroparity, and lower fecundity were K-selected.

The dichotomous classification has had much appeal, relating species life histories to environmental stability, invasion ecology, succession, and harvesting. But this theory has eclipsed evidence. In fact, even for simple populations such as bacteria and rotifers, experiments have failed to show expected responses of traits associated with r- and K-selected environments (Roff 1992). In addition, empirical evidence does not support the correlated sets of traits stipulated by r or K selection. As an example, we have already seen that large body size is often associated with small offspring size in marine teleosts (chap. 3, fig. 3.3). The persistence of the r-K dichotomy in the literature is particularly curious given early warnings by prominent ecologists that the classification was flawed and should be avoided altogether (Stearns 1977; Ricklefs 1979; Caswell 1982; Hall 1988; Roff 1992). But if not r- and K-selected, how should we expect life history traits to covary in association with complex life cycles?

Periodic Life Histories

Early observers noted a common suite of traits associated with commercially exploited fish-
es that did not fit the r-K dichotomy: relatively late maturation, iteroparity, high fecundity, small offspring size, and low early survival (e.g., Kawasaki 1980). In a groundbreaking multivariate analysis of life history traits for 216 marine and freshwater teleosts, Winemiller and Rose (1992) identified this same trait suite along with two others ordered along three dominant axes: (1) low fecundity, low early survival, and rapid maturation (opportunistic suite); (2) low fecundity, high early survival, and late maturation (equilibrium suite); and (3) high fecundity, low early survival, and late maturation (periodic suite). According to literature precedence (Wooton 1984), each suite was termed a *strategy*, implying selection for traits between three broad classes of environmental regimes (fig. 4.9). Opportunistic strategists exploit ephemeral, marginal, and stochastic environments with life history traits associated with short turnover times and rapid colonization. This suite differs from the r-selected suite in that batch fecundity is relatively low. A quintessential example is bay anchovy, which spawns scores of small batches of eggs that in sum amount to several times their weight over a single reproductive season. Equilibrium strategists capitalize on stable environments, investing higher parental resources on offspring size and care. Periodic strategists have relatively long generation times and high fecundities matched with periods of favorable environments and patterns of patchiness that affect early survival, respectively (see Finding Patches above). Periodic strategists thereby hedge against environmental uncertainty through long reproductive life spans and high fecundity, but at the expense of slow turnover. Most commercially harvested fishes tend to exhibit this strategy.

A triangular continuum of covarying life history traits thus emerged through a robust statistical analysis of hundreds of species. To generalize and explore this continuum further, the fecundity-offspring size data set presented in chapter 3 (fig. 3.3; 288 teleosts and elasmobranchs) was analyzed in a multivariate manner similar to Winemiller and Rose (1992). Size at maturation, batch fecundity, and offspring size were tested as principal vectors of the triangular life history continuum (fig. 4.10). Multivariate loadings on the first principal component (x axis) showed positive correlations between offspring and adult size and negative correlation between those two traits and fecundity, consistent with the equilibrium strategy. Factor scores for viviparous and oviparous elasmobranchs and viviparous teleosts were positively associated with this suite of traits. The second principal component (y axis) received positive loadings for adult size and fecundity. On the bivariate factor plot, a high density of scores for individual species was associated with small offspring size. Teleosts and viviparous *Sebastes* were included in this aggregation of scores and gave evidence of traits associated with periodic (large size at maturation, high fecundity, small offspring size) or opportunistic (small size at maturation, low fecundity, small offspring size) strategies. Note that in comparison to species associated with the equilibrium strategy, the distribution of scores between periodic and opportunistic strategy endpoints is more continuous.

Epitome species associated with the periodic strategy included ocean sunfish, Atlantic sturgeon, and commercial species such as Atlantic cod and striped bass. Opportunistic species included small teleosts: sheepshead minnow *Cyprinodon variegates*, mosquitofish *Gambusia affinis*, and Atlantic silversides. *Sebastes* spp. fell within the transition region along Factor 2, but trended toward periodic life history traits. Interestingly, this group showed little variation associated with Factor 1, indicating low variance in offspring size. Other transitional species included Atlantic herring, Pacific salmon *Oncorhychus* spp., and gafftopsail catfish *Bagre marinus*, the latter of which produces large eggs (~20 mm diameter) and trends toward equilibrium life history scores.

The triad of life history types thereby seems

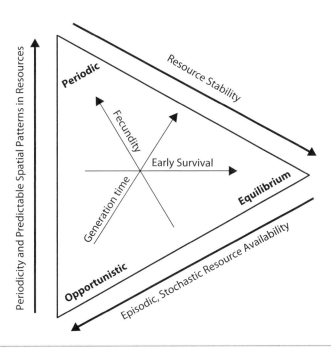

Figure 4.9. Life history types of fishes, representing ecological constraints and trade-offs between principal life history traits. Redrawn from Winemiller 2005, with permission.

robust to reproductive and taxonomic classifications. Separate multivariate factor analyses
were conducted on oviparous teleost and viviparous elasmobranch subsets. Loadings showed
similar trends among data subsets, with positive correlations with offspring size and adult
size on one of the two factors (periodic–opportunistic continuum) and a negative correlation
between offspring size and fecundity/adult size
on the other factor (equilibrium mode). For viviparous elasmobranchs, white and common
thresher sharks *Alopias vulpinus* epitomize the
equilibrium endpoint, but at the other extreme
within this continuum is the pygmy shark *Euprotomicrus bispinatus* (maximum size 25 cm)
with a relatively high fecundity (eight pups).
In targeted multivariate analyses of life history traits of elasmobranchs, Frisk et al. (2001)
classified all elasmobranchs as equilibrium
strategists, but Cortes (2000) uncovered three
life history modes, one corresponding to a peri-
odic-like life history strategy (large litter, moderate to high longevity, large adult size, small
offspring).

The current multivariate analysis comple-
ments others conducted on separate data sets,
comprising diverse fish fauna and provinces
(Winemiller and Rose 1992; Cortes 2000; Vila-Gispert et al. 2002; Tedesco et al. 2008; Mims
et al. 2010). Together these analyses present
compelling evidence that many or most fishes
exhibit to varying degrees a periodic life his-

tory type. Such species experience infrequent
cycles of favorable conditions for early survival
and broad spatial patchiness in early survival
conditions.

Within species and populations, life history
traits are also responsive to periodic selection
regimes. In American shad *Alosa sapidissima*,
Leggett and Carscadden (1978) found that longevity and the degree of iteroparity were positively correlated with latitude. They concluded
that high-latitude traits (higher longevity, age
at maturity, reproductive life span) were associated with higher variation in juvenile survival as a result of greater seasonality in environmental conditions. Although speculative,
shorter growth seasons during the first year
of life at higher latitudes (Conover and Present
1990) could decrease early survival at these latitudes if juveniles were to trade off higher consumption and growth against predation risk
(Lankford et al. 2001). In turn, selection could
favor adults with longer lifespans at higher
latitudes.

The Periodic Strategy and Migration

The periodic strategy is strongly linked to migration. The 10 highest scores associated with
the periodic suite (fig. 4.10, Factor 1) were all
highly migratory, capable of seasonal migrations exceeding 1,000 km (species include
ocean sunfish, Atlantic sturgeon, bluefin tuna,
Atlantic halibut *Hippoglossus hippoglossus*, yel-

Facing page

Figure 4.10. Life history trait associations for teleosts and elasmobranchs. (A) Factor analysis of adult
size, offspring size, and batch fecundity for 288 species of teleosts and elasmobranchs. Loadings for each
factor are presented for Factor 1 (*x* axis) and Factor 2 (*y* axis). End-member trait associations—periodic,
opportunistic, and equilibrium—follow those of Winemiller and Rose (1992). For different reproductive
groups, 67% confidence ellipses are shown. (B) Factor loadings (F1 and F2) are shown for separate reproductive groups (see fig. 3.3). Example species include *Mola mola* (ocean sunfish), *Acipenser oxyrinchus* (Atlantic sturgeon), *Thunnus thynnus* (Atlantic bluefin tuna), *Gadus morhua* (Atlantic cod), *Morone saxatilis*
(striped bass), *Clupea harengus* (Atlantic herring), *Oncorhynchus tshawytscha* (Chinook salmon), *Oncorhynchus keta* (sockeye salmon), *Sebastes paucispinis* (Bocaccio), *Sebastes dallii* (calico rockfish), *Anchoa mitchilli*
(Bay anchovy), *Menidia menidia* (Atlantic silverside), *Gambusia affinis* (mosquitofish), *Cyprinodon variega-
tus* (sheepshead minnow), *Awaous guamensis* (o'opu nakea [freshwater goby]), *Bagre marinus* (gafftopsail
catfish), *Euprotomicrus bispinatus* (pygmy shark), *Carcharodon carcharias* (white shark), and *Alopias vulpinus* (common thresher shark). See figure 3.3 for data sources.

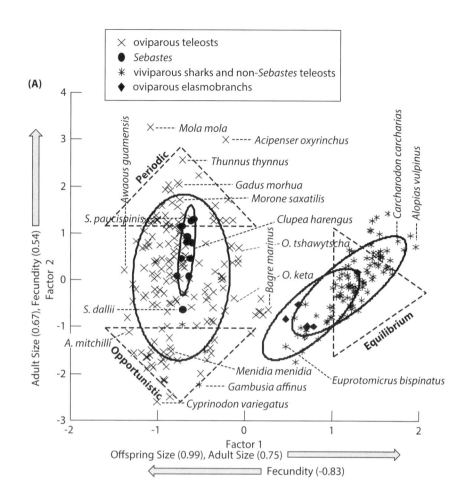

(A)

(B)

Trait	F1	F2
All (*n*=288)		
Fecundity	-0.83	0.54
Adult Length	0.74	0.67
Offspring Size	0.99	-0.04
Oviparous (*n*=166)		
Fecundity	-0.07	0.97
Adult Length	0.69	0.70
Offspring Size	0.86	-0.79
Viviparous (*n*=104)		
Fecundity	-0.27	0.96
Adult Length	0.91	0.37
Offspring Size	0.98	-0.08

lowfin tuna *Thunnus albacares*, cobia *Rachycentron canadum*, black drum *Pogonias cromis*, king mackerel *Scomberomorus cavalla*, Atlantic cod, and dolphinfish *Coryphaena hippurus*). These fish are all large with maximum sizes exceeding 1 m. Roff (1988) constructed a simple optimization model that equated migration to reproduction in an energetic framework. Gonad weight and energetic content increase with fish length, but so too do energy expenditures per distance traveled. Applying a series of allometric functions that described relationships between swimming metabolism, gonad weight, and body size, Roff found that larger fish disproportionately accumulated surplus energy stores that would equate to greater distances traveled. As evidence that reproductive energy may trade off against travel distances, higher-latitude American shad are larger and exhibit later maturation and lower batch fecundity but will migrate greater distances over their reproductive lifespan (Glebe and Leggett 1981). Further, within populations of migratory fish, there is generally an increase in distances traversed at larger sizes (Harden Jones 1968; Waldman et al. 1990; Pearcy 1992; Rjinsdorp and Pastoors 1995), and for species that exhibit partial migration (chap. 6), the resident portion of a population tends to be composed of smaller individuals relative to the migratory component (Hjort 1914; Jonsson and Jonsson 1993; Secor 1999). Increased migratory tendency with adult size also applies to equilibrium strategists where late maturation and large body size are associated with highly migratory species (e.g., white and common thresher sharks). In contrast, opportunistic species tend to be small bodied; inhabit shallow estuaries, embayments, or reefs; and exhibit curtailed seasonal migrations.

Through migration, dispersal of small offspring, and repeated reproduction the periodic strategist samples environments that vary in structured ways across wide temporal and spatial scales. Repeated spawning and associated migrations operate to sample low-frequency conditions that differentially influence early survival and growth (fig. 4.9). Among and within populations, variable migration patterns (i.e., juvenile vs. adult migrations; resident vs. migratory individuals) will sample habitats at different spatial scales (chap. 6). Within each reproductive season, offspring spawned weeks apart will differentially grow and survive depending on the timing of plankton blooms and other key resources. Many small larvae will disperse amidst concentrated patches of zooplankton as they begin to feed. All of these preceding traits are associated with a periodic strategy. Other life history end-members also sample environmental variance but not over as many spatial and temporal scales. The opportunistic strategist samples smaller-scale, and less structured, variance; the equilibrium strategist minimizes the influence of environmental variance through large offspring size, parental care, or attachment to stable habitats.

Schooling through Food Webs

Schooling can result in rapid exploration of favorable environmental or food web conditions that occur periodically. Alternatively, it can cause inertia to change when conditions become unfavorable. In this way, schools represent "self-organizing" systems (Ulanowicz 1997) in which individual school members self-organize by collectively maintaining school spacing, shape, and polarity (chap. 2, Schooling and Flocking). Further, schools retain information through the collective actions of individuals conserving seasonal migration patterns (Parrish and Edelstein-Keshet 1999). In keeping with the self-organization construct, individuals within schools can turn over, yet the school persists. New members from adjacent schools can contribute their experience, but only when new entrants occur in sufficient numbers will the polarity of the school be influenced. The dynamic interchange of members between schools led Lebedev (1969) to suggest that multiple school aggregations should be considered as fundamental entities that are responsive to changes within ecosystems and food webs. Similarly, Fréon et al. (2005) noted the fractal and dynamic nature of members among school nuclei, schools, school clusters, and subpopulations of school clusters. The flux of school membership within school aggregations can result in either stabilizing or destabilizing feedback cycles stimulated by changes in climate, the food web, or fishing. Schooling has adaptive attributes where information that benefits survival or growth can be rapidly transmitted through positive feedback among members, but the phenomenon can also result in apparent maladaptive responses when schooling feedbacks result in ecological traps. The contraction into a single fjord (Vestfjord) by the bulk of the Norwegian Atlantic herring population during winter, for instance, resulted in high levels of predation and other losses owing to poor water quality from the aggregation's collective respiration (Dommasnes et al. 1994; Corten 2001).

The extreme cycles of population growth in marine planktivorous fishes such as California sardine *Sardinops sagax* or Peruvian anchovy *Engraulis ringens* have long defied mechanistic explanations associated with climate, oceanography, food web, or fishery influences (see also chap. 7, fig. 7.5). For instance, large-amplitude low-frequency (decades to centuries) cycles in abundance have occurred over millennia for California sardines, long before industrial fisheries developed (Baumgartner et al. 1992). These population dynamics are commonly associated with large changes in spatial range (Schwartzlose et al. 1999; Petitgas et al. 2010). Bakun (2001) proposed that mobile planktivorous fish populations represented complex systems owing to their schooling behavior—that their dynamics cannot be directly linked to changes in the environment or food web without understanding how these changes are mitigated by schooling.

Schooling provides refuge from predation, particularly in exposed pelagic settings, but the vulnerability of school members varies with school size. Classically, three functional responses—linear, saturating, or sigmoidal —describe the effect of prey concentration on predation rates. Experimental and field observations support the view that some degree of

saturation is expected at large school size or density (predator swamping), which is consistent with a saturating or sigmoidal relationship. At small school size, a depensatory (sigmoidal) response occurs where predators continuously crop down the population of schooling fish to a low stationary abundance. But given a change in climate, food web structure, or migration pattern, a release from predation pressure can occur, allowing the population to "break out" and increase to a much higher abundance level (fig. 4.11; Bakun 2006).

Cycles of range expansion and contraction in many pelagic fishes correspond to abundance dynamics associated with predation and schooling. As proposed by Bakun (2001), periods of depressed abundance and limited spatial range by a schooling population occur as a result of efficient predation by co-occurring predators. With decreased abundance, however, schools within a population become increasingly independent from each other. A given school or school aggregation may adopt a more exploratory behavior and break out from its recent range, resulting in a release from predation or improved forage conditions (fig. 4.11). In essence, the school moves to a new size spectrum. Increased growth, survival, and fecundity by this exploratory segment can stimulate further range extension by exploratory schools. Recall that this is a self-organizing system with fluid membership and social transmission of spatial behaviors. Schooling can thus self-reinforce dominant behaviors (a.k.a. autocatalysis; Ulanowicz 1997). As an example, Parrish and Edelstein-Keshet (1999) describe the rapid inundation of fish aggregation devices (engineered flotsam set out by commercial fishermen to attract pelagic fishes) by large schools of tunas that rapidly overwhelm the initial attraction radius of the device. Autocatalysis can accelerate coupled changes in abundance and range expansion. Bakun (2001) suggests that at some point the school aggregate merges and becomes a homogenous hyperschool, resulting in the loss of previous migration behaviors. When food web or environmental conditions no longer support growth and survival, the hyper-

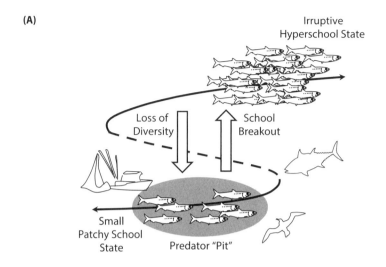

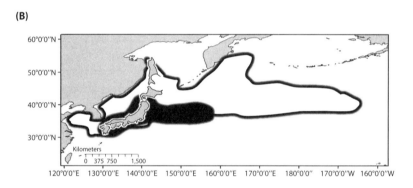

Figure 4.11. Irruptive outbreak of pelagic schooling fish as (A) conceived in the school-trap hypothesis and (B) observed in Japanese sardines. The school-trap hypothesis (Bakun 2001) is represented as a bifurcation diagram (transition between two schooling states), where predation is the forcing variable and abundance the state variable. A threshold change (improved recruitment, changed spatial behaviors, food web changes) is required for schools to break out from their predator "pit," after which irruptive growth and range expansion occur. The range of Japanese sardine *Sardinops melanostictus* in (B) shows periods of low (1950s, black regions) and high (1980s, white regions) abundance. Over this period the fishery expanded from <0.2 to 5.4 million t (Watanabe 2002). Drawn from information in Fréon et al. 2005.

school is now recalcitrant to change, resulting in a much more rapid decline than would have occurred had diversity in population segments been maintained.

The rapid expansion and contraction of Japanese sardine *Sardinops melanostictus* in the 1970s and 1980s represents a possible case study (Bakun 2001). In the initial phase of low

abundance, two population segments were recognized: a nearshore fast-maturing (age 1) group and a migratory slow-maturing (age 2–3) oceanic group. A single event or series of events stimulated a linked change in abundance and range expansion, resulting in an autocatalytic response in which the migratory group became overwhelmingly dominant. By the mid-1980s the population's range showed remarkable expansion beyond its typical historical distribution (fig. 4.11), extending into the Sea of Japan, shelf waters of Kamchatka, and the Kuroshio Current, ~3,000 km beyond its normal range. Bakun (2001) speculated that this superexpanded population encountered an ecological trap when conditions no longer favored its new distribution. During the period of range expansion, sardines apparently abandoned former inshore areas, which may have resulted in homogenization and inertia to environmental/food web changes and contributed to its rapid collapse during the late 1980s and early 1990s, perhaps owing to mortality during the early juvenile period (Watanabe 2002).

Although one might criticize the above conceptual model (termed the "school-trap" hypothesis by Bakun 2001) as relying solely on circumstantial evidence, the approach is of heuristic value. Such conceptual frameworks are often opposed by many alternative explanations and require cautious interpretation. On the other hand, they can lead to improved hypotheses, broadened perspective, and new avenues of research. For instance, telemetry is increasingly used in novel ways to infer networks of social interaction. Using fish-attracting devices ($n = 13$) surrounding Oahu (Hawaii, USA) as nodes of attraction for yellowfin tuna, Stehfest et al. (2013) showed paired associations by acoustically tagged individuals that persisted as long as 60 d. We should expect schooling to influence broad migration patterns and range dynamics of mobile fishes, yet few conceptual models link the two (Fréon et al. 2005). Bakun (2001) introduces schooling as a propensity for self-organization within fish populations. As defined by Ulanowicz (1997), this propensity does not represent an efficient ("bottom-up") mechanism of biological response, but rather a system with its own internal dynamics, which in this case are periodically reset by climate and other external forcing (including fishing). Phenomenological frameworks would seem to convey poor predictive abilities to forecast future changes in fish populations relevant to

fisheries or habitat management. Yet one can use emergent properties (or propensities) of populations to develop precautionary stewardship tactics, as introduced in the next section and in chapter 7 (see also Scheffer 2009).

The Storage Effect

Redundancy in complex systems is predicted to reduce trophic efficiency and production but increase stability and resilience (Ulanowicz 1997). Referencing the previous section, short-lived pelagic planktivorous fishes comprise the world's highest-yield fisheries but are prone to high fluctuations in abundance and range dynamics. In contrast, abundance and spatial dynamics of longer-lived fishes tend to be dampened with respect to dominant environmental cycles. Here iteroparity (multiple age-classes of adults) represents a source of redundancy, which confers stability to both yield and spatial dynamics through a process known as the storage effect. Coined by Chesson and Warner (1981), the storage effect represents the notion that within complex life cycles certain life stages are relatively invariant or insensitive to environmental perturbations. Think insect pupae, diapausing stages of zooplankton, dormant spores, or large adults in vertebrates. These life history stages dampen population responses to environmental vagaries and contribute to population stability and resilience. The storage effect applies to complex life cycles where part of the life cycle, the longer part, is buffered against environmental variability, and the other, typically shorter, phase is more tightly coupled to periodic environmental variations that only occasionally favor survival (Frank and Brickman 2001; Secor 2007).

The concept of the storage effect arose from resolution of a paradox on why coral reefs harbored high fish diversity. Sale's (1977) competitive lottery hypothesis postulated that new settlement on coral reefs by juveniles was vicarious, dependent upon a space opening up through predation or an environmental disturbance. Otherwise the reef was closed (saturated) to settlement by individual species. The random allocation of space to new juveniles maintained species coexistence and high diversity. A problem with the concept occurred when any small competitive advantages were introduced, say, owing to relative abundance, swimming, or condition. In such instances the system moved quickly to competitive exclusion by a single dominant species (Chesson and Warner 1981). Through a life table modeling approach, Warner and Chesson (1985) concluded that three conditions were necessary for stable coexistence of species in saturated reef habitats: (1) early vital rates and larval supply among species are variable across years, (2) adults survive over long periods of poor settlement, and (3) episodic recruitments (i.e., high rates of settlement) are matched by generation time. They termed this suite of demographic conditions the storage effect, where one life stage is driven by environmental variability and the other is buffered against competitive interactions and environmentally driven changes.

In fisheries science, the concept of a storage effect is implicit in how we consider the renewal process—the year-class phenomenon (Secor 2007). The role of particularly strong years for juvenile production (year-classes) in cycles of abundance for Atlantic herring and other marine fish populations was discovered by Hjort and his team a century ago (Hjort 1914). A single year-class (the 1904 year-class) of Norwegian Atlantic herring sustained a subsequent decade of stock abundance and yield to commercial fisheries. A strong year-class can stimulate the next cycle of population abundance. Thus scientists and managers have long recognized that longevity and age structure can counteract protracted periods of poor juvenile production (Beamish and McFarlane 1983; Longhurst 2002; Berkeley et al. 2004). Still, age structure is often viewed as the consequence of variations in year-class success, rather than as an emergent structure related to the propensity for ecological redundancy and stability. This propensity is given some evidence by recent discoveries of

increased longevity for many marine fishes (fig. 4.12; Secor 2007).

The storage effect can be visualized as part of the size spectra attributable to the adult period (fig. 4.3). Log_{10} mass units scale to trophic levels (predators typically take prey 1%–5% of their mass; Kerr and Dickie 2001; Beaugrand et al. 2003), and one observes that adults "hold position" within the food web, converting their stored mass to reproduction rather than substantially increasing their size. While the larva and juvenile will traverse four to five orders of magnitude in mass within the first year of life, the adult stage ranges less than tenfold in mass and can persist there for years to decades. In simple terms, iteroparity and reproductive longevity increase the number of years that a cohort of adults samples conditions favorable for early survival. The benefit in dampened variability (stability) over a generation will be a negative geometric distribution defined by the standard deviation and proportional to $\sqrt{(1/N)}$, where N is the number of adult age-classes. Note that this reasoning can also apply to a semelparous species with overlapping generations such as the Japanese eel or Chinook salmon, where multiple age-classes spawn each year.

For Japanese eels, the early life phase varies in concert with oceanic conditions, whereas the relatively long juvenile continental growth phase is insulated from environmental variations. Fluctuations of glass eels (young juveniles) are coupled to oceanic conditions (e.g., El Niño, temperature, primary production) experienced during the leptocephalus stage. In contrast, the yellow eel (late juvenile) stage is buffered from the environment to a greater extent because of its larger size and behavioral and physiological attributes that adapt to seasonal and spatial changes in the environment. For eels, the storage effect is determined by the duration of the yellow eel phase and how well it is matched with intervals between oceanic conditions that favor recruitments (larval production). Overlapping generations of eels are critical for the operation of the storage ef-

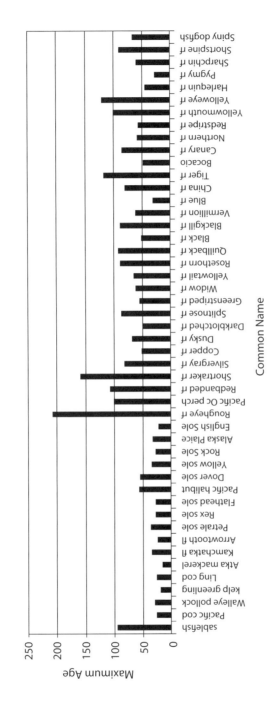

Figure 4.12. Maximum ages reported for Alaska and British Columbia demersal teleost fishes (Munk 2001). Abbreviations are as follows: fl, flounder; rf, rockfish. Redrawn from Secor 2007, with permission. (Note that a regrettable error occurred in the original figure, where the species axis was reversed.)

fect. Multiple year-classes of silver (adult) eels buffer population responses to oceanic conditions because they span long-term environmental oscillations. The degree of overlapping year-classes depends upon a diversity of maturation rates in the population, which in turn is related to the variable growth conditions (diversity of habitats) that yellow eel experience (chap. 7, American Eels: When Does Diversity Matter?).

A second interesting case study is the commercially fished North Atlantic spiny dogfish *Squalus acanthias*, a late-maturing (~12 y for females) small shark with a long gestation time and low fecundity (2–15 pups; Nammack et al. 1985). Despite expectations for slow changes in population growth rates and high susceptibility to fishing, this species experienced a large population expansion, the result of a decadal shift in its food web (Fogarty and Murawski 1998; Link and Garrison 2002). The species has a longevity >30 y that enables it to persist and respond to long-term cycles in the environment and fishing removals. Remarkable persistence has also been ascribed to barndoor skate *Dipturus laevis*, a species thought to be extirpated in parts of its range because it was no longer observed in surveys (Casey and Myers 1998). Yet it emerged again in recent survey data and even gave evidence of moderate population growth (Gedamke et al. 2009). Similar to spiny dogfish, longevity in this elasmobranch hedged against long-term cycles of environmental and anthropogenic influences (Frisk et al. 2002).

High rates of exploitation can truncate age distributions either through cumulative mortality or selection of large and old individuals. Either way, loss of older individuals can result in loss of migration pathways, habitat contraction, and diminished generational sampling of periodic conditions favorable for replacement (Petitgas et al. 2010). Conservation of age structure can be promoted through reducing fishing mortality or through fishing size limits. Age diversity indices can serve as performance metrics in stock assessments complementing conventional reference points and are reviewed

further in chapter 7 (Diversity in Age Structure: The Storage Effect).

Summary

In marine food webs, turnover and production rates are inversely related to organism size. Further, the small plankton upon which fish production depends is transient and patchily distributed. To exploit this resource, mating systems of many teleosts cause large numbers of small larvae to disperse. The production period and dispersal patterns of larvae broadly overlap scales of periodicity and patchiness of prey resources, but predation levels in these planktonic food webs are extremely high. Offsetting high levels of larval mortality are high fecundity, diverse spawning behaviors, and iteroparity (or overlapping generations), which typifies most teleost fishes. A second dominant reproductive mode (e.g., elasmobranchs and teleosts with parental care) minimizes early mortality by placing offspring into stable environments or into a different segment of the food web where predation mortality is less.

Size spectra were used to depict how fish life cycles scale to production in marine ecosystems. Cohorts of larvae are most productive when they (1) find themselves at the crest of a seasonal production wave of plankton and (2) are able to "surf" in front of the breaking wave of larger predators. But for many species, it will be impossible to offset early mortality losses, and only during the juvenile period does the cohort begin to gain biomass. Owing to increased swimming performance and sensory capabilities, the juvenile can exhibit greater discrimination in selecting habitats that favor growth and survival. Settlement, the movement from a pelagic habitat to a demersal one, is one common example of a shift between food webs during the juvenile period, but one that initially can engender substantial predation risk as small juveniles enter novel habitats. Understanding fitness trade-offs in selecting habitat during the juvenile period and other stages of life has been central in past explanations on the evolution of complex life cycles, but it entails assumptions of maximizing reproductive rates and stationarity.

The emergent properties of complex life cycles run counter to the view that fitness is maximized for a set of equilibrium conditions. Rather, variance and diversity in life cycles emerge as patterns that offset periodic variation and patchiness in marine ecosystems. For instance, compelling evidence for the match-mismatch hypothesis is not due to the matchup between peaks in larval and prey production, but rather to the degree of overlap (variance) between them. Evidence increasingly points to the prevalence of the periodic life history strategy (high fecundity, low early survival, large adult size, iteroparity) in marine fishes; this mode is sufficiently common to warrant avoidance of the tired and flawed typology of r- and K-selected species. The periodic strategy represents a suite of traits and behaviors that filter dominant sources of ecosystem variance into more modulated long-term population responses, thereby contributing to persistence and resilience. Fishes that epitomize the periodic strategy are highly migratory.

Not all emergent patterns associated with life cycles are stabilizing. Bakun and colleagues introduce schooling by mobile planktivorous fishes such as sardines and anchovies as a self-organizing system with its own internal dynamics. Cycles of range contraction and expansion attend periods of boom and collapse for these productive species, the possible result of autocatalytic behaviors and inertia arising from schooling dynamics. These large cycles in biomass of intermediate-level consumers alter entire size spectra and associated trophic efficiencies in upwelling and shelf ecosystems.

An emphasis on emergent properties associated with complex life cycles, rather than deductive explanations, limits the predictive frameworks that might otherwise address management and conservation agendas. Here and elsewhere precautionary tactics are promoted, conserving traits and behaviors that contribute to population productivity, persistence, and resilience. The storage effect concept emphasizes conservation of age structure. Age structure serves as a surrogate for cumulative events in a fish's life that contribute to diverse demographic outcomes. As a result of their migrations, large and old members of a population may uniquely sample environmental conditions with implications to subsequent generations through transmission of migration pathways, unique spawning behaviors, or energetic investments to offspring.

Segue

Our descriptions and explanations of complex life cycles remain incomplete. How does the cycle renew itself? Or, conversely, do patterns of dispersal or migration recycle at all? A dominant view held for fishes, birds, and other vertebrates is that the life cycles of populations are closed. This assumption was made repeatedly in this chapter in terms of selection for life history traits (e.g., iteroparity in periodic strategists implies a closed life cycle). On the other hand, more open cycles of renewal were apparent for school aggregations of planktivorous fishes. Increasingly, traditional viewpoints on closed life cycles of marine fishes are not well accommodated by the sheer diversity of migration and habitat use patterns observed within species and within populations. Chapter 5 will address mechanisms and related concepts on what leads fish back to their natal (parental) habitats. Additionally, we will explore what happens in more open populations, where not all individuals find their way back home.

5   Population Structure

Closed and Open Life Cycles

Migration is an important adaptation to environments that vary in space and time.
—Harrison and Taylor (1997)

Finding Their Way back Home

Spawning migrations sometimes, but not always, return marine fishes to where they started life. An apparent dichotomy in fish migration theory is whether a spawning adult will return to its place of origin or spawn elsewhere. This conundrum applies not so much to the fish itself but to the investigator who wishes to generalize life cycles, migration pathways, and population structure given the fact that not all individuals will end their life cycle where they began. That these wayward individuals are often in the minority might lead us to discount their behaviors as itinerant. Life cycles in so many instances seem well adapted to each population's unique environmental circumstance. Further, reproductive isolation, often revealed through genetic approaches, requires population members to find their way back home across many generations. Although past students of fish migration recognized the importance of open life cycles (e.g., Hjort 1914; Cushing 1962a; Harden Jones 1968), these views have been ignored, partly in support of developments in fisheries assessment science that required assumptions of closed populations (Smith 1994; Cadrin and Secor 2009). During recent decades, however, scientific and management assumptions have relaxed, allowing closed populations to become more open. Associated conceptual developments include those on connectivity, metapopulations, spatially explicit stock assessment models, and broadened management reference points such as stability and resilience. Still, before entering into a fuller treatment of semiclosed and open popu-

lations, it is important to recognize that many recent discoveries still point to the remarkable abilities of marine fishes to find their way back home. Five examples follow.

Homing on a Fixed Itinerary: Fraser River Sockeye Salmon

Among marine fishes, Pacific salmon spawn at moderately high altitudes, leaving oceans for clear and shallow fluvial ecosystems elevated hundreds of meters above sea level. The constancy in the timing of their appearance in these inland habitats and the ease of their capture and observation during spawning migrations have promoted salmon as the model species for concepts related to natal homing (Quinn 2005). Fraser River sockeye salmon are among the best-studied populations (Gilbert 1915; Burgner 1991). In the timing of spawning runs, specificity of spawning sites, and migration of juvenile parr to unique rearing habitats, sockeye salmon epitomize life cycle closure centered on local adaptation. Demersal spawning habitats—stream bed or lake beach—can be mere hundreds of meters apart, yet spawners utilizing either habitat are unique in their time of arrival, their body morphology (lake spawners are deeper bodied), their mating systems (territorial and satellite males), and the timing of emergence and subsequent migration behaviors of their young. Further, they are genetically discrete despite the potential (1) to mate with spawners in adjacent streams or lakes and (2) to migrate and spawn elsewhere in the Fraser River watershed by intermingling with unrelated schools. That a fish spends most of its life ranging thousands of kilometers throughout the Alaska Gyre (fig. 5.1) and then returns and spawns at a fixed site on a fixed itinerary is remarkable. It exemplifies how life cycle closure can cause populations to be tightly coupled to specific environments—in this case, the highly seasonal and structured habitats of high-elevation aquatic ecosystems (foothill and montane zones).

Most sockeye salmon will spend the majority of their lives at sea, occurring broadly throughout the subarctic Pacific Ocean (albeit with basin warming they are increasingly distributed northward with recent invasions into the Arctic Ocean; Babaluk et al. 2000). Salmon originating from the northwest Pacific, Alaska, Siberia, and Japan migrate seasonally in wide arks across thousands of kilometers (fig. 5.1) but maintain varying degrees of regional separation (Burgner 1991). Still, within a given ocean domain, overlap among populations is considerable. For instance, 16 maturing fish captured in the same net in the center of the Alaska Gyre were subsequently recovered in 10 different spawning rivers (fig. 5.1; Neave 1964).

For returning Fraser River salmon, Vancouver Island blocks the way (fig. 5.1). Historically, adults looped around the southern tip, returning via the Juan de Fuca Strait. In more recent warmer El Niño years, salmon predominately returned from the north through the Queen Charlotte Strait (Groot and Quinn 1987). Summarizing, sockeye salmon bound for the Fraser River will have (1) spent 2–4 y at sea, (2) dispersed broadly in the Alaska Gyre, (3) mixed extensively with unrelated salmon, and (4) circumnavigated Vancouver Island by alternate routes. Yet they consistently enter the river each summer as "runs"—several-week pulses of fish—destined to diverge further within the watershed and spawn at discrete locations and times as ~90 separate spawning populations (Beacham et al. 2004). The orderly arrival of Fraser River adults from diverse marine starting points and their naïve transit around complex coastlines implicates precise orientation, navigation, and time-keeping abilities (Quinn 2005; chap. 2, Navigation Maps: Inherited, Imprinted, or Learned?).

The final push up-river is a one-way trip. Many months of final feeding and fattening at sea have provisioned the salmon for this difficult migration against flows 200–10,000 $m^3 s^{-1}$, over distances 100–1,100 km, and up elevations 10–1,200 m (Crossin et al. 2004; Eliason et al. 2011). Energetic expenditures during upriver migrations have selected morphological, energetic, and reproductive traits that differ among spawning populations. In comparison to the downriver Weaver Creek population (distance = 161 km; elevation = 10 m), females within the high-elevation Chilko Lake population (distance = 629 km; elevation = 1,158 m) are smaller, more streamlined, and begin their migrations with one-third greater stores of somatic energy and twofold greater lipid reserves, but one-third fewer oocytes (Crossin et al. 2004). Adults destined for Chilko Lake migrate faster at 33 km d^{-1}, in comparison to a speed of 20 km d^{-1} for those traveling to Weaver Creek (Crossin et al. 2007). The manner in which temperature influences swimming energetics (cardiac performance and aerobic scope for activity) also varies significantly between these populations (Eliason et al. 2011). Energetic and morphological traits of other spawning populations within the Fraser River are similarly ordered by upriver migration difficulty (Crossin et al. 2004). Sockeye salmon spawn in a greater range of aquatic habitats than other salmons, including rivers, tributary creeks, interlake channels, springs, and lake beaches (Burgner 1991). Once on spawning grounds, adults build and defend nests (redds), mate, and spawn over curtailed periods (typically <3 weeks) during fall months.

The migrations of adult salmon have important carryover effects to their young. A long winter separates fall spawning from the spring, when embryos (termed alevins) will swim up into the water column and begin to feed. Spawning sites stepped across 1,000 m of elevation differ substantially in winter temperatures over which embryos incubate. Higher-elevation populations spawn weeks earlier than lower-elevation ones to accommodate the longer incubation times (Brannon 1987). During

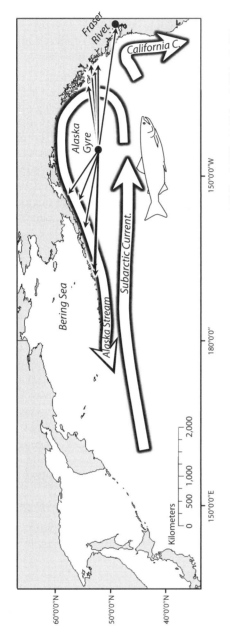

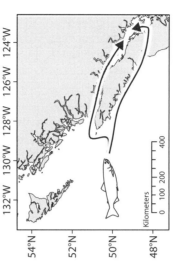

Figure 5.1. Sockeye salmon originating in the Fraser River and elsewhere in British Columbia, Canada, undertake seasonal migrations within the Alaska Gyre prior to their return migrations. A single tagged sample within the gyre (filled circle) resulted in returns of spawning run fish from 10 tributaries throughout Alaska and British Columbia (Neave 1964). Inset shows how those destined for the Fraser River navigate alternative routes around Vancouver Island depending on climate conditions.

the next spring, offspring emerge from redds and transition into early stage juveniles (fry) that undertake population-specific dispersal behaviors, causing them to arrive into lake rearing habitats shared by several other populations. The schedule of oceanic and upriver migrations by adults is apparently selected on the basis of their carryover effects on embryo incubation and juvenile rearing conditions that are still many months ahead in the future (Miller and Brannon 1982; Quinn 2005). Failed reproduction in salmon attends departure from these life cycle schedules. Beginning in 1996, for reasons unknown, Weaver Creek and other low-elevation populations began participating in much earlier runs, exposing adults to superoptimal temperatures during upriver migrations and high en route mortality, resulting in population declines (Cooke et al. 2004; Crossin et al. 2007; Eliason et al. 2011). Possible causes include direct and indirect effects of oceanic conditions, climate (regional warming), disease, and hydroelectric flow regulation that influence run times and upriver migrations (Robards and Quinn 2002; Cooke et al. 2004; Hodgson et al. 2006; Miller et al. 2011).

In the Fraser River, 90 separate itineraries lead generations of salmon to occupy the same set of incubation and rearing habitats. In this way, "Homing creates reproductive isolation and allows for local adaptation in life history traits" (Stewart et al. 2003). Reproductive isolation across generations can lead to genetic drift, which is measured through neutral (nonselective) genetic or phenotypic traits. Based on analysis of microsatellite DNA and other nuclear loci, classification success of 13,000 Fraser River spawners was 60% for each spawning population ($n = 47$), 79% for each juvenile rearing lake ($n = 25$), and 90% among major spawning runs ($n = 4$; Beacham et al. 2004). Reproductive isolation is not absolute but modified by a population's ecological history and degree of straying (Hendry et al. 2000; Stewart et al. 2003). Deliberate introduction of sockeye salmon to Washington Lake (Washington, USA) prior to 1945 provided a unique opportunity to evaluate the number of generations required to generate detectable reproductive isolation using DNA microsatellites (Hendry et al. 2000). A spawning group initially established itself in a tributary, and then within a 10- to 20-y period this same group colonized a beach spawning habitat ~7 km distant from the tributary. Significant genetic divergence was observed 13 generations after the beach spawning group established itself. Thus adaptation to specific spawning sites led to reproductive isolation within timescales pertinent to ecological processes (<100 y). Of course, in this and other instances, both straying and homing are required to establish new populations (Cury 1994; Quinn 2005). The topic of straying is reserved for later.

Homing a Sea Apart: Celtic Sea Atlantic Herring

In their consistent use of sets of demersal spawning sites over narrow ranges of dates, Atlantic herring might be imagined as a fully marine fish version of Pacific salmon. Unlike the more robust salmon fry, the herring larva that emerges from coastal and offshore spawning beds has little control over its dispersal fate. The Downs herring stock, for instance, originates from spawning beds in the English Channel, a highly advective environment with dominant northeasterly flow through the Channel into the North Sea. Winter-spawned larvae experience cold water and will not transform into juveniles for ~5 months. During the first 70 d, strong currents cause larval displacements of 250–500 km into the North Sea (Dickey-Collas et al. 2009). By May, late-stage larvae (similar to European plaice young; see chap. 4, fig. 4.6) enter shallow waters of the Wadden Sea some 1,000 km distant from the spawning grounds. But Atlantic herring also spawn in estuaries, fjords, small seas, and bank systems where stable frontal systems act as dispersal barriers that retain larvae (Iles and Sinclair 1982; Heath et al. 1987). Thus, in comparison to sockeye salmon, where numerous breeding populations correspond to a large set of spawning sites that are nearby to nursery habitats, population richness in Atlantic herring is related to the diversity of hydrographic features that transport larvae to, or retain larvae within, nursery habitats (Harden Jones 1968; Sinclair 1988).

The locations of spawning habitats that are consistently occupied across generations are well known by fishermen who exploit concentrations of ripe herring (Iles and Sinclair 1982; Stephenson et al. 2009), but how natal homing operates remains largely unknown and controversial. How do larval herring, particularly those transported far from spawning grounds, find their way back home? A well-demonstrated example of natal homing occurs for Celtic Sea Atlantic herring, which spawn on shelf banks along the southern coast of Ireland. During winter, most larvae will be swept northeastward into the Irish Sea over their first month of life. Once in the Irish Sea, they mix extensively with juveniles that originated there from local fall spawning (Brophy and Danilowicz 2002). Because no spawning occurs in the Irish Sea during winter, Brophy et al. (2006) were able to differentiate winter-spawned (Celtic) versus fall-spawned (Irish and Celtic) larvae by assigning hatch dates through otolith microstructure analysis (fig. 5.2). They found that despite extensive mixing of populations within the Irish Sea as juveniles, winter-spawned larvae consistently returned to the Celtic Sea as adults to spawn. The ability of the larva to perceive and store information related to its place of origin within the Celtic Sea, despite its rapid displacement into nonnatal habitats and subsequent intermingling with unrelated fish, is remarkable and defies current understanding. Still, the phenomenon cannot be denied. In an early summary of herring population studies, Harden Jones (1968) generalized this pattern of larval drift and adult return as the migration triangle (chap. 1, fig. 1.1). Natal homing by Celtic Sea herring conforms well to the Harden Jones concept of life cycle closure.

Homing an Ocean Apart: Atlantic Bluefin Tuna

Like salmon and herring, spawning schools of Atlantic bluefin tuna in the Mediterranean have experienced a long history of exploitation, and their consistent summertime occurrence in particular regions (the Balearic, Tyrrhenian, and Aegean Seas) has traditionally been interpreted as natal homing. Yet another group of bluefin tuna spawns an ocean apart in the Gulf of Mexico. Collections of larvae confirm spawning locations in the Gulf of Mexico and Mediterranean Sea, but like the Celtic Sea herring, bluefin tuna larvae are quickly transported away from where they are spawned, limiting the opportunity of larvae to imprint to local oceanographic conditions. Traditional tagging studies showed that adult bluefin tuna will seasonally traverse the entire North Atlantic Ocean but do so at low frequency (<10%; Rooker et al. 2007). Further, Gulf of Mexico adults are much larger and older on average than Mediterranean spawners (maturing at a >180-cm versus >100-cm curved fork length; Nemerson et al. 2000). For Atlantic bluefin tuna, restricted spawning in the Gulf of Mexico and Mediterranean Sea, life history differences in adult spawners, and concentrated fisheries in shelf waters have supported separate management units on either side of the mid-Atlantic 45°W meridian since 1982 (Fromentin and Powers 2005; Rooker et al. 2007). Recent discoveries support natal homing for bluefin tuna but also expose flaws in current management assumptions (Taylor et al. 2011).

An aggregation of large bluefin tuna that occurs off the coast of North Carolina in the United States has received concentrated study, with hundreds of fish receiving archival or pop-up electronic tags (Block et al. 2005; Walli et al. 2009). Of these hundreds, scores were relocated, and from these, daily geolocations were estimated. Similar to the sockeye salmon tagged in the open ocean, bluefin tuna departed from North Carolina shelf waters in diverse directions: some eventually migrated into the Gulf of Mexico or Mediterranean Sea during spawning seasons; others, despite being of a presumed adult size, occurred elsewhere during spawning seasons. Long-term records demonstrated year after year return of several adults to the Mediterranean Sea during spawning seasons (Block et al. 2005). Although such behavior is

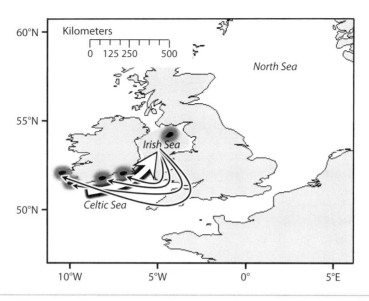

Figure 5.2. Atlantic herring larvae originating in the Celtic Sea (spawning areas are black) are advected into the Irish Sea, where they intermingle (light gray) with locally produced larvae and juveniles. Otolith microstructural analysis demonstrated that despite this mixing, they exhibit natal homing (solid arrows) to Celtic Sea spawning grounds (Brophy et al. 2006).

strongly suggestive of life cycle closure, this inference is uncertain because the natal origin of these fish was unknown.

Stable isotopes in otoliths were recently used to separate fish originating from juvenile nursery habitats in the Mediterranean Sea and the US Atlantic Shelf (Gulf of Mexico origin). The study indicated >95% natal homing to either population (Rooker et al. 2008). For the North Carolina aggregation, a modest fraction (~25%) of this aggregation originated from the Mediterranean Sea (Secor et al. 2013). Thus tracer studies together with telemetry studies provide strong evidence that Mediterranean-origin fish are important components of US fisheries. Strong mixing between the two populations means that US fisheries have likely long depended upon Mediterranean production (Taylor et al. 2011), yet the United States has limited influence on allocation decisions about the Mediterranean stock because mixing is not included in how the species is assessed or managed.

Homing and "Self-Recruitment" in Reef Fishes

For many reef fishes, natal homing is viewed from the perspective of a biphasic life cycle: (1) larval dispersal and (2) juvenile-adult residency. After the newly transformed juvenile has settled on a reef, it remains there (or in the immediate vicinity) as an adult and does not undertake extensive migrations that would cause it to reproduce elsewhere. Reefs often occur in open advective environments, and, lacking an adult homing migration, the presumption has been that most larvae will ultimately reproduce on nonnatal reefs after dispersing some distance. This hypothesis was tested in fringing reefs of Lizard Island, part of the Great Barrier Reef (Coral Sea, Australia), where ~10 million Ambon damselfish embryos were marked by short-term incubations in tetracycline that permanently stained their otoliths (Jones et al. 1999). The field experiment was no trivial effort: over a 90-d period, clutches of ~2,000 embryos at a time were placed in bags containing the dye at six sites surrounding the island's periphery (~4 km diameter). Over that period, the marked embryos hatched and mixed with hundreds of millions of other unmarked damselfish larvae as they dispersed, driven by prevailing currents. Young juveniles settling onto the fringing reefs of Lizard Island were captured in light traps and examined for a tetracycline mark (n = 5,000). Using classic mark-recapture estimators (chap. 2, Tagging: Connecting the Dots), the small number of positive recaptures scaled up to rates of juvenile return to Lizard Island, termed *self-recruitment*, ranging between 15% and 60% (table 5.1).

The recapture of a mere 15 marked damselfish juveniles by Jones et al. was transformative, eventually prompting a whole subdiscipline termed self-recruitment (see Between Closed and Open Populations: Connectivity below). As Jones et al. (1999) remarked, "These 15 recaptures are extremely significant as, given the potential for oceanic dilution and mortality, recapturing any marked larvae was previously considered impossible." Indeed, this needle-in-the-haystack study is among the most cited studies in marine ecology (>600 citations in 2013, according to Google Scholar). Substantial proportions of self-recruits have now been demonstrated for other reef fishes and invertebrates with dispersive pelagic larvae (Cowen

and Sponaugle 2009). The discovery of significant self-recruitment in reef fishes has led to (1) new views on how larvae actively migrate and (2) novel oceanographic modeling tools to track larval dispersal fates (chap. 3, Alignment of Larval Dispersal with Mating Systems). Self-recruitment is critically important to strategies of spatial management and conservation, such as marine protected areas and marine refuges (Halpern and Warner 2003; Botsford et al. 2009), which in large part has stimulated growth of this subdiscipline.

More fundamentally, however, self-recruitment is largely redundant with concepts of hydrographic containment of life cycles developed by Harden Jones (1968), Cushing (1975), and Sinclair (1988). The chief difference seems to lie in the expectation of natal homing: for temperate and boreal species with dispersive larvae, life cycles have long been conceived as closed, whereas until recently the converse expectation existed for tropical reef species with dispersive larvae. To explore this distinction further, the results of Jones et al.'s study on damselfish were compared to a similar larval mark-recapture experiment in a more hydrographically contained temperate estuarine ecosystem (Secor et al. 1995). Here the saltfront results in a strong density discontinuity that retains eggs and larvae (chap. 3, fig. 3.10). For

Table 5.1. Self-recruitment estimated from two larval mark-recapture experiments

	Damselfish scenario 1	Damselfish scenario 2	Striped bass
Number marked	10,000,000	10,000,000	6,540,000
Population estimate	2,000,000,000	500,000,000	54,300,000
Proportion of recaptured sample marked (M)	0.003	0.003	0.18
Proportion of population marked (E)	0.005	0.02	0.12
Self-recruitment (M/E)	0.6	0.15	1.49

Notes: Striped bass data from Secor et al. (1995); damselfish data from Jones et al. (1999). The damselfish study required assumptions concerning the proportion of the wild population that were marked as embryos, bracketed by two scenarios. Population estimates apply to the time of release, assuming similar survival rates between marked and unmarked embryos and larvae.

the Patuxent River Estuary, self-recruitment estimates for larval and juvenile recaptures exceeded 100% (table 5.1). Because self-recruitment cannot exceed 100%, the estimate likely contained errors arising from sampling error and differential survival between marked experimental and unmarked wild larvae. Still, the high degree of hydrographic containment observed in this ecosystem is in keeping with a longer tradition of concepts of hydrographic containment and closed life cycles in temperate/boreal systems. We will return to the subject of self-recruitment and connectivity as they relate to metapopulation theory later in the chapter.

Natal Homing in Elasmobranchs

White sharks are pan-global highly migratory predators that are incidentally captured and observed in subtropical, temperate, and even boreal ocean environments. Assumed to be shelf inhabitants, electronic tagging has shown long and rapid transoceanic migrations. A single large shark traveled from South Africa and Australia and back (>20,000 km) in <9 months (Bonfil et al. 2005), but, until recently, whether such movements represented seasonal homing migrations was unknown. Further, regular seasonal concentrations of white sharks are centered off California, South Africa, and Australia: might these be separate populations, each exhibiting a degree of life cycle closure? In a cross-technology study deploying telemetry and genetics, Jorgensen et al. (2010) found that, like birds of two worlds, white sharks undertook repeated seasonal migration between (1) coastal California, (2) the Hawaiian Archipelago, and (3) a broad subtemperate offshore region (18°N–30°N) termed the "white shark café" (fig. 5.3). "Café" was an apt description based on the concentration of males in this region, which are presumed to feed and await mating opportunities with females that occasionally visit the area. That sexes differ in migrations (termed *differential migration*) is also avian-like. Despite seasonal migrations that span thousands of kilometers, individual sharks showed precise homing to specific habitats, sometimes

within 1 km of where they had been recorded a year prior. The repeated seasonal recurrence of individuals within the same local habitats off California was particularly compelling evidence for life cycle closure and possible natal homing, as these coastal areas serve as nursery habitats (Jorgensen et al. 2010). Further, mitochondrial DNA, a conservative marker of reproductive isolation, showed no evidence of mixing between the California aggregation with those sampled off South Africa and Australia. Critical aspects of the white shark's life cycle remain to be discovered: Is the café the only location where adults mate? How do juveniles first undertake adult migration circuits? Definitive proof of natal homing will require juveniles to be tracked until first spawning. Still, given the difficulty of studying this large and rare species, the discovery of avian-like breeding and feeding migrations is remarkable.

Studies on natal homing in elasmobranchs are surprisingly rare, given the ease with which telemetry and tracer approaches can be applied owing to their accessibility and large size (Guttridge et al. 2009b). Further, strong site fidelity to functional habitats (feeding, pupping, and mating site fidelity) has been well described for sharks (Dicken et al. 2007; Speed et al. 2010). A noteworthy exception is a long-term study on lemon shark *Negaplrion brevirostris*, centered in the Bimini Islands (Bahamas, Caribbean Sea), that has uncovered important evidence for philopatry. Using microsatellite genetic tags specific to individual females, Feldheim et al. (2004) observed nursery site fidelity by 75% of pregnant females, which returned according to their 2-y gestation cycles. Again, natal homing remains to be confirmed for this population, but that females returned to pupping habitats up to five times during the course of the study is compelling indirect evidence.

Life Cycle Closure

Obstinate Nature: Imprinting

Natal homing within a population provides safe harbor: "From one generation to the next, individuals are risk averse and avoid the experi-

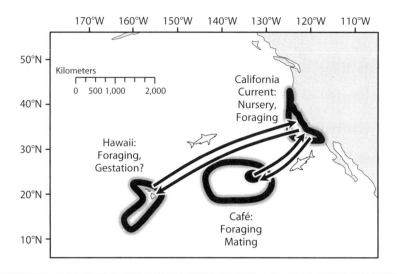

Figure 5.3. To-and-fro avian-like migrations of northeast Pacific white sharks discovered through telemetry. Fish reside along the California coast from August to February and then seasonally migrate elsewhere, either to Hawaiian waters or to a region known as the "white shark café," a hypothesized mating
habitat (Jorgensen et al. 2012a). Here male sharks concentrate in a smaller region (black), which females
intermittently visit. Adapted from Jorgensen et al. 2012b.

ence of new reproductive environments. Fidelity to the native site is strong and individuals
are obstinate in not considering alternative reproductive sites" (Cury 1994). Philopatry serves
as an assurance system for the generational
continuity of (1) mating systems, (2) favorable
spatial regimes for young, and (3) ontogenetic
and seasonal migrations to nonreproductive
habitats. Mating systems in dispersive marine
ecosystems occur in localized regions to ensure fertilization success (chap. 3, Alignment
of Larval Dispersal with Mating Systems). Exploration of new spawning habitats would in-
cur risk to the individual in not finding mates
or locating suitable incubation conditions (e.g.,
temperature, substrate, flow) for the developing embryo (Hendry et al. 2004a). Philopatry
represents a safe harbor for the conservation of
spawning times and locations, which favor hydraulic delivery of young to nurseries (Harden
Jones 1968) or retention (Sinclair 1988). Within
a philopatric (closed) population, related individuals undergo ontogenetic and seasonal migrations that are selected for specific sets of

habitats favoring survival, growth, and reproduction during the member's entire life history.
In this way, natal homing provides a starting
point with carryover effects on subsequent migration routes and habitats, which individuals
will use over their lifetime.

The architecture of philopatry might be considered in a hierarchical sense (chap. 1, table
1.1). Philopatry (a.k.a. breeding philopatry), the
use of the same spawning habitat by generation
after generation, can occur only when individuals within a generation return to their site
of origin to spawn—otherwise known as natal homing. For iteroparous fishes, consistent
return from one spawning season to the next,
or spawning fidelity, maintains natal homing
(fig. 5.4). This hierarchy follows from imprinting, the most commonly invoked mechanism
controlling philopatry in marine fishes (Cury
1994). During an early genetically controlled
period, larvae or juveniles sense and memorize
environmental stimuli, distinctive for their
site of natal origin (Stabell 1984). Imprinting
then provides a mechanism for natal homing

and subsequent spawning site fidelity (for iteroparous fishes), which over generations leads to philopatry and population structure owing to reproductive isolation. Because imprinting is irreversible, it produces a rigid life cycle that is unresponsive to changing ecosystem conditions (Cury 1994). Straying therefore becomes a necessary release from closed life cycles (Harden Jones 1968), a topic discussed further below.

For Fraser River sockeye and other Pacific salmon populations, young juveniles imprint on stream (or lake) odors. Compelling evidence for the "olfaction hypothesis" comes through transplant studies and experiments on Pacific salmon in Lake Michigan, where coho salmon *Oncorhynchus kisutch* were successfully imprinted as juveniles and induced to return as adults to natal streams that were deliberately dosed with artificial odors (Hasler and Scholz 1983). Such odors have a limited ambit of detection (<50 km) even in high-flow environments. Still, over the scale of their detection, odors can govern the final incursions to particular spawning sites, a critical guidance system giv-

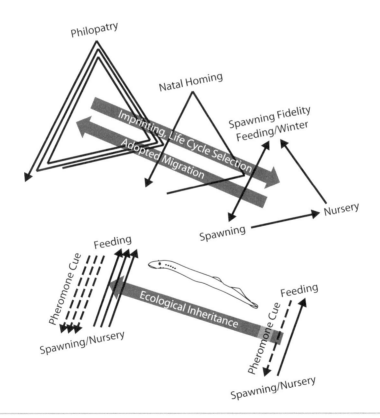

Figure 5.4. Depictions of philopatry, natal homing, and spawning fidelity as migration triangles between principal habitats. Under the imprinting/life cycle selection concept, philopatry selects for natal homing and spawning fidelity. Under the adopted migration hypothesis, philopatry is an emergent property of spawning fidelity and natal homing. The lower panel depicts the pheromone hypothesis for sea lamprey, where pheromones produced from ammocoetes larvae cause multigenerational use of the same spawning grounds. Note, however, in this system of ecological inheritance, individuals do not necessarily return to the spawning ground where they originated. The dashed arrows indicate that the same spawning ground is used by the species over generations.

en the close proximity of adjacent spawning sites among discrete populations. Although one often thinks of fish imprinting to a unique geolocation, other fishes—particularly pelagic ones—likely cue to geomagnetic fields or particular oceanographic features that vary year to year in their location (Cury 1994; Teo et al. 2007; Brochier et al. 2009).

Imprinting remains an incomplete explanation for natal homing in marine fishes; its generality limited by the relatively few taxa and systems studied. Further, some evidence shows that genetics can play a role in encoding where fish return. In several salmon transplant studies, a portion of returning adults arrived in spawning tributaries from which brood adults were drawn, rather than the tributaries into which the offspring of those adults had been introduced (Quinn 2005).

Hydrographic Containment: Member-Vagrant Hypothesis

For Atlantic herring and Atlantic bluefin tuna, imprinting is not so easily understood as in salmon, but if we accept such a mechanism (say, imprinting to geomagnetic fields early in life), natal homing should conserve ontogenetic and seasonal migration patterns over generations. In the member-vagrant hypothesis, Sinclair (1988) gave priority to natal homing as a means for populations to exploit oceanographic features that would favor the retention of larvae across generations. These retention areas range from saltfronts in estuaries to coastal current systems to entire oceanic gyres (chap. 3, Alignment of Larval Dispersal with Mating Systems). But regardless of the feature's size and attributes, mating behaviors that utilize such larval retention features were postulated to arise through natural selection, serving to keep populations from intermingling and improving the demographic fates of larva, which would otherwise be prone to loss from dispersal. Sinclair termed this *life cycle selection*: similar to what has been described above for salmon, spawning at a particular time and place preserves a legacy by which populations occupy a specific set of habitats. Life cycle selection for regions

of larval retention was simulated in a "homing evolutionary model" in which larval transport of Peruvian anchovy *Engraulis ringens* was simulated over hundreds of generations to determine which times and places would lead to highest retention in the Humboldt upwelling ecosystem (Brochier et al. 2009). Emergent spawning patterns matched well with observed spawning times and places.

Life cycle selection and population structure should cause limited attrition (vagrants), including (1) embryos and larvae lost to the population through diffusion and advection and (2) juvenile and adult strays to other populations. "The very existence of a population, we argue, depends on the ability of larvae to remain aggregated during the first few months of life" (Sinclair 1988). Harden Jones (1968) and Cushing (1990) had also argued that hydrographic containment (life cycle closure within a bounded oceanographic setting) was necessary for the perseverance of populations, but Sinclair (1988) and Iles and Sinclair (1982) provided a more explicit mechanism that linked natal homing to the then-emerging view that ocean fronts and other density features played a major role in controlling the spatial ecology of larvae (chap. 3, Alignment of Larval Dispersal with Mating Systems). Intriguing corollaries supporting the member-vagrant hypothesis included positive correlations between (1) population richness (number of populations) and number of larval retention areas among diverse marine species, and (2) the size of larval retention areas and adult abundance for Atlantic herring populations and other species (Page et al. 1999; Hay et al. 2009; fig. 5.5).

The member-vagrant hypothesis arguably represents an exaggerated view on the role of philopatry and natal homing in defining patterns of migration in marine fishes, particularly as imprinting and other mechanisms of natal homing remain largely unknown. Current evidence that natural selection leads to the perseverance of closed life cycles comes largely from neutral markers of lineage, which as an emergent property of past genetic drift provides only indirect support for population

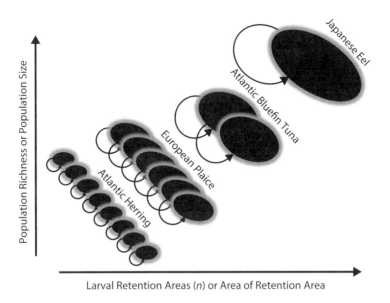

Figure 5.5. Depiction of increased population richness with number of larval retention areas (black) under the member-vagrant hypothesis (Sinclair 1988). The size of each retention area scales with physical area and population size. Arrows represent natal homing.

structure. Increasingly, however, functional alleles have provided insights on ecological and behavioral drivers of population separation under the rubric of biocomplexity research, reviewed further in chapter 6 (Cod Ecomorphs and Biocomplexity).

Recalcitrant Nature: Adopted Migration and Tradition

A contesting explanation to hydrographic containment posits that social transmission of migration causes persistent life cycles (McQuinn 1997a). This adopted migration (or entrainment) theory postulates that juveniles learn life cycle circuits through their association with experienced individuals during periods of spatial overlap (fig. 5.6). Behavioral propensities to learn and school cause individuals to adopt migrations and life cycles through their interactions with older, experienced individuals who themselves adopted the migration pathways of previous generations. When such interactions do not occur, exploratory juveniles can establish novel migration circuits, some-

times leading to irruptive population growth, as postulated for Japanese sardines and other schooling fish (chap. 4, Schooling through Food Webs). The adopted migration theory contrasts with imprinting in reordering the hierarchy of behavioral properties from (1) breeding philopatry (life cycle selection) → natal homing → spawning site fidelity to (2) spawning site fidelity → natal homing → breeding philopatry (tradition; fig. 5.4). A principal tenet of the adopted migration theory is that there occurs a labile period during which interactions with experienced individuals in schools or other aggregations lead to entrainment of individuals into a life cycle (fig. 5.6; Cushing 1962a; Guttridge et al. 2009b). Such life cycles can persist over generations, causing individuals to return to natal habitats and eventually leading to reproductive isolation and associated genetic drift.

Atlantic herring are central in debates on life cycle closure (Dickey-Collas et al. 2009), and just as we can find evidence for imprinting in Atlantic herring (Brophy et al. 2006; Stephenson et al. 2009), there are other instances

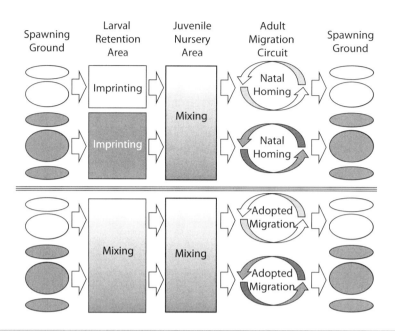

Figure 5.6. Two central explanations for closed populations in marine fishes. The top panel represents larval retention, imprinting, and natal homing; the bottom panel depicts social transmission of adult migration circuits. White and gray shading indicate spatial separation of individuals between life history stages; graded shading between white and gray indicates mixing of individuals. Reprinted from Secor 2010, with permission.

where evidence supports adopted migration. Along the coast of Norway, a large migratory coastal population of spring-spawning herring interacts with small local populations residing in fjords and bays. Within the Lurefjord, Johannessen et al. (2009) uncovered evidence (based on otolith microstructure and morphometric traits) that the large coastal population crossbred with the local Lurefjord herring population. Despite crossbreeding and the potential for larvae to disperse from the fjord, a resident population persisted as what Johannessen termed a "cultural" population. Age structure analysis indicated that the fjord population comprised a fraction of the coastal spawned individuals, which adopted a resident behavior.

Schooling and other aggregation behaviors provide the opportunity for social transmission and learning of migration behaviors. Unlike birds, marine fishes do not associate with their parents during their initial migrations

and must enter adult migration circuits later in life (chap. 2, Navigation Maps: Inherited, Imprinted, or Learned?). Social transmission was an early hypothesis (Cushing 1962a), competing with other mechanisms such as instinct, kinotaxis, or serendipity. Most marine fishes begin to aggregate and school during the late larval and early juvenile periods (chap. 3, The Active Larva). As we have already seen, schooling has a strong influence on navigation through consensus, leadership, and dynamic information exchange. The efficient control of schooling behavior, particularly for aggregating pelagic species, provides a propensity for group attachment to more experienced conspecifics (chap. 2, Information Systems).

Learning from experienced congeners involves behavior copying and can lead to longlasting traditions (Giraldeau 1997). A classic experiment on social transmission of migration behavior involved manipulating the interactions between newly settled juvenile and well-established French grunts on reef patches (Helfman and Schultz 1984). Traditional diel migrations between the reef and seagrass habitats were adopted by naïve juveniles that were transplanted in the presence of experienced adults, but no such behaviors occurred in the absence of experienced conspecifics. But this experiment presents an incomplete picture of social transmission (Dodson 1988). Did the juveniles merely school with experienced fish, or were behaviors learned such that they in turn could entrain newcomers? Would such behaviors persist over generations? As already reviewed (chap. 2, Navigation Maps: Inherited, Imprinted, or Learned?), fish can learn and memorize habitat locations and migration routes (e.g., Brown and Laland 2003). That these cognitive capacities might be socially facilitated seems quite feasible based on the avian literature, but definitive field evidence remains regrettably elusive for marine fishes. Further, how the emergent properties of schooling behaviors play out over seasons and life cycles is unknown. To more fully understand the association between individual aggregation behaviors and migration, one would need longer-term studies directed at higher-order aggregations, such as systems of interacting schools (e.g., Stehfest et al. 2013).

A corollary of adopted migration theory is that transmission of beneficial behaviors would be more rapid for social than asocial individuals. In opposition to this view, naïve guppies *Poecilia reticulata* in the laboratory tended to associate more with experienced fish that showed less oriented behaviors than those that exhibited oriented movements toward forage rewards. The naïve fish could not keep pace with these best-performing fish (Swaney et al. 2001). Thus poor teachers might disrupt migration systems that depend on learned behaviors. Swaney et al. (2001) suggested that learning beneficial behaviors may be optimal in populations showing a continuous range of performance, where individuals learn from others that have slightly better performance. The propensity for individuals to aggregate with similar-sized individuals could promote continuous information transfer from slightly larger higher performers to smaller learners (McQuinn 1997a; Guttridge et al. 2009a; Johannessen et al. 2009).

The historical aggregations of spawning cod off Newfoundland (chap. 3, Five Mating Systems) provide circumstantial evidence for adopted migration. After spawning, adults departing for shelf feeding habitats are led by large scouts (Rose 1993). When scouts are diverted, echograms indicated a shift in the entire school's behavior. Joining the migrating herd were younger members that did not participate in spawning but were now comingling with adults in their shelfward migrations. Rose (1993) proposed that these juveniles are entering the adult migration circuit through behavioral entrainment. In contrast to juveniles adopting adult behaviors, Huse et al. (2002) showed that younger and more numerous cohorts can affect the migration of older fish. For Norwegian herring, the ratio between numerous juveniles and experienced older fish (numerical dominance of juveniles) explained major changes in migration patterns to wintering habitats in herring. Numerical dominance of naïve fish can then lead to innovation and colonization of new habitats, which if favorable might initiate novel life cycle patterns. Thus accelerated changes in spatial behaviors might occur when one population segment undergoes rapid growth relative to other segments (Secor et al. 2009).

If individuals of differing natal origins do not intermingle during the behavioral labile period, closed cultural populations can result. Robichaud and Rose (2004) proposed that genetic differences among small coastal Atlantic cod populations resulted from the lack of overlap between migration pathways. Thus genetic isolation can arise through tradition without any innate homing propensity. Isolation of breeding populations and genetic drift would be compromised by invasion of strays, regardless of whether the isolating mechanism was imprinting or social transmission.

The sudden appearance or disappearance of aggregations of herring and other marine fishes has been explained by positive feedbacks associated with efficient social transmission of colonizing behaviors and negative feedbacks associated with lost traditions, respectively (Corten 2001; Óskarsson et al. 2009; Petitgas et al. 2010). The intergenerational conveyance of migration behaviors can be disrupted when fishing affects age structure or severely depletes aggregations of experienced fish (Huse et al. 2010; Petitgas et al. 2010). As a population declines, behavioral entrainment into its migration circuit becomes less efficient, and entire life cycles can be lost. For instance, intensive high-seas fisheries off Brazil and Norway in the 1960s depleted Atlantic bluefin tuna aggregations that abruptly disappeared in the 1970s. Fonteneau and Soubrier (1996) speculated that these "frontier" fisheries occurred on unique population segments. Once lost, conservatism of migration pathways would curtail reinvasion of these historical regions (Petitgas et al. 2010; Fromentin et al. 2014). Alternatively, social transmission can be disrupted when strong year-classes of naïve juveniles "flood" experienced adults, causing adults to entrain into the novel behaviors of juvenile fish (Huse et al. 2010).

Similar to natal homing, the exclusive dependence on adopted migrations would lead to a system recalcitrant to environmental change. For instance, Norwegian herring in recent decades are densely aggregated in the same few coastal fjord systems during winter, despite conditions that are superoptimally warm and densities that result in high predation and hypoxia (Dommasnes et al. 1994; Corten 2001). Adopted migration behaviors would be expected to produce lagged responses to climate change in accord with the frequency with which juveniles establish novel migration pathways. Traditional use of habitats would be expected to persist, even as those habitats become suboptimal. Thus, similar to imprinting, adopted migrations are expected to produce nonlinear responses (nil or lagged responses, positive or negative feedbacks, and carryover effects) between environmental change and migration behaviors.

Telemetry approaches could provide direct evidence for adopted migrations, but so far

studies have not been specifically designed to test such behaviors. In a single test, Svedäng et al. (2007) tracked movements by Atlantic cod between the Kattegat (east of Denmark) and North Seas. Because individuals released together at similar locations and dates did not undertake similar seasonal migrations, Svedäng et al. concluded that social transmission of migration behaviors was not occurring. Contrary to this result, other tagging studies often show that individuals tagged in proximity to each other are recaptured at similar times and places (Secor 1999; Robichaud and Rose 2001, 2004; Jorgenson et al. 2010; Stehfest et al. 2013). What is needed in experimental tagging and telemetry studies is information on the social history of released fish. Elements of adopted migration theory could be directly tested by manipulating social interactions among fish prior to their release.

Conservatism of Nursery Locations through Ecological Inheritance

Multigenerational use of the same spawning and nursery habitats can occur through ecological niche expansion (chap. 1), in which adults and offspring alter environments in a manner that causes future generations to use those same areas. Such a mechanism was initially proposed for anadromous salmons, where juveniles released pheromones or other chemical attractants that cued subsequent runs of salmon to utilize nearby spawning habitats (Nordeng 1971; Larkin 1975). This so-called pheromone hypothesis was subsequently disproved (Hasler and Scholz 1983; Brannon and Quinn 1990), who demonstrated that natal homing in Pacific salmon occurred through imprinting (see Obstinate Nature: Imprinting above). But another anadromous species, the sea lamprey Petromyzon marinus, is a strong candidate for the multigenerational conservation of spawning and natal habitats through ecological transmission (fig. 5.4). Sea lamprey are a parasitic species that spend most of their adult life roving for hosts in marine or Great Lake ecosystems, but, similar to salmon, they return to upstream habitats and spawn large sticky eggs on cobble, gravel, and sand (Clemens et al. 2010). Their offspring, termed *ammocoetes larvae*, find their way to soft substrate and burrow for many years, filter feeding and giving off olfactory cues that lead adults to spawn in proximate locations (Li et al. 1995; Vrieze and Sorensen 2001). The attraction of adults to the same spawning habitats over years and decades is related to the modification of that habitat's chemical environment by larvae. Thus conservatism of spawning and nursery locations in sea lampreys occurs absent natal homing (Rodriguez-Munoz et al. 2004; Waldman et al. 2008; Spice et al. 2012). Indeed, management measures to control lamprey numbers in ecosystems where they are judged detrimental emphasize biocides targeting nursery habitats (Slade et al. 2003; Hansen and Jones 2008). Conversely, efforts to restore sea lampreys depend on conserving and founding new ammocoetes seed banks (Neeson et al. 2012; Ward et al. 2012).

Between Closed and Open Populations: Connectivity

Connectivity, broadly defined as the exchange of individuals within networks of local populations or habitats, has emerged as a field of inquiry across all classes of migrating animals. The consequences of connectivity will depend not only on the number of individuals that move among populations, but the question at hand. Genetic connectivity describes classic metapopulation investigations and focuses on population and metapopulation extinction rates. Demographic connectivity emphasizes shorter timescales and includes effects of straying on population growth (ecological connectivity), viability (conservation connectivity), and production (harvest connectivity; terms from Lowe and Allendorf 2010). Similar to our definition for migration (chap. 1, Classifying Migration), connectivity entails changed ecological status.

A chief focus across these types of connectivity is on dispersal by species exhibiting biphasic life cycles, such as demersal marine reef fishes and invertebrates (Gaines et al. 2007; Jones et al. 2009). These life cycles involve sedentary juvenile and adult stages and dispersive larvae. In such systems, local populations are defined by the reef habitats where juveniles and adults occur, and these habitat patches are connected through larval dispersal. Reef networks are connected through larval dispersal, where each patch serves as both a source and recipient of migrants (Roberts 1997). The degree of connectivity between patches is influenced by (1) the distance between patches; (2) the duration of larval dispersal discounted by mortality; (3) ocean circulation features contributing to advection, diffusion, and retention; and (4) larval behaviors that modify their transport (Cowen and Sponaugle 2009). Previous views held that reefs were repopulated by a ubiquitous pool of highly dispersed larvae, limited in their settlement patterns by available habitat (e.g., the lottery hypothesis; Sale 1982). But it is now known that larval dispersal distances are modified by survival (dispersal kernels; fig. 5.7) that are often insufficient to effectively connect reef habitats that are hundreds of kilometers distant (Kinlan and Gaines 2003; Cowen et al. 2006). New empirical estimates, albeit small in number, indicate that self-recruitment can exceed 50% (Jones et al. 1999, 2009; Almany et al. 2007).

Both modeling and empirical approaches have focused on how the arrangement of patches, regional circulation features, and capacity of larvae to disperse influence whether a given patch is colonized, grows, declines, or becomes vacant. In an ambitious modeling domain that included the entire Caribbean Sea, Cowen et al. (2006) released millions of virtual larvae designed to emulate the monthly spawning by diverse life history types (snapper [Lutjanidae], goby [Gobiidae], and damselfish [Pomacentridae]) from 260 sites distributed about the basin. The larvae were subjected to oceanic transport through a high-resolution three-dimensional circulation model fit to the regional seabed and tuned to 5 y of wind data. Importantly, larvae were simulated with three levels of vertical behaviors that would favor their settlement in nearby reef habitats at either 5,

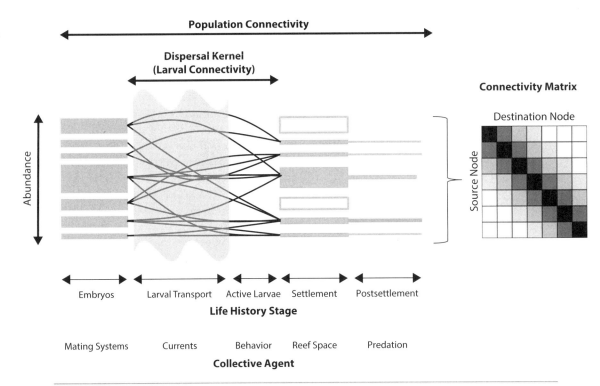

Figure 5.7. Population connectivity in a biphasic life cycle. Mating systems inject sets of embryos from reefs (rectangles) into semiopen systems. Cohorts of larvae then "sample" spatial and temporal variance in food web and oceanographic conditions, some of which are favorable for larval dispersal (thin lines) to reef patches. The larval dispersal kernel, a function of larval numbers and their transport and demographic fates, results in varying settlement rates across reef patches (Cowen and Sponaugle 2009). Presented on the right is a dispersal kernel (connectivity matrix) for source and destination reefs. In this example, self-recruitment to reefs is indicated by the higher densities of settlers (darker shading) along the diagonal.

15, or 30 d into their dispersive phase. Results showed that some degree of self-recruitment by larvae was needed for patch replenishment throughout the Caribbean basin. That self-recruitment varied substantially among reefs indicated a strong role for oceanographic forcing in controlling patch gains and losses. Interestingly, their analysis suggested that species like snapper, which are long lived (≥15 y) exploit less frequent circulation features that favor both self-recruitment and immigration to more distant reefs, an example of the storage effect (chap. 4, The Storage Effect).

Might self-recruitment through limited lar-val dispersal apply more broadly to marine fishes? European plaice spawn in major regions of the North Sea, exposing their offspring to strong hydrodynamic forces. As reviewed in chapter 4 (Ontogenetic Habitat Shifts), plaice nurseries are sandy shoal habitats in tidal embayments and estuaries that fringe much of the North Sea (fig. 5.8). Using a circulation model, Hufnagl et al. (2013) tracked virtual larvae under the assumption of passive dispersal. Predicted settlement patterns indicated a high degree of self-recruitment between major spawning regions and adjacent nursery habitats (fig. 5.8). Offspring produced in bank regions exposed to stronger flows or longer transport (e.g., Dogger Bank and the German Bight) experienced more diverse settlement fates.

The question of self-recruitment leads to well-trod ground: the migration triangle, hydrographic containment, and the member-vagrant hypothesis. Thus we see that population structure scales between Pacific salmon having no dispersive larval stage and numerous populations, schooling tunas and sardines with limited population structure and homing to dynamic environmental cues over large regional scales, and Japanese eels with larvae dispersing throughout much of the North Pacific Gyre, supporting a single panmictic population (fig. 5.5). Whether termed self-recruitment, population integrity, or unit stocks, the concept of closed populations has been a guiding concept in marine fish ecology, one deemed necessary, particularly in the management of exploited populations. A recent challenge, however, is to recognize the more open nature of populations.

Open Life Cycles

Homogenized Life Cycles: The Basin Model

An idealized open life cycle would represent the composite of individual behaviors guided only by their points of origin and environmental circumstance, and leading to spatial distributions characterized by physical forces and habitat quality alone. For embryos and larvae where diffusion predominates, field observations often show a dense centroid surrounded by serially declining abundances (Heath 1992; Bradbury et al. 2003. New embryos should emanate from a patch as they grow into the larval period—the shape of the centroid and the rate of peripheral dispersal governed by diffusion and advection (chap. 3, Embryo Dispersal; fig. 5.7). Under a scenario where adults select spawning areas that are favorable to offspring, we might expect concentrated patches of embryos and larvae to track highest-quality habitats conditioned on their points of origin.

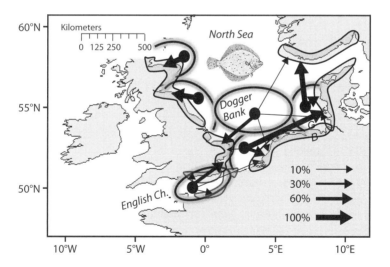

Figure 5.8. Larval connectivity between principal spawning regions (ellipses) and nursery areas (shaded coastal regions) for North Sea plaice, estimated by Hufnagl et al. (2013). Arrow widths indicate degree of connectivity. Note the more diverse or distant destinations for larvae spawned in the English Channel, Dogger Bank, and German Bight (GB). Modeled after Hufnagl et al. 2013 and used with permission. Plaice illustration from the Fishery Resources Monitoring System: http://firms.fao.org/firms/en.

This view accords with Sinclair's member-vagrant hypothesis, where larvae entrained into and concentrated within physical retention features experience higher survival and reproductive success than those that do not.

Active larvae, juveniles, and adults select habitats on the basis of their capacity to sense and respond to structural features and environmental gradients. If individuals are uniform in their ability to select habitats, then members of a population will initially select highest-quality habitats, but as these become occupied, members will sequentially move into lower-quality habitats so that, in aggregate, individuals are receiving the same level of per capita benefit over the range of habitats occupied. Thus an individual juvenile fish may be responsive only to several hectares of its immediate physical and social environment, yet when we combine its behavior with that of others, the emergent property can be described as density-dependent habitat selection—with highest concentrations centered on the highest-quality patches of habitat.

Whether governed by diffusion or active habitat selection, the spatial distribution of open populations can be simplified to a single idealized behavioral response across all members of the population. In his review of animals and plants, Brown (1984) noted that the spatial distribution of densities within populations followed a Gaussian distribution (see fig. 5.10) and suggested that highest abundances were centered in (1) the middle of a population's range and (2) over the highest-quality habitats. The decaying tails of the Gaussian distribution represent spatial gradients of ever-diminishing habitat quality. Diffusion of passive particles (embryos and early stage larvae) shares similar statistical properties (Denny 1993). Under the scenario of no individual variation in spatial behaviors, a population should equitably distribute itself according to habitat resources so that density is a perfect surrogate for habitat quality.

The classic model of density-dependent habitat selection is the ideal free distribution (Fretwell and Lucas 1970). The best-quality habitat patches are first filled, and as competitive interactions increase, individuals move freely to alternate patches of sequentially lower-quality habitat regardless of prior occupants. The result is that across the population's range, individual survival, growth, and reproductive prospects are identical. An elegant depiction of the ideal free distribution is MacCall's (1990) basin model, where patches are represented as basins, the depth of which is related to habitat quality (contributions to reproductive success) and the surface related to carrying capacity (fig. 5.9). The topography of the basin's floor represents habitat quality across the population's range, the basin's shores are the population's range, and the volume that fills the basin is population abundance. Basin sills represent impediments or thresholds to dispersal or regions of poor habitat quality. Through a diffusion-like process, individuals fill each basin and assort themselves so that they optimize their reproductive values (MacCall uses per capita growth rate, $(dN/dt)/N$, to index habitat quality). MacCall added realism to the basin model by incorporating territorial behaviors (ideal despotic distribution; Fretwell 1972) and "viscosity" (inertia) to individual movements, causing a deformed surface of higher abundances centered on the deepest portions of the basin (fig. 5.9). The basin model represents a depiction of a series of stable abundance states where each basin defines a local attractor (chap. 7, Resilience Theory). Thresholds are overcome by collective agencies including population growth, ontogenetic habitat shifts, personality modes, predator breakouts, and partial migration (chap. 6).

Three general patterns relate geographic distribution with density: (1) the geographic range stays constant, and density increases proportionately with abundance; (2) the density stays constant, and range increases proportionately with abundance; and (3) the basin model, where both range and density change with abundance (fig. 5.10; Petitgas 1998; Shepherd and Litvak 2004). Discriminating among these alternatives requires simultaneous information on range, density, and abundance,

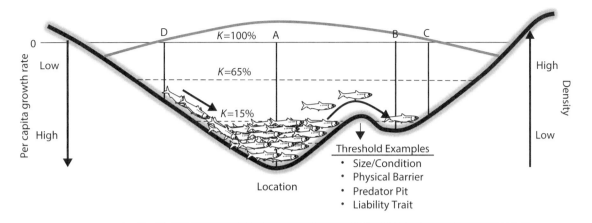

Figure 5.9. MaCall's (1990) basin model depicting changes in spatial range with increased density. Density changes in a way that equitably distributes habitat resources among individuals (ideal free distribution). Habitat resources are depicted as basins, the depth of which relates to reproductive fitness. Local attractors will cause individuals to gravitate toward the deepest segments. The level of infilling is representative of carrying capacity (K). In this depiction, habitat values are ordered A > B > C > D. Note, however, that barriers or thresholds can reorder the manner in which habitats are utilized. Under density-dependent habitat selection, B and C would be simultaneously occupied. Behavioral agencies can deflect the surface (gray line) and include territoriality, homing, or threshold changes such as predator breakouts by sardine populations (chap. 4, Schooling through Food Webs).

as well as habitat quality, which has curtailed rigorous tests for density-dependent habitat selection. For the most part, only corollaries of density-dependent habitat selection have been examined. For the heavily fished Scotian Shelf (Canada), Shackell et al. (2005) observed expected range contraction in cod with declining abundance, but densities in core areas also declined. Also in Atlantic Canadian waters, yellowtail flounder *Limanda ferruginea* exhibited increased habitat selectivity at lower densities, a principal corollary of the basin model (Simpson and Walsh 2004). For populations of anchovy and sardines, simultaneous increases in range and density were associated with increased population abundance (Barange et al. 2009). Using genetic markers, Lecomte et al. (2004) concluded that range expansion and contraction of Pacific sardine *Sardinops sagax* and northern anchovy off California (USA) occurred for single contiguous populations rather than through coalescence and division of local subpopulations. In submersible observations of

continental shelf waters, Sullivan et al. (2006) related local densities of demersal juvenile fishes to different types of structured habitat and showed that at high densities individuals were less discriminating among habitats.

Concepts related to the ideal free distribution are exaggerations because they depend on functional relationships between habitat quality and density that change only with abundance. Changed habitat rankings in time, individual attributes, and movement ecology are largely discounted. Although some have argued that they are valuable as null models against which to test more complex spatial patterns of abundance (e.g., Smogor et al. 1995; Petitgas 1998), it is nearly impossible to distinguish the principal types of density-dependent habitat selection, or to disentangle them from underlying population, environmental, and life history dynamics (fig. 5.10; Shepherd and Litvack 2004). For instance, how do we know whether an expansion is driven by an increased spatial extent of habitat or population growth? Con-

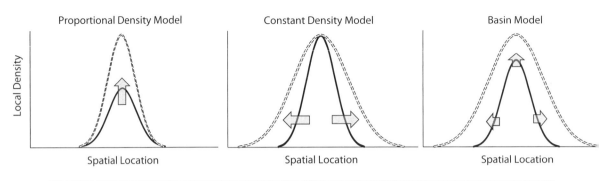

Figure 5.10. Three general concepts of density-dependent habitat selection. Modified from Shepherd and Litvak 2004, with permission.

trasting these alternatives would require detailed time series on habitat quality, density, and abundance. In terms of the topic at hand, density-dependent habitat selection cannot be feasibly scaled up to migrations and life cycles of marine fishes because it oversimplifies collective behaviors by a common diffusion-like property.

Still, from the standpoint of state-space models and movement ecology (chap. 2, Rules of Aggregation), density-dependent habitat selection is one of several simple models that can be tested against short-term movements. For Pacific skipjack tuna *Katsuwonus pelamis*, a species widely distributed throughout the tropical and subtropical Pacific Ocean, Sibert et al. (1999) parameterized movements on the basis of diffusion as modified by advection (biased random walk). Advection in this case indexed directed movement, which in some instances could be driven by density. A differential equation—simultaneously solving for diffusion, advection, and mortality terms—was fit to ~4,000 tag-recapture events across 10 broad Pacific Ocean regions and four seasons. Model outcomes—more dispersed distributions during certain seasons and regional variation in the direction of movements—agreed well with observed recoveries and past studies on distributions, providing support for the application of simple physical models in understanding short-term changes in fish distributions (see also Humston et al. 2000, 2004).

Fish of a Different Feather

As we more precisely track individuals over major phases of their life, unexpected behaviors are increasingly common. With increased observations of "misplaced" fish has come the realization that divergent migrations in fact represent structured modes of collective behavior with important consequences to population and metapopulation dynamics (Secor 1999, 2004). Rare observations such as a large tuna captured in an estuary (Murdy et al. 1997) or a freshwater carp captured in marine waters (Barraclough and Robinson 1971) have long been noted as anomalies defying simple explanation. The incidence of the odd fish at the odd place and time, as with birds (chap. 2, Losing Their Way), may be the result of navigational or orientation errors or oceanic flows that cause unusual transport of larvae. Alternatively, such modes may be "built into" populations through genetic and behavioral propensities in which personality, metabolism, and schooling lead to alternate migration behaviors. As Robichaud and Rose (2004) note, "There appears to be a cod for each environment, with diverse forms at once exploiting and being molded by physical oceanographic and biological properties of the marine ecosystem."

First, let us return to the phenomenon of larval fish dispersal that frequently results in misplaced individuals and large advective losses (Hjort 1914; Houde 1989; Kim and Bang 1990). As reviewed previously for Japanese eel leptocephali, lost larvae represent a type of ecological overhead, an apparently wasteful system required to ensure transport of sufficient numbers of larvae to juvenile nursery habitats. Retentive features are often "leaky" or not present at all, resulting in a range of spatial fates for pelagic embryos and larvae (Campana et al. 1989; Cowen et al. 2006; Saenz-Agudelo et al. 2011). Another example of lost larvae occurs for tropical butterflyfishes *Chaetodon* spp., which are transported from their subtropical latitude spawning sites to US temperate waters as a result of their entrainment into the northward flowing Gulf Stream. From this shelf current, larvae transform and migrate into temperate estuaries, grow during summer and fall months, but then, unable to migrate south over long distances, they die en masse as temperatures plummet each winter (McBride and Able 1998). Similar doomed dynamics have been observed for larval cohorts of Atlantic herring and Atlantic menhaden, which enter temperate estuaries in the late fall but subsequently experience lethal winter mortality (Townsend et al. 1989; Lozano et al. 2012). These wayward eel, butterfly fish, herring, and menhaden larvae—what Sinclair (1988) termed *vagrants*—say something about the efficiency with which larvae are transported to suitable juvenile habitats. But might this type of ecological overhead serve another purpose?

Roamers

As an entrée into more structured migration behaviors within populations, consider roamers Roaming successively leads individuals away from aggregations or common migration pathways. In late-maturing fishes, juvenile roaming behavior is particularly prevalent. Juvenile striped bass, for example, leave their natal freshwater habitats and roam widely, migrating hundreds of kilometers into nonnatal estuaries and coastal waters over their first

several years of life (Secor and Piccoli 1996; Pautzke et al. 2010; Able et al. 2012). Conversely, juvenile barramundi *Lates calcarifer* originate in marine coastal waters off northern Australia and roam into communicating estuaries, mangrove creeks, swamps, rivers, and billabongs (McCulloch et al. 2005; Walther et al. 2011). Interestingly, many juveniles that enter these seasonally dynamic habitats become resident adults that fail to navigate their way back to reproduce in coastal waters (Milton and Chenery 2005). Thus juvenile roaming behaviors are expected to have strong carryover effects on adult growth and reproduction.

Roaming behaviors may be moderately common. In their review of Atlantic cod tagging studies throughout the species range, Robichaud and Rose (2004) classified 20% of tagged groups (28 of 138 groups) as roamers (their term was "dispersers"). They described roamers as those that dispersed broadly over large shelf areas, with low site attachment or low rates of annual return to similar areas. Jørgensen et al. (2008) contrasted roaming adult cod to ones that exhibited natal homing, comparing simulated populations that exploited near (750 km) and distant (1,500 km) spawning areas along the Norwegian coast. For the modeled population of homers, the distant spawning ground was hardwired for a component of the population regardless of adult size and condition. For the population of roamers, only large and well-conditioned adults occupied the distant spawning ground. Interestingly, under scenarios of high losses due to fishing, the distant spawning area was completely lost for the homing population but persisted, albeit at a much reduced level, for the population of adult roamers. Examples of roaming associated with range extension and fish size or longevity include black drum *Pogonia cromis* spawning at the northern extent of their range in the Chesapeake Bay (Jones and Wells 1998) and red bass *Lutjanus bohar*, which extend their range into deeper waters off the Great Barrier Reef (Marriott et al. 2007). The trend of moving into deeper waters with adult size was noted long ago for flatfish and other demersal fishes (Heincke 1913).

Fish with Personalities

The term *roaming* implies a specific behavior, and indeed a primary cause attributed to roaming is an individual's personality or, alternatively, its behavioral syndrome. Behavioral syndromes are well established in fishes, measured as the consistency (repeatability) of individuals to perform certain behaviors under provocative laboratory conditions, and can develop early in life (Fuiman and Cowan 2003; Biro and Stamps 2008; Conrad et al. 2011). Syndromes include exploring individuals, which show a propensity to explore novel situations. Bold individuals explore regardless of risk. Aggressive types are measured by their success in acquiring limited resources, and social individuals demonstrate attraction and alignment with conspecifics. Behavioral syndromes are not orthogonal to one another. As might be expected, overlap occurs between bold, exploring, and aggressive types, and these types tend not to be social. Level of activity (locomotive performance, active metabolism) is yet another ascribed personality type. These syndromes fit well within the movement ecology explanatory framework (chap. 1, Movement Ecology: A Series of Individual Movement Phases) in that they motivate movement behaviors, albeit conditioned on environmental and social circumstance (Conrad et al. 2011). Through artificial selection and crossbreeding experiments, behavioral syndromes have been shown to be heritable and hypothesized to be under pleiotropic control (multitrait control by a single gene or gene complex), where syndromes are switched on or off on the basis of environmental, developmental, or social interactions (Conrad et al. 2011). Still, behavioral types represent propensities, which are conditioned on past social and environmental experiences (Dingemanse et al. 2009; Nakayama et al. 2012). Further, there is evidence that an individual's flexibility to change personality may itself be heritable within populations (Sih and Bell 2008). This conditional aspect in the expression of personalities likely manifests considerable variation in aggregate behaviors observed in natural

populations (B. B. Chapman et al. 2011a; Conrad et al. 2011).

A handful of mesocosm and field studies are beginning to demonstrate how personality types are relevant to roaming, straying, and resident movement behaviors at collective scales. Individuals are assayed for their performance and then released into natural or seminatural conditions. Small annual tropical stream Trinidad killifish *Rivulus hartii* were ranked according to boldness (assayed in the laboratory), marked, and then released into both artificial and natural streams (Fraser et al. 2001). In both systems, boldness was positively correlated with dispersal distance 24 h later. Similarly, in an artificial stream—a series of five pools separated by flowing raceways—individual mosquitofish *Gambusia affinis* assayed to be bold or exploring types showed greater dispersal between pools than did fish showing propensities to shoal (fig. 5.11; Cote et al. 2012). Behavioral types have also been studied by collecting groups of fish showing different behaviors in the field. For instance, Edeline et al. (2005) measured higher locomotive performance and greater thyroxine levels (active behavioral type) for European eel juveniles that had moved greater distances upstream in the field.

In simple terms, behavioral types lead to propensities to stay put or head out. Summing such movements for many individuals leads to a skewed distribution, similar to movements associated with the area-restricted search behavior (chap. 2, Rules of Aggregation). But in this case the skewed distribution is due to variability between rather than within individuals. The majority of individuals tend to stay put, with a smaller fraction undertaking longer-distance exploratory behaviors (Paradis et al. 1998; Fraser et al. 2001). Behavioral types, particularly those that underlie exploratory behaviors, are hypothesized to be key attributes that allow for straying, colonization, and metapopulation dynamics (Fraser et al. 2001; McMahon and Matter 2006; Cote et al. 2010; Conrad et al. 2011; Hamann and Kennedy 2012). Consistent with individual propensities underlying leptokurtic distributions are the skewed distributions of

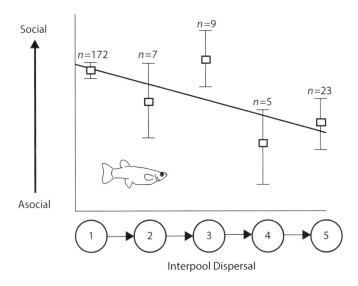

Figure 5.11. Experimental mosquitofish that exhibited asocial personalities showed greater dispersal among connected pools. Individuals were assayed for consistency in social (propensity to shoal) and asocial behaviors (propensity to exhibit active, bold, and exploratory behaviors) and then released into a series of connected pools. Redrawn from Cote et al. 2010, with permission.

recoveries in tagging studies for marine fishes, exemplified in a large multidecadal study where over 1.5 million Pacific herring were tagged off British Columbia (Hay et al. 2001). Of those at large for longer than 1 y, >90% were recaptured in the region of release followed by a rapid decline of recaptures in adjacent regions (fig. 5.12).

Beyond the skewed distributions of individual movements, collective agencies continue to intervene as we scale up to aggregate migration behaviors such as homing and straying. Schooling is a catalyst for straying, whether straying results in colonization of new habitats or invasion into a different population. For instance, consider an individual Atlantic bluefin tuna captured well south of its normal range in the South Atlantic Ocean. This unusual movement will have trivial consequences at the population or metapopulation level unless others follow the same path. Just as interacting movement rules lead to complex schooling behaviors, it is the interaction between behavioral types that can allow for rapid transmission of exploring

behaviors. The complementary nature of bold/ exploring types with retentive/social types corresponds to simultaneous novel and conservative behaviors within populations that lead to straying and new homing traditions. Behavioral types are but one cause that underlies the propensity to stay or leave. Other attributes already reviewed include size, sex, growth rate, social transmission of behaviors, imprinting, navigational error, and larval dispersal.

The Necessity of Straying

The complement of homing and straying represents a key emergent property of complex life cycles that are both closed and open. Imprinting as a mechanism of natal homing would lead to maladaptive rigidity if homing was not complemented by straying. One can easily conceive of how past climate change, tectonic events, and other ecosystem changes would lead to extirpation of populations that relied upon a single life cycle. In tandem, however, natal homing and straying provide a powerful synergism permitting marine fishes to capitalize on novel

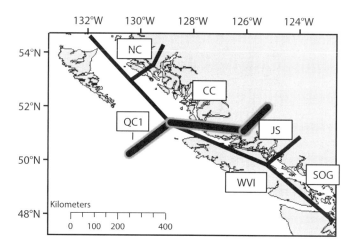

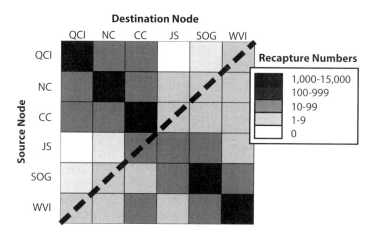

Figure 5.12. Regional recoveries of ~1.5 million Pacific herring tagged in British Columbia, Canada, depicted as a connectivity matrix. The matrix shows a strong tendency for shorter individual movements, where recaptures are highly skewed toward the region of tagging (here shown as management units; Hay et al. 2001). Regions are abbreviated as follows: QCI, Queen Charlotte Island; NC, North Coast; CC, Central Coast; JS, Johnstone Strait; SOG, Strait of Georgia; WVI, Western Vancouver Island. Note that the connectivity matrix indicates a north–south regional separation shown by the bolder line on the map and by the dashed line in the connectivity matrix. Most recaptures occurred at 1 y after tagging but ranged between <1 and 10 y. Data from Hay et al. 2001 for tagging periods 1936–1967 and 1979–1992.

reproductive habitats within a single generation (Cury 1994; Hendry et al. 2004a). Metapopulations depend on straying: even low rates of straying (<5% per generation) structurally contribute to long-term persistence of groups of subpopulations against environmental and demographic variability (Hill et al. 2002).

Straying also represents a form of ecological overhead. It is manifestly wasteful because in many instances strays will fail to produce progeny (Tallman and Healey 1994, and references therein). Still, on occasion, straying fish will successfully colonize new habitats or invade adjacent populations. In these instances, mechanisms associated with natal homing (imprinting, entrainment) lead to fixation by progeny to a new life cycle. Thus natal homing causes conservatism that allows populations to capitalize on long-term stability in conditions that favor replacement, and straying when coupled with natal homing allows rapid adaptation to changing ecosystem conditions when they arise. In this dual system of inertia and innovation, we might expect that the degree of natal homing or straying is related to the stability of habitats required for offspring survival (Fraser et al. 2001; Quinn 2005).

Straying rate estimates are rare in the marine fish literature, mostly available for salmons, where strays are more readily enumerated. Even in these instances, straying rates are subject to unavoidable bias. Straying represents those individuals spawning in nonnatal habitats. Thus, if we do not sample with equal effort every place that strays occur, our estimate is biased. Observing that 95% of juvenile fish, tagged in the natal system, return as adults to that system would imply low rates of straying, but without comprehensive sampling in all other possible spawning habitats, this estimate alone is inaccurate. Also, straying rates are variable for the same population, dependent on changes in habitat, migration corridors, and population status (Rieman and Dunham 2000). Indeed, a central issue in predicting metapopulation dynamics is the dependency of straying on local population dynamics (Gotelli 1991; see Metapopulation Theory below). Few "true" estimates occur. In a recent study using otolith strontium stable isotope tracers, Hamann and Kennedy (2012) reported straying rates from comprehensive sampling across natal systems of Chinook salmon in Idaho (USA). Compared to literature values reported for Pacific salmon, straying rates were moderately high, ranging from ~50% for systems separated by ≤1 km to ~10% for systems separated by >10 km. Otolith profile analysis of strays showed that when strays were juveniles (parr), they exhibited high rates of movement across the watershed. Males also tended to show higher rates of straying than females.

Other reported straying rates range from <10% for Pacific salmon, American shad *Alosa sapidissima*, Atlantic cod, and Atlantic bluefin tuna to >50% for shelf and reef fishes (Tallman and Healey 1994; Wright et al. 2006; Rooker et al. 2008; Walther et al. 2008; Tobin et al. 2010; Saenz-Agudelo et al. 2011). In well-studied Pacific salmon, straying is prevalent among populations and differs between species, sexes, and hatchery and wild fish, all indicating an underlying genetic propensity to stray (Quinn 1983; Quinn et al. 1991; Candy and Beacham 2000; Jonsson et al. 2003; Hendry and Stearns 2004). Straying rates for 120 groups of salmon were estimated for a New Zealand stream where Chinook salmon was deliberately introduced. Straying rates ranged widely, 0%–37% (mean 12%), varying with season of release and age of the returning adult (Unwin and Quinn 1993; Hendry and Stearns 2004). That salmon strayed most frequently to nearby systems suggested that navigation error played a role (Unwin and Quinn 1993). Straying rates in salmon tend to increase with duration of marine life, which further supports navigation error as a contributor to increased straying (Quinn et al. 1991; Unwin and Quinn 1993; Jonsson et al. 2003). School entrainment (Jonsson et al. 2003), unpredictable stream flows (Unwin and Quinn 1993), and catastrophic habitat loss (Leider 1989) have also been associated with higher rates of straying in Pacific salmon.

Self-recruitment from studies on larval connectivity in reef systems can be considered the complement of straying. Reported rates vary substantially, depending on oceanography and life history circumstances (Cowen and Sponaugle 2009). For the panda clownfish *Amphiprion polymnus*, large differences in self-recruitment (0%–27%) were observed along Papua New Guinea that corresponded to differences in reef exposure to coastal currents (Saenz-Agudelo et al. 2011). Another species of clownfish, *A. percula*, showed much higher natal homing levels, 60% (Almany et al. 2007). The latter study used a precise tool to evaluate natal homing (otolith stable isotope labeling of progeny), which is difficult to compare with other less precise life cycle tracer approaches (e.g., Patterson et al. 2005; Saenz-Agudelo et al. 2011). Regardless of method, a principal constraint will be sampling strays in a representative manner in nonnatal systems, which explains why few rigorous dispersal kernels exist in the literature despite rapid development of modeling and empirical approaches to estimate larval connectivity (e.g., Jones et al. 2009; Hufnagl et al. 2013).

In shelf systems, interest in conserving nursery habitats and ranking their importance have led to numerous otolith tracer studies focused on where adults originated as juveniles (Beck et al. 2001; Kraus and Secor 2005). Where representative sampling occurs, these same studies can generate estimates of straying; i.e., do adults return to spawn in regions overlapping or proximate to natal habitats? In a pioneering application of otolith tracers, Thorrold et al. (2001) estimated a 60% to 81% return by young adult (age 2) weakfish *Cynoscion regalis* to natal estuarine regions. This was a remarkable result given that adults are widely distributed in shelf waters during winter in US Atlantic waters with no evidence of spatial genetic heterogeneity (Cordes and Graves 2003). Similarly, Gillanders (2002) estimated that 89% of adult porgy *Pagrus auratus* in New South Wales (Australia) fisheries originated from local estuaries. For Scottish shelf waters, whiting *Merlangius merlangus* had substantially lower estimated natal homing rates—1%, 32%, and 73% (Tobin et al. 2010)—than several local populations of co-occurring Atlantic cod, which ranged from 91% to 100% (Wright et al. 2006). Across much broader regions surrounding South America, blue whiting exhibited high regional retention (≥90%) from juvenile to adult stages in shelf ecosystems separated by Patagonia (fig. 5.13; Niklitschek et al. 2010).

Estimates of straying and its complement natal homing remain rare for marine fishes, limited by (1) the capacity of tagging studies and life history tracers to precisely hindcast the natal origin of adults; (2) representative sampling of all potential natal sources and waypoints for larvae, juveniles, and adults; and (3) lack of information on survival differences between straying and homing groups. Currently, there is no way to bring all these considerations into play in rigorous empirical studies, so current estimates must be viewed as preliminary. Still, these initial estimates are critical in guiding simulations (e.g., Cowen et al. 2006; Hufnagl et al. 2013) as well as in developing spatially explicit models. Such models provide important insight into how migration ecology influences population and metapopulation dynamics.

Homing and straying as complementary behaviors were highlighted by Harden Jones (1968): "A return to the parent spawning ground allows a successful ground to be fully exploited, but the maintenance of a cadre of spawners on other grounds, and some straying provides a measure of flexibility to meet changing condition." Many have represented Harden Jones's legacy as a strong argument for natal homing and life cycle closure, emblemized by the migration triangle (Leggett 1984; Secor 2002a), but in this quote he provided a prescient prescription for metapopulation thinking that was decades ahead of its time.

Metapopulation Theory

Populations connected by even small levels of migration exhibit emergent properties that are understood through metapopulation dynamics. Stacey et al. (1997) define metapopulations as complex systems of geographically or ecologically isolated populations but with sufficient exchange of individuals to influence the demographic or genetic fate of those populations. The metapopulation concept arose from the view that populations occupy discrete spaces or patches, separated by less suitable habitats that are occasionally traversed in networks of exchange. In the most idealized construct, metapopulations function as a means to reestablish empty patches and persist so long as colonization exceeds extinction rates among the patches. In Levins's (1969) classic model habitat, patches are scored on the basis of whether they are occupied. In this way, ephemeral populations blink recurrently on and off in habitat patches over generations, but a pool of coloniz-

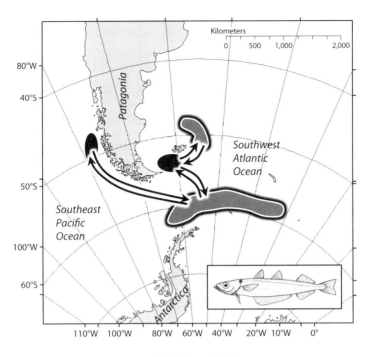

Stock Integrity Index		
Tracer	**Southeast Pacific**	**Southwest Atlantic**
Otolith Trace Metals	1.00 (0.001)	0.90 (0.15)
Otolith Stable Isotopes	0.91 (0.06)	1.00 (0.001)
Parasites	1.00 (0.001)	1.00 (0.001)
Integrated Index	0.96 (0.03)	1.00 (0.001)

Figure 5.13. Evidence for separate life cycles between populations that spawn in either the southwest Atlantic or southeast Pacific Oceans despite a high level of mixing in a region north of the Antarctic Peninsula. Spawning areas and feeding areas are black and gray, respectively. Using multiple life cycle tracers, Niklitschek et al. (2010) detected strong stock integrity, with 1.0 indicating complete separation (natal homing) and 0.5 indicating no separation. Standard error is given in parentheses. Redrawn from Niklitschek et al. 2010, with permission. Illustration of southern blue whiting by T. Ayling (Ayling and Cox 1982) and made available under an agreement through Wikimedia.

ers stabilizes the overall metapopulation (Harrison and Taylor 1997). Hanski (1997) provided four conditions for metapopulation dynamics to be of consequence. Local breeding populations are (1) spatially discrete, (2) susceptible to extinction, (3) networked through colonization, and (4) show independent population dynamics driven in part by habitat heterogeneity.

The metapopulation concept stands in contrast to traditionally held views of population structure in marine fishes, where exchange acts against population persistence (e.g., life cycle selection; Sinclair 1988). On the other hand, the two views could be considered complementary but to operate at different timescales, where metapopulation persistence occurs over longer timescales than population persistence. On an ecological timescale, colonization operates against the genetic integrity of the population, but over longer periods, persistence of the larger metapopulation is favored as colonization offsets local extinction. Interplay between extinction and colonization rates, modeled as a function of patch occupancy, also provides perspective on how spatial distributions arise (Gotelli 1991). By allowing these rates to depend on patch occupancy, a rescue effect emerges where immigrants from adjacent patches reduce the probability of patch extinction. Alternatively, by allowing immigration rates to be independent of patch occupancy, a "propagule rain" occurs across all patches (Gotelli 1991). Scenarios of how colonization might be independent of patch densities are easily conceived for pelagic spawning marine taxa with dispersive offspring, consistent with oversampling and storage effect collective agencies. Thus, at the metapopulation level, larval dispersal as well as juvenile and adult straying are manifest sampling behaviors. Owing to the high fecundity of most marine fishes, even low rates of straying can produce substantial opportunities to sample new environments or contribute to adjacent subpopulations.

To evaluate environmental-metapopulation feedbacks, local populations have been given unique demographic and exchange parameters in structured metapopulation models (Gyllenberg et al. 1997). Internal dynamics attached to the local population (birth and death rates) as well as degrees and types of exchange are stipulated, and metapopulation dynamics are audited as the aggregate response. Over generations, colonization across a set of patches (fig. 5.14; colonization kernel) is a function of (1) the likelihood of dispersing given the local state of the population and (2) the probability of successfully colonizing an unoccupied patch. Metapopulation responses can be simulated for patches that vary in size and environmental quality and for functional responses such as density-dependent dispersal. In a model of

habitat occupancy by butterflies among ~1,500 meadow patches, Hanski et al. (1995) estimated colonization rate as a function of patch area and interpatch distance. From this universe of networked patches, they sampled subsets (metapopulations) and evaluated the relationship of habitat occupancy versus potential colonization rate, expecting a single positive response. What they observed, however, were two stable states, with intermediate colonization rates supporting either high or low levels of patch occupancy. Gyllenberg et al. (1997) argued that introduction of structural realism is critical to evaluating the feedback between the environment and metapopulations, as relatively small changes in colonization rates can lead to threshold shifts in patch occupancy, such as observed in the Hanski et al. study.

In its ideal form, addressing only patch occupancy, metapopulation theory has been applied to Pacific salmon, Atlantic cod, and Atlantic herring, for which managed stocks are composed of local spawning aggregations. For herring, we have reviewed how social transmission can lead to intergenerational patterns of loss and colonization of spawning grounds (Corten 2001). Similarly, strong natal homing in Pacific salmon and Atlantic herring as a result of imprinting and hydrographic retention would cause local populations to be susceptible to overexploitation and ecosystem change (Cooper and Mangel 1999; Hill et al. 2002; Stephenson et al. 2009). For northern cod (chap. 3, Five Mating Systems), Smedbol and Wroblewski (2002) proposed a metapopulation composed of a few large shelf and shelf-edge populations and numerous bank and inshore bay spawning aggregations. Heavy offshore exploitation resulted in loss of both shelf and bank aggregations, suggesting a Levins-type metapopulation structure (chap.3, fig. 3.6). Note that because cod undertake seasonal migrations that cause spawning groups to intermittently mix, patches take a special meaning, defined by spawning fidelity rather than lifetime occupancy. This type of seasonal mixing implies that extinction of a local population could go unnoticed if groups of fish were harvested, surveyed, and assessed at a broader metapopulation scale (Petitgas et al. 2010). Under no fishing pressure, Smedbol and Wroblewski (2002) predicted extremely low probabilities of local extinctions. The recent period of intense directed fisheries on spawning aggregations has brought metapopulation considerations to the fore for Atlantic cod.

How to recover lost subpopulations of northern cod is uncertain. Consistent with the res-

cue effect, one approach targets restoration of the large offshore spawning aggregation to promote density-dependent recolonization of lost spawning aggregations (Rose et al. 2000). Another promotes recovering spawning aggregations upstream of targeted recolonization sites and depends on the likely dispersal trajectories of offspring (Smedbol and Wroblewski 2002). Reestablishing depressed or lost spawning ag-

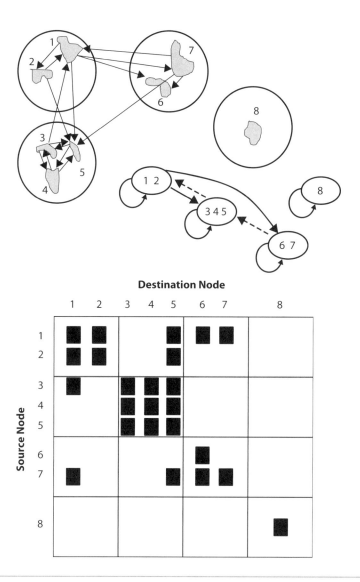

Figure 5.14. Levels of exchange across multiple spatial scales result in a networked patchy metapopulation, diagrammed three different ways.

gregations would entail fishing regulations that vary seasonally and regionally in accordance with assessments of demographics and exchange rates of linked populations (Stephenson et al. 2009). Later in this chapter we review such spatially explicit stock assessments.

Reef Systems as Structured Metapopulations

More elaborate metapopulation models depict the demographic give-and-take among connected populations through source-sink dynamics. In highly networked systems of populations and habitat patches, we may not expect to observe empty patches or local extinction. Source-sink dynamics occur between more and less productive habitat patches, where immigrants from more productive patches subsidize less productive ones (Pulliam 1988). If emigration is a positive function of local production, then stochastic variation in the supply of emigrants results in a rescue effect to the overall metapopulation, in which local population surpluses and deficits are offset by exchange rates (Stacey et al. 1997). Here as in the straying-homing synergy, the independence of local populations (asynchrony) coupled with a low rate of exchange between populations can lead to stability and persistence as emergent properties. Olivieri and Gouyan (1997) reference this synergy as the "metapopulation effect."

For marine metapopulations, the general definition articulated by Kritzer and Sale (2004) seems appropriate: a network of semiclosed populations that are connected but with each showing a degree of independent dynamics (fig. 5.15). Empirical and modeling studies on marine fishes have tended to shift away from the metapopulation focus on patchiness, genetic exchange, and local population extirpation. Although fragmentation is an important issue in reef and other coastal ecosystems, marine habitats are more homogenous and advective than terrestrial ones (chap. 4, Marine and Terrestrial Food Webs: How Different?), leading to increased focus on demographic exchange among biological entities (local populations) rather than on extinction rates attached to particular patches (Figueira and Crowder 2006). A

common feature among marine fishes not well accommodated by patch-occupancy models is that breeding populations mix during some period of their life cycle. Kritzer and Sale (2004) presented a continuum of population/metapopulation structures: a single population that is distributed among habitat patches but fully interbreeds → local populations connected by limited exchange rates → local populations with no exchange (fig. 5.15). In chapter 6, partial migration is introduced as a framework to explain how groups of fish move from one level of metapopulation structure to another.

Reef-associated fishes would seem ideal candidates for consideration of local populations centered on patches, connected through exchange by their dispersive offspring. The structural nature of these habitat patches has encouraged the use of closed fishing areas in locations centered on patches or networks of patches to sustain or enhance adjacent regional fisheries (Crowder et al. 2000; Jones et al. 2009). The effectiveness of this management approach will depend on whether and how much population replenishment occurs inside and outside of marine protected areas: whether enhanced growth, survival, and reproduction owing to spatial protection spills over as dispersive individuals (typically larvae) recruit to adjacent reef patches (fig. 5.16). If each protected patch includes a single closed population that self-recruits, then no spillover is expected. In such instances a unique population or assemblage will be maintained if replenishment occurs within the protected area (Almany et al. 2009). Alternatively, we might expect—particularly for pelagic spawning fish—that populations would not be closed. If larvae disperse and settle onto patches outside of the protected area in a structured manner (say, because of their proximity and dominant current systems), then spillover effects would be favored for only a subset of adjacent reefs (fig. 5.16; Crowder et al. 2000; Cowen et al. 2006). Such a system would have metapopulation attributes (Kritzer and Sale 2004). Finally, consider offspring that are more widely dispersed—a propagule rain. Enhanced reproduction in the protected patch would lead

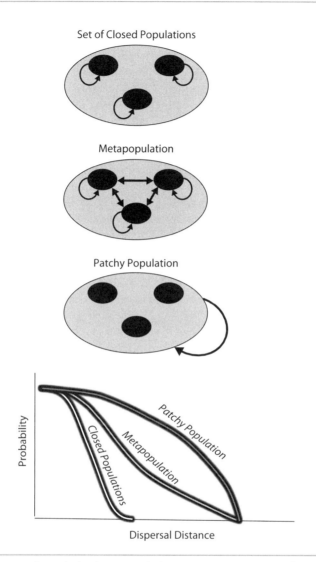

Figure 5.15. Continuum of population/metapopulation structures. Larger ovals represent geographic bounds of interest (e.g., a species range, a large marine ecosystem). Black circles represent groups or populations, and arrows indicate demographic exchange between self-reproducing populations. At one extreme are sets of closed populations with no exchange, and at the other are groups (contingents) that are not reproductively isolated and exhibit high levels of exchange. Redrawn from Kritzer and Sale 2004, with permission.

to spillover benefits that were more homogenously distributed across adjacent patches. These few scenarios of spillover from marine protected areas underscore metapopulation considerations in conservation strategies for reef species, which include patch size, number, spacing, and overall network structure (topology; Almany et al. 2009; Kininmonth et al. 2010).

The arrangement of reef patches and their relative production will contribute to source-sink dynamics within metapopulations. Populations on source patches will be stable and produce a surplus of offspring that recruit to other patches, whereas those in sink patches will require subsidy from elsewhere to persist (Pulliam 1988). Larval connectivity informs how patches are connected and their classification as sources or sinks (Figueira and Crowder 2006). But, as Almany et al. (2009) point out, these systems should not be viewed as stationary, and source and sink patches can reverse their roles over time. In the reserve system of coral reefs in the Great Barrier Reef Marine Park, protected reef patches generally conformed to expectations of increased probability of self-replenishment and spillover (fig. 5.17; Almany et al. 2009). Protected reefs are biased much larger (>1 km²) than unprotected reefs, enhancing self-recruitment and subsidy. Spacing between protected patches favors spillover; ~80% of intervals were within 20 km, which is within the range of larval dispersal distances for many endemic species (Jones et al. 2009).

Nonstationarity in patch quality and dispersal trajectories, poorly sited reserves, displaced fishing (i.e., fishing that would have otherwise occurred within the reserve), and other anthropogenic activities can confound conservation aims attached to marine protected areas and in some instances may result in more harm than good (Crowder et al. 2000). Further, conservation objectives are often complex, including (1) maintenance and enhancement of genetic diversity, assemblage composition, and ecosystem services; (2) research on unexploited populations, communities, and the reefs they occupy; and (3) promoting the stability, persistence, and resilience of populations and metapopulations. Because of the numerous goals attached to marine protected areas, performance indicators and assessments are not straightforward, resulting in a high volume of science, synthesis, and critical review (Crowder et al. 2000; Houde et al. 2001; Polunin 2002; Lubchenco et al. 2003; Palumbi 2003; Sale et al. 2005; Jones et al. 2009; Steneck et al. 2009).

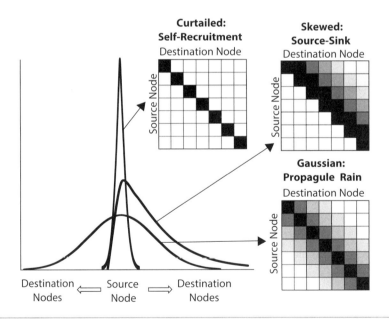

Figure 5.16. The goal of enhanced fisheries from marine protected areas depends on whether dispersive individuals (typically larvae) spill over into adjacent unprotected areas. Presented here are larval connectivity matrices associated with self-recruitment, source-sink, and propagule rain dispersal outcomes.

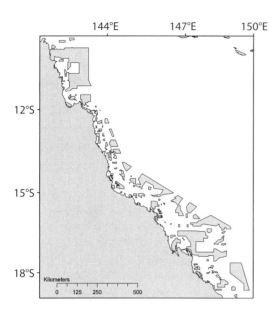

144°E   147°E   150°E

12°S

15°S

18°S

Kilometers
0   125   250   500

Figure 5.17. Location of 164 no-take reserves within the Great Barrier Reef Marine Park, Queensland, Australia. Note varied sizes of, and distances between, protected areas. A general prediction is that increased spillover is expected for larger and more closely spaced protected areas.

Taking Stock of Population Thinking

The Unit Stock Concept

Prior to metapopulation and connectivity frameworks, marine fish populations were defined spatially according to the unit stock concept, a traditional simplification that allowed fisheries scientists to compare regional exploitation rates against natural replacement rates for particular groups of fish (Secor 2013). The unit stock concept has received criticism for its traditional adherence to closed population assumptions. On the other hand, the stock concept, if defined operationally, is still a useful construct for the practical demands of spatially managing exploited populations.

For hundreds of years, fishermen, naturalists, and scientists have speculated about how groups of fish are geographically bounded. In the early 19th century, the philosopher Chong (1814) noted cycles of Pacific herring abundance in two regions along the Korean Peninsula that were offset from each other. Further, herring from either region exhibited substantial differences in vertebral counts. These differences led Chong to conceive that migrations were limited so that groups of herring were spatially discrete, each exhibiting a unique cycle of abundance. Later, in the North Sea, regional morphometric differences in Atlantic herring were quantified by Heincke (1898), which he ascribed to curtailed migration by local populations (races in the late 19th-century lexicon). This finding supplanted earlier ideas of a single large panmictic group of Atlantic herring interacting with fisheries throughout the North Sea. Similarly, Gilbert (1915) used fish scale morphometrics of Fraser River sockeye salmon to support separate races between spawning tributaries. Chong, Heincke, Gilbert, and other early observers of local population structures argued that the collective movements of individuals, organized as races, should make them differentially vulnerable to fishing and environmental influences. For instance, Heincke believed the siting of a whale rendering plant had local effects on a single proximate race of herring (Smith 1994). Gilbert described how a canyon collapse into Hells Canyon River Gorge impeded spawning runs and led to the catastrophic loss of a race of Fraser River sockeye salmon.

Building on this concept of local influence, early fisheries scientists sought to understand how fishing, climate, habitat loss, pollution, and hatchery augmentation—forces that are bounded geographically—influenced the internal dynamics of certain groups of fish. This issue led to two central questions. (1) Which group of individuals is affected? (2) Having identified that group, how are its internal dynamics (survival, growth, reproduction) affected? In an early quantitative treatment on geographic scales of influence, Dahl (1909) refuted the premise that cod hatcheries effectively augmented catches in individual Norwegian fjords.

He did so by showing that the within-group dynamics across several years (measured by juvenile abundances in fjords) were not related to how many cod were stocked into each fjord. Rather, cod showed correlated dynamics across fjords, which indicated broader regional-scale population dynamics (Secor 2004). In a related example, Hjort and Lea (1914) discovered the dominant year-class phenomenon, which emerged from the age structure of Atlantic herring. A critical observation was that the same dominant year-class was apparent regardless of where herring schools were sampled up and down the Norwegian coast, suggesting that climate and fishing influences were integrated by a larger group of fish, one that ranged across the entire Norwegian Sea.

The stock concept allowed for the definition of discrete entities so that aggregate group demographics could be audited against removals by fishing (Cadrin and Secor 2009). Once groups were practically identified by when and where they were fished, demographic models such as Russell's catch equation could be applied to balance fishing rates against rates of natural renewal. Such groups were termed *unit stocks*, emphasizing their discrete internal responses to fishing and other outside forces. Early definitions were decidedly agnostic about why groups might exhibit unique demographic responses and emphasized practical considerations such as accessibility to fisheries (Russell 1931; Cope and Punt 2009) and homogeneity of demographic attributes (Waples et al. 2008). These definitions persist. The United Nations Food and Agriculture Organization defines stock as a "sub-set of one species having the same growth and mortality parameters, and inhabiting a particular geographic area" (Sparre and Venema 1998). But natural renewal in fisheries stock assessment follows theories of density-dependent replacement rates that occur within biological populations (i.e., stock recruitment and surplus production models). Therefore a common prerequisite in identifying stocks during the past 50 y has been to establish reproductive isolation, particularly through use of molecular markers (Reiss et al. 2009).

New appreciation for structured populations and metapopulations strongly indicates that narrow definitions of discrete homogenous populations are unlikely to be correct and can lead to biased (liberal or conservative) stock assessment results (Cadrin and Secor 2009; Petitgas et al. 2010; MacCall 2012). Fishing and other external drivers occur at multiple scales such that groups other than populations are often the most affected (fig. 5.18). Further, the internal dynamics within populations and other groups are not homogenous nor, with a nod to metapopulations, are they exclusively "internal." An alternative is to define stocks by the geographic extent of fishing or other influences of interest. Here stock is being operationally defined; that is, by the question we are asking or by other practical considerations such the method employed (Booke 1981; Secor 2004). Operational definitions need not be arbitrary. Rather, the question at hand specifies the relevant ecological entity, which can range from species to brood (fig. 5.18). Conservation biology, for example, typically focuses on the persistence and recovery of species, metapopulations, and populations. The Distinct Population Segment of the US Endangered Species Act spans these levels of organization but also considers political realities by including governmental boundaries (US Fish and Wildlife Service 1996; Ford 2004). Similarly, the US Magnuson-Stevens Fisheries Conservation Act defines stock as a species, subspecies, geographical grouping, or other category of fish that lends itself to management as a unit (National Oceanic and Atmospheric Association [NOAA] 2007). Operationally, then, a stock can be variously defined by its ecological, technical, recreational, economic, or fishery attributes.

Bounded Fisheries

Government policies that prevent overfishing of stocks include stipulation of thresholds related to an overfishing rate and an overfished abundance level as well as harvest target levels that relate not only to maximum sustainable yield but also to uncertainty and risk of future stock depletion. Further, in overfished

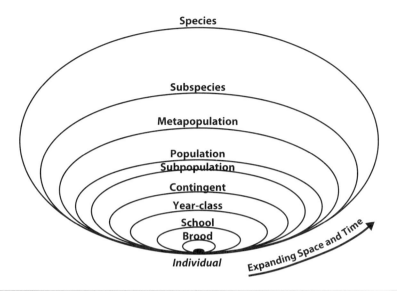

Figure 5.18. Levels of ecological organization relevant to the unit stock. From Secor 2013, with permission.

stocks, policies require fishing controls that enable timely recovery of biomass. As a result, the health of many fished stocks has improved (Worm et al. 2009; Hutchings et al. 2010; NOAA 2012). In numerous instances, however, overfished stocks have failed to recover, with a principal cause being misspecified stock structure (Hutchings et al. 2010; Murawski 2010; Petitgas et al. 2010; Kritzer and Liu 2013). As the basis for effective fisheries assessment and management, delineation of unit stocks requires three criteria: (1) identification of the stock, a group of fish with homogenous internal dynamics and limited exchange with other stocks; (2) evaluation of the stock unit area, the geographic boundaries associated with the seasonal movements and concentrations (habitats) of that stock; and (3) long-term stability of the stock and its boundaries (Begg et al. 1999; Cope and Punt 2009; Link et al. 2011).

How stocks might be defined by their spatial behaviors, distribution, and demographics can be at odds with how fishermen interact with groups of fish (Secor 2013). The principal way in which humans historically interacted with fishes was through systems of customary

tenure such as in the coastal villages of Japan, where hereditary rights were conferred to individual families and villages to fish in certain coastal regions (Kada 1984; Kalland 1984). Over the past 150 y, fisheries have become increasingly industrialized, leading to increased vessel travel, more efficient searches, and much improved capture techniques. Expanded fishing ranges have led to conflict and the need for spatial management among fishermen, communities, and governments. Layered systems of zonal management now occur globally. As one example, in US Atlantic Ocean waters, nearshore state fisheries (3 mi from shore) are regulated through the Atlantic States Marine Fisheries Commission, and those occurring 3–200 mi from shore are regulated by US Regional Management Councils. Beyond the 200-mi Economic Exclusive Zone, fisheries are regulated according to the United Nations Convention on the Law of the Sea through regional fishery management organizations such as the Northwest Atlantic Fisheries Organization and the International Commission for the Conservation of Atlantic Tunas. Thus zonal management scales from local harbors to international compacts.

Assessing where fish are caught across management boundaries is complex and expensive. Fishery management boundaries, whether termed practical management units, statistical areas, or unit stock areas, are often mismatched with the spatial ambits of populations and other groups of fish. A particularly vexing issue is mixed stock fisheries. Mixed stocks occur when stock separation is incomplete and groups overlap in their spatial range for periods of time. Where fisheries occur on mixed stocks, sustainability of those fisheries depends upon the internal dynamics of each stock, weighted by their degree of overlap. Because stock demographics and levels of mixing are difficult to simultaneously assess, they are often ignored, resulting in false apparent trends in fishery assessments.

Consider the mixed stock issue for Atlantic bluefin tuna. The Gulf of Mexico population was depleted 30 y ago (~30% of historical levels; see chap. 7, fig. 7.2). The Mediterranean population is approximately tenfold more productive and supports higher fishing rates. But under likely stock mixing between the two populations, catches in the Mediterranean Sea and northeast Atlantic Ocean will disproportionately affect the Gulf of Mexico population, which has been managed under a long-term but unsuccessful rebuilding plan (Taylor et al. 2011). Stock depletion is a common attribute of mixed stock fisheries, in which the less productive stock receives higher proportional removals; in these instances, management controls should guard against overharvest of the subordinate stock (Ricker 1958; Nehlson et al. 1991).

Distributional shifts in groups and populations of fishes are common, the result of changes in climate and ocean states, population irruptions, overfishing, colonization, and food web changes. Mapping these distribution shifts onto unit stock boundaries shows how apparent trends from stock assessments can be misleading. Link et al. (2011) provided a set of likely scenarios where shifted stocks were considered respective to a fixed unit stock boundary; they are expanded here (fig. 5.19). First, consider transient shifts due to environmental forcing.

As an example, climate warming might cause a north temperate stock to shift its distribution northerly across a boundary. In this instance, the original unit stock area will show stock extirpation, whereas the adjacent area—if already occupied by the same species—will show increased abundance. The shifted distribution of Georges Bank winter skate *Leucoraja ocellata* during recent decades (Frisk et al. 2008) is an example of this phenomenon. Contraction of a stock that originally straddled a management boundary to one or the other side of that boundary can manifest a similar result. One example was contraction of the shelf-distributed northern stock of Atlantic cod to small coastal stocks (Rose et al. 2000; Petitgas et al. 2010). If the shifted distribution causes species expansion to new areas, then developing fisheries can emerge, as was recently observed for albacore *Thunnus alalunga* and bluefin tuna in the northeast Atlantic Ocean (Dufour et al. 2010). Depending on the type of shifted pattern, assessments can result in alternatively overly optimistic (e.g., high abundance for a contracted coastal cod stock; novel stocks of albacore tuna) or pessimistic (e.g., vacated Georges Bank winter skate) portrayals of stock abundance (Link et al. 2011).

Stock Biocomplexity and Spatially Explicit Stock Models

Recent agendas to rebuild depleted and collapsed stocks, conserve endangered species, and protect critical habitats have expanded the traditional definition of stock to consideration of multiple population structures (e.g., contingents and metapopulations). Such structures can confer important yield, stability, and resilience functions (Secor 1999; Hilborn et al. 2003; Ruzzante et al. 2006; chap. 7, Diversity in Spatial Structure: The Portfolio Effect). In a series of case studies on collapsed and depleted marine fish populations, Petitgas et al. (2010) argued that recovery strategies that focused exclusively on rebuilding biomass in overfished populations were likely to fail. Instead, the spatial and behavioral architecture of those populations must be recovered as well. Conserving biomass

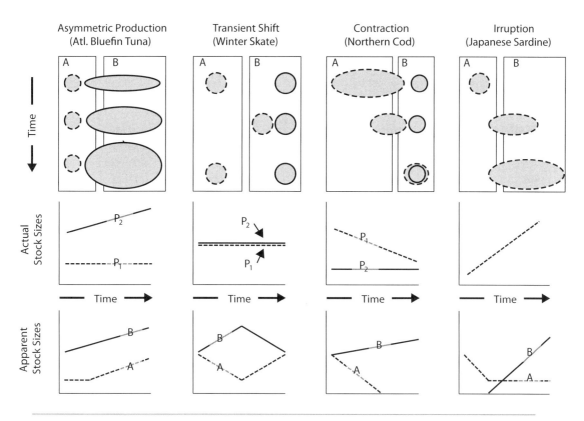

Figure 5.19. Diagrams illustrating the interplay between shifts in stock structure and resulting trends in actual and apparent stock abundances. The top panels show shifting stock dynamics where two populations (P_1 and P_2) are outlined in solid or dashed lines. These stocks occur in two stock unit areas represented by adjacent boxes A and B. The middle panels show actual stock trends in abundance for the two stocks, and the bottom panels show apparent abundance dynamics occurring within the two stock unit areas. Examples of shifting stocks include asymmetric production, transient shift, contraction, and irruptions. Figure originally modified from Link et al. 2011 and reprinted from Secor 2013, with permission.

among multiple components serves as bet hedging against catastrophic losses of entire stocks (Smedbol and Stephenson 2001; MacCall 2012) and resilience of populations to climate change (Hilborn et al. 2003; Kerr et al. 2010a). Ocean observing systems, telemetry, and molecular and other natural tags can all serve to identify these key structural/behavioral entities in support of conservation measures (chap. 2, Analysis of Movements and Migration; Manderson et al. 2010), but they will require expanded and costly research and monitoring programs.

Asynchrony among semiclosed populations produces emergent properties that can both stabilize and destabilize metapopulations, sometimes termed *stock complexes*. Simulation models that use realistic demographic rates and scenarios of exchange among interacting stocks allow one to ask when complex stock structure matters. In age-structured simulations (specifying different growth rates and stock recruitment functions for each subpopulation), Kerr et al. (2010b) showed that misspecifying stock structure in Gulf of Maine cod could lead to underestimates of productivity. Independent dynamics of coastal subpopulations contributed to higher yield in the overall shelf metapopulation. In a similar exercise,

Kell et al. (2009) observed that the consequences of lumping rather than splitting population components of British Isles herring caused an assessment model to yield optimistic predictions on the level of overall fishing rates and probability of recovery following depletion.

An advantage of simulation modeling is the flexibility in accommodating multiple types of information and levels of structural organization (Kerr and Goethel 2013). Three principal classes of modeling spatial structure are (1) simple spatial heterogeneity where fishing, mortality, and production rates differ regionally for the same stock; (2) movement models, which follow the demographic fates of individuals as they move between regions that vary; and (3) multistock or metapopulation models that assume some degree of reproductive isolation (Cadrin and Secor 2009). Although the current section emphasizes the latter class, the three classes principally serve to specify types of data and modeling approaches, and can be combined in overall assessments of how movement behaviors, natal homing, and stock production influence stock performance against regional differences in fishing mortality, selectivity, and catchability (Goethel et al. 2011; Taylor et al. 2011). Still, other frameworks can bring in oceanographic influences, such as those leading to retention or advection of young. In a simulated metapopulation of 10 cod subpopulations (Heath et al. 2008), model outcomes suggested that population structure was associated with oceanography in portions of the North Sea, but elsewhere in the North Sea and off western Scotland, population structure required a degree of natal homing by first-spawning adults.

For a simulated two-component herring metapopulation, Secor et al. (2009) evaluated the emergent properties—yield, stability, and persistence—for connectivity scenarios of straying and behavioral entrainment (adopted migration). Under cycles of environmental forcing, density-dependent straying tended to stabilize populations while density-dependent entrainment often led to extirpation of one of the spawning groups. Higher levels of connec-

tivity, regardless of type, tended to increase the degree of synchrony between the two groups and decrease metapopulation yield and stability. Most types of connectivity caused dominant year-classes to be distributed across the groups, destabilizing individual populations that would have otherwise benefitted through the accumulation of spawning stock biomass (chap. 4, The Storage Effect). Thus the simulated herring metapopulations required some degree of connectivity for long-term persistence, but this may involve a trade-off in terms of lost yield and stability. Small levels of connectivity (~5%) caused negligible influence, but at higher connectivity levels the internal dynamics of subgroups were disrupted, causing the entire stock complex to destabilize.

Management strategy evaluation is a simulation framework where alternative operating models (premises about population and metapopulation behaviors) are tested against their influence on assessment and management outcomes (fig. 5.20). Spatially explicit stock assessments integrate stock composition information into an assessment framework. Such applications show promise (Cadrin and Secor 2009; Goethel et al. 2011; Link et al. 2011) but remain rare. Recently, Taylor et al. (2011) used telemetry, conventional tagging, catch, and otolith chemistry data to construct seasonal and age-specific matrices of regional stock movements by Atlantic bluefin tuna. These movements were integrated into a statistical catch-at-age assessment model to estimate past abundance trends and fishing rates. The model used maximum likelihood parameter fitting, which permitted simultaneous use of diverse types of data to support critical movement estimates. The efficacy of past management based on the assumption of no mixing was evaluated by comparing (1) the range of historical exploitation rates emulated under realistic scenarios (operating models) of mixing with (2) current accepted overfishing thresholds and targets.

Spatially explicit stock assessment models coupled with management strategy evaluation hold particular promise in evaluating the role of stock complexity on yield, stability, and resilience outcomes. Balanced against this prospect is the understandable reluctance by managers and stakeholders to accept predictions from simulations. Continued development of spatial simulations in stock assessments will depend on (1) surveys and research that corroborates the behavior of spatially explicit population models; (2) evaluating how error propagates through such models; and (3) building stakeholder involvement in operating models, including development of key scenarios and interpretation of model outputs (Miller et al. 2010; Kerr and Goethel 2013). Operating models can be tested against an ever-increasing database on marine fish movements from tagging and other approaches (Goethel et al. 2011).

Summary

The general theory of life cycle closure via natal homing has received strong support in empirical and modeling studies of the past several decades. The weight of tagging, telemetry, and life cycle tracer evidence supports a strong propensity toward natal homing (natal return of an adult) in diverse marine fishes. Examples reviewed in this chapter include (1) the 90 separate itineraries of Fraser River sockeye salmon; (2) natal homing by Atlantic herring, bluefin tuna, and reef fishes in apparent opposition to dispersive flows that sweep larvae away from natal regions; and (3) the avian-like seasonal fidelity by white shark to feeding, mating, and pupping habitats that are thousands of kilometers distant from each other. On the other

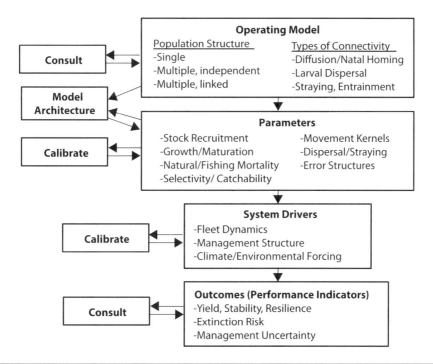

Figure 5.20. Example of the operation of a management evaluation strategy, the goal of which is to explore outcomes of scenarios that are plausible and represent key, but poorly understood, aspects of the system. These aspects often center on spatial behaviors of stocks and fisheries, as shown in this specific example. Consulting entails goal-setting and interpretations with interested parties as an iterative process. Calibration references model precision and degree of realism. Shown here is but a subset of possible models, parameters, drivers, and outcomes.

hand, with the exception of Pacific and Atlantic salmon, tests for natal homing remain rare. Natal homing and breeding philopatry have been conceived as an assurance system for the continuity of (1) mating systems; (2) favorable spatial regimes for young; and (3) ontogenetic and seasonal migrations among sets of habitats that support feeding, growth, refuge, and other critical functions.

Imprinting to natal cues (odor, geomagnetic fields) remains a dominant explanation informed by studies on Pacific and Great Lakes salmon, but direct evidence for imprinting is sparse for marine fishes in general. Multiple systems of inheritance likely come into play in explaining elements of life cycle closure. These systems include (1) genetic inheritance unrelated to imprinting demonstrated by salmon transplant studies; (2) cultural inheritance evinced by sudden (within-generation) shifts in distributions and spawning habitats for Atlantic herring and other pelagic fishes; and (3) ecological inheritance, or the pheromone plumes of lamprey ammocoetes larvae, which establish and maintain adjacent spawning grounds. The accepted framework—that natal homing and spawning site fidelity are selected as attributes that give rise to philopatry—remains a central explanation for life cycle closure, but alternatives are not easily discounted. The adopted migration hypothesis posits that life cycle closure is an emergent property of spawning site fidelity without the requirement for natal homing. Similarly, the pheromone hypothesis for lampreys leads to consistent spawning habitat use generation after generation without the requirement of natal homing or spawning site fidelity. A central unanswered question in the control of life cycle closure is how juveniles entrain into adult migration circuits. This mystery remains an empirical challenge but will likely see resolution in the coming decades with improved observational capabilities, increased attention to behavioral aspects of migration ecology, and public agendas that focus on protection of critical nursery habitats and spillover from marine protected areas.

In contrast to boreal and temperate marine fishes, populations of tropical reef fish had been presumed to be relatively open, with natal homing by dispersive larvae unlikely given the arrangement of reefs and the low likelihood of retention. A somewhat audacious larval mark-recapture experiment demonstrated that many larvae in fact did return to the reef island from which they originated. Larval connectivity studies now show a continuum of exchange rates between reefs (patches) that relate to reef sizes and distances, ocean circulation features, larval behaviors and vital rates that modify their transport, and life history attributes that can stabilize levels of connectivity and self-recruitment. Larval connectivity studies are a powerful means to link the oceanographic fates of larvae to where they originated, resulting in emergent population and metapopulation structures. Population connectivity more broadly considers dispersal throughout a generation, synonymous with migration ecology.

Open life cycles begin with the notion that a marine fish cannot remain where it was spawned throughout its life history. Most marine fishes must migrate through food webs as they grow, necessitating some degree of dispersal. The null model of dispersal is diffusion or a random walk, which for short periods of time (days to months) can serve as a basis for collective movements, such as the dispersion of haddock larvae or migratory skipjack tuna. More elaborate null models such as the basin or state-space models present elegant vehicles by which simple movement parameters can be fit to distributional, tagging, and telemetry data. But such movements cannot be feasibly scaled to longer-term migrations that define life cycles. Importantly, populations are not uniform in constituent individual behaviors; roamers, strays, and personalities intervene.

The synergy between natal homing and straying simultaneously confers stability in replacement rates while allowing rapid exploitation of new opportunities or a refuge from catastrophe. Metapopulations result from straying. Classic metapopulation models designed to track genetic exchange and extinction risk have been modified to incorporate increased spatial realism and demographic interactions over ecological timescales. Empirical and modeling studies of demographic exchange between populations have attracted particular interest in marine fishes where fishing, climate, habitat degradation, and other impacts differentially affect local populations, but at the metapopulation level their influence is mitigated by the type (e.g., self-recruitment, source-sink, propagule rain) and degree of population exchange. When subpopulations are conceived of as habitat patches (reef subpopulations) connected through demographic exchange, a metapopulation effect can be empirically measured or modeled to support management objectives for marine protected areas, such as conservation of diversity, community structure and function, and enhancement of adjacent fisheries and recreational opportunities through spillover effects.

Marine fish populations are not strictly closed, recognized in the past decades through developments in (1) metapopulation theory; (2) differential, partial, and adopted migration concepts; and (3) threshold changes in population integrity, including range shifts, irruptions, and collapses. The state of being simultaneously both open and closed confounds simple attempts used in fisheries management to define populations as unit stocks. On the other hand, noteworthy advances have been made in assessment science and fishery management despite the past need to make simplifying assumptions related to closed population structures. Further, by operationally defining stocks according to the management problem at hand, stewardship objectives can be better aligned with available information. Nevertheless, misspecified stock structure is a leading explanation as to why fished stocks fail to recover or otherwise respond in expected ways. Spatially explicit stock models and related simulation approaches such as management strategy evaluation are effective means of evaluating alternate population structures and the risk to management objectives caused by misspecified stock structure.

Segue

In chapter 4, we reviewed how features of marine ecosystems were accommodated by periodic life history strategies, which balanced the risks and rewards of starting life small and reproducing in a transient environment. In chapter 6, partial migration is erected as the spatial equivalent of this life history type, a set of collective agencies through which safe harbors and new seas are simultaneously exploited over multiple scales—from cohorts to species.

6 Propagating Propensities

The preoccupation with individuals performing spectacular feats of homing and orientation and the tendency to dismiss individuals not performing as expected may have hampered our understanding of the migration process.
—Quinn and Brodeur (1991)

Partial migration provides a behavioral turntable from which exclusive migratoriness and sedentariness can easily and rapidly be reached . . . through selection and related microevolutionary processes.
—Berthold (1999)

Collectively, individual patterns of migratory behavior make up ecologically important, population-level patterns of partial migration.
—B. B. Chapman et al. (2011b)

Conditional Migrations

Migration is conditional: conditional on individuals' physiological, sensory, and genetic capacity to imprint, explore, and adapt; conditional on habitat quality and resource availability; conditional on size, maturity, and gender; and conditional on social interactions—population density, learning, tradition, and mix of personality modes. That migration is not canalized as a genetic trait or fully determined by the environment directs attention to interacting propensities, which give rise to populations that are rarely limited to a single life cycle or a single set of directed migrations (e.g., the classic migration triangle). For marine fishes, the exigencies of reproducing in a dispersive environment, advection of young, and transiting through size-structured food webs are met through alternate life cycles that cause populations to exhibit both closed and open attributes. These alternate life cycles exploit different sets of marine ecosystems and are to varying degrees responsive or recalcitrant to changes in climate and environmental forcing. The theory of partial migration accommodates such complex systems of conditional migration. Partial migration has been broadly recognized in salmons, birds, terrestrial vertebrates, and insects, but it has seen limited application to marine fishes. The goal of this chapter is to show how the phenomenon of partial migration, if underlain with propensities rather than specific mechanisms, is fundamental to (1) understanding the evolution of migration in marine fishes and (2) scaling individual movement propensities to homing, range expansion, colonization, metapopulation structure, and speciation.

Partial migration is perhaps best known for birds of "two worlds" (e.g., temperate songbirds), where some individuals stay year round in their spring breeding habitats, whereas others migrate to lower latitudes during winter months (Lack 1944; Berthold 1996; Alerstam and Hedenstrom 1998). Early theoretical treatments sought to explain partial migration through evolutionary stable strategies —modeling the resistance of a simulated population composed of a certain fraction of migratory individuals against invasion by a population composed of only resident or migratory individuals (Lundberg 1987; Kaitala et al. 1993). An enduring question, however, has been the relative role of genetics versus environment in the evolution of partial migration. A single model bird species, the blackcap warbler, gives unique insight into the heritability of traits associated with partial migration. More recently, studies on birds, fishes, and other taxa have focused on phenotypic plasticity as a cause for divergence in migration behaviors, but this too cannot fully explain the many types of partial migration observed in marine fishes. In the following sections, (1) Arctic charr is introduced as a model species exhibiting many attributes typical of partial migration; (2) partial migration theory and phenotypic plasticity are presented as explanatory frameworks for understanding the evolution and maintenance of divergent migration behaviors within populations; and (3) partial migration is considered more broadly as an emergent property. Following this chapter's summary, I revisit some of Harden Jones's central premises in a recapitulation of themes developed in previous chapters to present partial migration as a unifying concept of marine fish movement and migration behaviors.

The "Problem" of Sympatric Ecotypes: Arctic Charr

Arctic charr exhibit partial migration through distinct ecomorphological types that vary in size and color as well as feeding, mating, and migration behaviors. In the same Arctic watersheds, dwarf and larger resident forms comingle with still-larger migratory oceangoing charr. The small dwarf contingent matures early and tends to be male. The larger resident form occupies a separate trophic niche and matures later, and the anadromous oceangoing contingent is principally the largest females. These three forms recur again and again in thousands of populations throughout the species range, a remarkable example of parallel ecotypic variation (Schluter and Nagel 1995; Jonnson and Jonnson 2001). Coexisting ecomorphs could be maintained through assortive mating, each keeping to its kind. Successive invasions could account for separate populations of the three ecomorphs. Alternatively, they might represent phenotypic plasticity—modalities in physiological, morphological, and behavioral responses to varying environmental conditions (a.k.a. polyphenism). Through a series of common garden and transplant experiments, Nordeng (1983) proved that both phenotypic plasticity and genetics play a role (fig. 6.1). When ecomorphs were mated with others of the same type, offspring developed into all three ecomorphs. But significant intergenerational carryover effects occurred: a slight propensity for crosses between dwarf forms to produce more resident offspring and crosses between anadromous adults to produce more migratory offspring. Still, early rearing conditions had a larger influence with higher food levels associated with increased resident behaviors. In transplant experiments across the Norwegian Sea, progeny from a northern population released into a southern river system again developed into all three ecotypes as adults, but the southern native population did not originally produce the oceangoing anadromous contingent. Subsequent genetic tracer and crossing experiments have confirmed that sympatric ecotypes arise through phenotypic plasticity (albeit sometimes followed by genetic divergence) rather than through simultaneous invasions of resident and migratory genotypes (Adams and Huntingford 2004; Adams et al. 2008). Nordeng also observed that migration behaviors were reversible: resident individuals can adopt anadromous behaviors later in life, and in rare instances migratory charr become resident (fig. 6.1; see also Näslund 1990; Radtke et al. 1996; Babaluk et al. 1997).

Arctic charr exhibit partial migration attributes common to marine fishes and birds: (1) partial migration often manifests itself as two or more ecomorphological types, which occupy differing trophic niches (a.k.a. resource polymorphism; Smith and Skúlason 1996); (2) early rearing conditions carry over into lifetime differences in migration behaviors; (3) contingents are not genetically discrete, but their occurrence and relative frequency are heritable; (4) migration behaviors covary with maturation rates and reproductive success; and

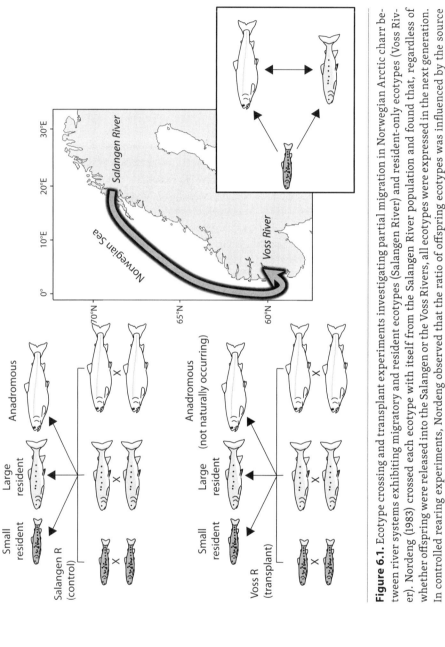

Figure 6.1. Ecotype crossing and transplant experiments investigating partial migration in Norwegian Arctic charr between river systems exhibiting migratory and resident ecotypes (Salangen River) and resident-only ecotypes (Voss River). Nordeng (1983) crossed each ecotype with itself from the Salangen River population and found that, regardless of whether offspring were released into the Salangen or the Voss Rivers, all ecotypes were expressed in the next generation. In controlled rearing experiments, Nordeng observed that the ratio of offspring ecotypes was influenced by the source population and early feeding conditions. Within generations, ecotypes can exhibit differential migration (switching) as shown in the inset. Figure concept from Adams 1999 and adapted from Klemetsen 2010, with permission.

(5) contingent behaviors are not necessarily permanent—they can reverse themselves over an individual's lifespan.

Partial Migration Writ Narrow: Genetic Controls

Bird species have been classified by the types of migration they exhibit (obligate and facultative, resident, mainly migratory, mainly summer visitor, etc.), resulting in a somewhat idiosyncratic lexicon not dissimilar to that used to classify marine fishes (see chap. 1, appendix). Berthold (1999) proposed a unifying concept, postulating that partial migration was ancestral to most if not all classes of migration behavior, and in many instances partial migration traits remained latent within the genotypes of most species (table 6.1). Supporting this view is the ubiquitous occurrence of partial migration among temperate bird species and the potential for partial migration to occur in less studied tropical and boreal species (Lack 1944; Berthold 1999; Newton 2008), as well as variation in the fraction of migratory individuals within the same population, which can change rapidly over generations.

Berthold and colleagues studied heritability of traits associated with partial migration in blackcap warbler, a model temperate bird species. In the laboratory, blackcaps exhibit restlessness (a.k.a. Zugunruhe; chap. 2, The Elusive Sixth Sense: Magnetoreception), a trait representing the propensity to migrate. Importantly, this trait can be observed, measured, and manipulated in the laboratory. By selectively mating either sedentary or migratory adults (depending on degree of restlessness observed), a population that initially exhibited intermediate levels of restlessness descended into purely sedentary or migratory lines within three to six generations (fig. 6.2; Berthold 1996, 1999). Offspring of crosses between wild populations exhibiting high levels of restlessness (partially migratory population from Germany) and sedentary behaviors (resident population from the Canary Islands) showed intermediate levels of restlessness. High rates of heritability, 0.37–0.46, associated with restlessness in blackcaps were observed for a range of crossing and artificial selection experiments (Berthold and Pulido 1994, Pulido and Berthold 2010).

In his model for genotypic control of avian migration, Berthold (1999) emphasized two points—(1) natural selection for populations that are resident, migratory, or partially migratory can occur over ecological timescales owing to high heritability of migration syndromes, and (2) partial migration is regulated through a threshold response to a trait such as early restlessness. An important implication of the theory is that phenotypes associated with partial migration remain latent in the genome under most selection regimes. Otherwise, selection over evolutionary time would lead to genetic canalization of migratory or sedentary genotypes and loss of partial migration. The threshold response was further generalized by Pulido (2011, reviewed below) as a means by which genes associated with migration behavior can remain latent in their expression over evolutionary time.

Regrettably, there is no model marine fish similar to blackcap that supplies compelling evidence for genetic control of migration. Fish ecologists are still searching for a threshold trait such as restlessness that can be readily measured and manipulated across generations. Still, substantial progress has been made in understanding basic elements of partial migration in model species such as Arctic charr, Atlantic salmon *Salmo salar*, threespine stickleback, European eel, and white perch *Morone americana* (Dodson et al. 2013). These studies indicate important interactions between genetic and environmental control of partial migration, perhaps overlaying the deeper genetic structure revealed by Berthold and his colleagues.

Phenotypic Plasticity

Over evolutionary and ecological time, nonstationary or periodically varying marine ecosystems necessitate plastic responses by organisms—responses that may or may not be

Table 6.1. Central postulates proposed by Berthold (1999) on the evolution of partial migration and related evidence for marine fishes

Berthold's postulate	Evidence for marine fishes
Migration and partial migration occurred in the most ancestral species	Like birds, elasmobranch and bony fishes exhibited early radiation of traits associated with increased efficiency in locomotion and development of vertical position control, both highly adaptive for migration (Long 1995). Large-bodied fishes occur early in the paleontological record, which also implicates early migration behaviors. Extant taxa with long evolutionary histories exhibit partial migration (e.g., Carcharhinidae, Acipenseridae, Salmonidae, Anguillidae, Moronidae).
As an adaptive and successful life history strategy, partial migration became ubiquitous	Electronic tagging and tracer studies show diverse migration behaviors within populations consistent with partial migration. Recent review papers on partial migration document an ever-increasing number of diverse taxa (Jonsson and Jonsson 1993; Secor 1999; Kerr et al. 2009; Secor and Kerr 2009; B. B. Chapman et al. 2012). From a broader perspective still, B. B. Chapman et al. (2011b) asserted that "the most common type of migration is known as 'partial migration.'"
Partial migration is maintained through its genetic latency; migratory and sedentary lineages can be reversed	The most compelling evidence for genetic latency of partial migration is from crossing experiments on Arctic charr (Nordeng 1983), but transplant experiments of Arctic charr and other salmonids also indicate latency in modes of migration behavior (Quinn 2005). The rapid expansion and changed behavior of invasive fishes introduced from initially resident populations also implicates the genetic latency of partial migration.
Partial migration is controlled by a threshold trait	Early size, growth, condition, and behavioral thresholds have been documented for a range of taxa—principally Salmonidae but also Anguillidae, Osmeridae and Moronidae (see text for examples).

Note: The postulates are modified from the five listed in Berthold's synopsis, which placed greater emphasis on the phylogenetic origin of migration and less emphasis on underlying mechanisms.

adaptive in terms of improved growth, survival, and reproduction. Phenotypic plasticity describes "the capacity of a single genotype to produce different phenotypes to varying environmental conditions" (Pfennig et al. 2010). Such plasticity is not without cost, requiring genetic, sensory, developmental, and behavioral capacities to respond to environmental cues (Pigliucci 2001). Further, a fully plastic response would interfere with early development and metamorphosis and preclude genetic selection and evolutionary change (Price et al. 2003; Ghalambor et al. 2007). Rather, phenotypic plasticity is genetically governed through

the norm of reaction, in which the responsiveness of phenotypes to environmental variation is regulated (Pigliucci 2001). Thus, in some parts of an Arctic charr's range, environmental conditions are curtailed in such a way that the norm of reaction does not express a migratory form. Yet if these fish are transplanted to another environment, the migratory phenotype would emerge, latent within the population's genotype (fig. 6.1). For transplanted fish, an intermediate level of phenotypic plasticity allows an evolutionarily rapid shift toward a new adaptive peak—that associated with the migratory ecotype (Price et al. 2003). But predicting

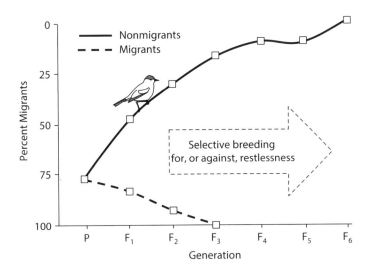

Figure 6.2. Artificial selection for restlessness (migratory behavior) in blackcap warblers. The experiment by Berthold et al. (1990) was initiated with parental (P) generation of pairs of breeding warblers from a partially migratory population (*N* = 267; southern France). Selection for restlessness led to a fully migratory population within three generations; selection for sedentary behavior resulted in a fully resident population in six generations. P, parental generation; F, filial generations. From Berthold et al. 1990, with permission.

the phenotypic outcomes of a given population is complex. Reaction norms are collective agencies, the result of genetic architecture, physiological and behavioral constraints, and past environmental and biotic selection (Pigliucci 2001).

Developmental Plasticity

Plasticity in migration behaviors in animals such as Arctic charr is conspicuous: ecomorphs are associated with either a resident or migratory behavior. These two modes are thought to arise as threshold responses related to the conditional state of the individual early in development. In Arctic charr and other salmonids, early growth or feeding conditions favor one or another ecomorph (Morita et al. 2000; Klemetsen et al. 2003; Olsson et al. 2006). The threshold occurs at a switch point in ontogeny and depends on a "liability" trait (e.g., a particular size or condition), the expression of which determines whether one or the other migratory trajectory is adopted (Páez et al. 2011; Dodson et

al. 2013). Restlessness in Berthold's blackcaps certainly qualifies as a liability trait. Several liability traits have been proposed for fishes. In Arctic charr, Atlantic salmon, and other species, higher early growth has been linked to residency (Nordeng 1983; Metcalfe et al. 1990; Morita et al. 2000; Kraus and Secor 2004b), but in other populations of Atlantic salmon, the opposite has been documented—fast growth leads to migratory behavior (Jonsson and Jonsson 1993; Hutchings and Myers 1994a; Thériault and Dodson 2003). Factors other than growth—such as body size, basal metabolic needs, and behavioral dominance—can also lead to conditional responses (Metcalfe et al. 1995; Morinville and Rasmussen 2003; Olsson et al. 2006; Kerr and Secor 2009; Dodson et al. 2013). Central hypotheses related to liability traits and conditional responses include (1) deviation in early metabolic trajectories cause high consumption demand in some cohorts, leading to migration; (2) cohorts of young fail to reach a certain body size and remain resident; and (3) selection (Darwinian fitness) favors the maintenance of alternative reproductive behaviors, which are associated with migratory and resident forms (reviewed by Jonsson and Jonsson 1993; Secor 1999; B. B. Chapman et al. 2011b; Dodson et al. 2013).

Effects of early growth rate on partial migration have been documented for Chesapeake Bay white perch, an estuarine-associated species that spawns in freshwater tidal regions. Larvae are retained in this same region, but then some but not all juveniles migrate to brackish waters. The propensity of juveniles to migrate is associated with slow larval growth (Kraus and Secor 2004b). Larval growth rates are in turn influenced by weather conditions that affect the timing of larval production (Kerr and Secor 2009). As a result, the ratio of migratory to resident individuals varies year to year according to weather conditions (fig. 6.3; Kraus and Secor 2004b). Following early migration by slower-growing larvae, growth improves during the juvenile period, suggesting that the conditional response depends on metabolic trajectories, where migration is compensating for a "bad start" (Metcalfe and Monaghan 2001; Kerr and Secor 2010).

In Lake Biwa (Japan), partial migration of the small smelt-like fish ayu *Plecoglussus altivelis* is related to a body size threshold, where larger juveniles are capable of overcoming flow and migrating from natal lake habitats to upstream feeding areas (fig. 6.4A). Dwarf residents remain behind in the lake, but during the next summer they spawn earlier and their young enjoy a longer growth season, forming the next generation of migrating stream ayu. In this way, partial migration alternates between generations: offspring of migratory spawners become resident and those of resident spawners become migratory (Tsukamoto et al. 1987).

Growth rate and body size thresholds can be both positively and negatively related to the propensity to migrate. This apparent inconsistency has been associated with evolutionary trade-offs between growth, survival, and reproduction. In their review of the evolution of partial migration in fishes, Jonsson and Jonsson

(1993) consider age-dependent trade-offs related to survival and fecundity. They contend that migration incurs costs to survival and delays reproduction, but these costs are balanced against higher growth and maximum achieved size in the nonnatal habitat. They proposed that rapid early growth and large juvenile size could support either early migration or early maturation, but not both. Dodson et al. (2013) advanced a model of consecutive thresholds, an initial threshold specifying early migration and a second one related to early maturation (see fig. 6.5B). In theory, resident and migratory behaviors can be evaluated by their relative fitness or reproductive success, but a more difficult challenge is predicting why and under which circumstances both tactics should coexist.

Simulation modeling of "mixed evolutionary stable strategies" has been used to predict equilibrium ratios of alternate (mixed) life history types by balancing their reproductive contributions in a stationary population. The stability of a given equilibrium ratio (partial migration) is evaluated against other possible mixture ratios (Kaitala et al. 1993). As an example, Gross (1985, 1991) hypothesized that disruptive selection (genetic selection against intermediate forms) resulted in male coho salmon that were either dwarf resident jack or large migratory hooknose ecotypes (fig. 6.4B; see also chap. 3, Sperm Competition). Partial migration represents the stable mixture of these two reproductive tactics, dependent upon their relative frequency and reproductive success (Gross and Repka 1998). Residency and early maturation traded off against limited access to defended females. For migratory males, large size and female access traded off against delayed maturation and ocean mortality (Hutchings and Myers 1994a; Hendry et al. 2004b). Growth rates or size thresholds associated with frequency-dependent selection are sensitive to changes in rearing habitats, which can disrupt frequency-dependent equilibriums (Gross 1991). Indeed, this result calls attention to a central tenet of the theory: mixed evolutionary stable strategies can only persist in stationary systems (Holt and McPeek 1996).

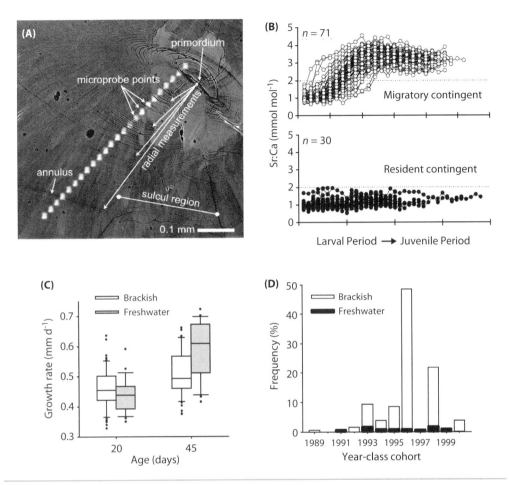

Figure 6.3. Features of partial migration for Chesapeake Bay white perch. (A) Otolith microanalysis links a series of Sr:Ca measures from the earliest part of the larva's life to peripheral regions of the otolith corresponding to juvenile and adult periods. For the larval and early juvenile period, assayed points are related to daily increments. (B) Otolith Sr:Ca profiles represent early dispersal into brackish water (>2 mmol mol⁻¹) or residency in freshwater. (C) Larval growth rates for brackish and freshwater contingents are shown as box-and-whisker plots. The freshwater (resident contingent) exhibited significantly higher growth rate over the first 45 d after hatching. (D) Classified contingents vary according to year-class strength and environmental conditions. Note that in some years (drought years) only the resident contingent occurs. Reprinted from Kraus and Secor 2004b, with permission.

Partial migration responses need not be bimodal. For life history types in male Atlantic salmon, Thorpe (1987) postulated three conditional responses dependent on juvenile growth rate: (1) fast-growth juveniles mature early and remain resident, (2) intermediate-growth juveniles remain resident for a relatively short period and migrate early in life, and (3) slow-growth juveniles remain resident for a longer period and migrate later in life (fig. 6.5A). These same modes are consistent with Dodson et al.'s (2013) concept of consecutive migration/maturation thresholds (fig. 6.5B). In contrast to disruptive selection, this system of conditional thresholds is associated with stabilizing selection (selection against extremes) for several

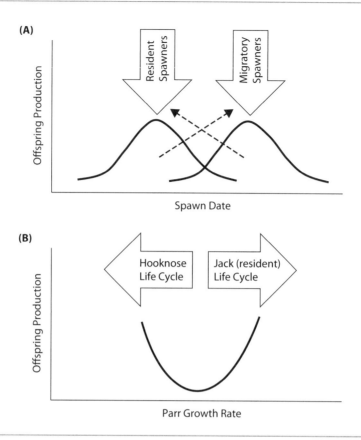

Figure 6.4. Two types of partial migration driven by spawning behaviors in (A) Lake Biwa ayu and (B) Pacific coho salmon. (A) The resident lake contingent of ayu spawns earlier than the migratory river contingent; its offspring enjoy a longer growth season, and they develop as the next generation's migratory contingent (Tsukamoto et al. 1987). (B) The mating success of coho salmon hooknose (migratory) and jack (resident) ecotypes is related to their relative size and frequency (Gross 1985). From Secor 1999, with permission.

modes of migration associated with juvenile growth rates. Stabilizing selection for partial migration is supported in this species by modes in presumed liability trait distributions (small-resident, large-migratory) that are stable across generations and, in Atlantic salmon, highly heritable thresholds in this trait (Páez et al. 2011).

Threshold Model

In quantitative genetics, the expressed genotype results from alleles that are additive in their contribution. This mechanism seems at odds with emergent partial migration behaviors, which are bimodal or multimodal in na-

ture, and directs attention to threshold models, where some continuously distributed liability trait is divided into discrete phenotypic outcomes (fig. 6.6). The liability trait is linked to a suite of other traits that determine migration outcomes. This trait (restlessness, early growth, size, dominance) is represented by its propensity (here termed *liability*) to cause an individual to undertake one of several discrete behaviors (Roff 1996; Pigliucci 2001). For partial migration, consider a normal distribution of a population's liability trait under stationary environmental conditions. Extremes in the liability trait will result in populations that are composed of either migrants or residents, and

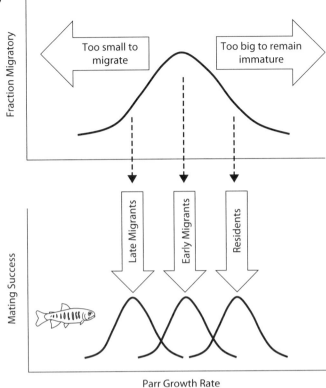

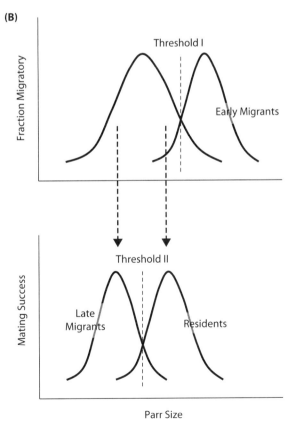

Figure 6.5. Multimodal partial migration responses to mating success in Atlantic salmon. Stabilizing selection for conditional responses to early growth results in multiple ecotypes. Scenario (A) considers simultaneous thresholds for three modal groups; scenario (B) considers consecutive thresholds during the life history. (A) from Secor 1999; (B) from Dodson et al. 2013.

partial migration is predicted across a range of intermediate phenotypes that traverses a migration threshold (fig. 6.6). The threshold condition for the liability trait is controlled by an underlying norm of reaction. But at intermediate levels of liability trait expression the population exhibits partial migration and is particularly sensitive to the environment. For this receptive range of trait expression, intergenerational control operates through phenotypic plasticity (fig. 6.6).

Two interesting consequences of the threshold model, highlighted by Pulido (2011), contribute to the long-term latency of partial migration. First, because a large range of genotyp-

ic variation is expressed as a single behavioral outcome, a portion of trait variance remains insensitive to natural selection (fig. 6.6). Plasticity in the phenotype is lost, and a given phenotype (e.g., resident behavior) is in fact underlain by substantial genetic diversity. This phenomenon, termed *genetic assimilation*, curtails loss of genotypic variance associated with changing selection regimes (Pfennig et al. 2010; Pulido 2011). Thus resident and migratory genotypes can be maintained in a population, which expresses only one of the two behaviors. This latent genetic variation can be released rapidly in novel environments, where a different threshold might prevail (McGuigan and Sgró 2009).

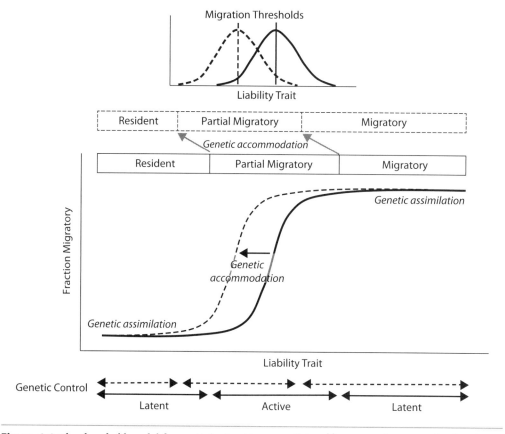

Figure 6.6. The threshold model for partial migration, specifying (1) a norm of reaction for the phenotypic expression of migration (top plot), (2) latency (genetic assimilation), and (3) genetic accommodation under a new selection regime (dashed lines). See text for details. Figure concept adapted from Pulido 2011.

The second dynamic relates to phenotypic plasticity expressed near the migration threshold, where both migratory and resident contingents coexist, and migration is most sensitive to changes in the environment through developmental plasticity (Pigliucci 2001). In nonstationary settings, this phenotypic plasticity can be self-reinforcing, causing shifts in the migration threshold, and should these be adaptive, genetic accommodation occurs through changes to the underlying reaction norm (Ghalambor et al. 2007; Crispo 2008). Because the liability trait conveys to a larger suite of traits determining migration outcomes (sensu the migration syndrome; Dingle 1996), changes in the reaction norm can result in rapid and nonlinear changes in population behaviors (Pfennig et al. 2010; Dodson et al. 2013).

Although individuals are responding to changes in the environment through their plastic response to encountered environments, they are also collectively responding through the population's norm of reaction. When new environments favor divergence in phenotype, wholesale population changes can occur as selection operates on the norm of reaction over multiple generations (Price et al. 2003; Pfennig et al. 2010). In support of the potential for rapid selection, Páez et al. (2011) reported surprisingly high heritability in Atlantic salmon associated with size-dependent thresholds for partial migration. Early schooling behavior in fishes may also act as a liability trait. In recent multigenerational common garden experiments, significant genetic inheritance was detected in threespine stickleback schooling behaviors for populations that were initially migratory or resident (Greenwood et al. 2013). Further, functional molecular markers corresponded with schooling behaviors and lateral line attributes associated with schooling. This experimental system for threespine stickleback seems quite promising in terms of examining how schooling as a liability trait is responsive to novel environmental settings.

A particularly compelling example for rapid changes in threshold responses occurs in in-

stances of colonization. As an example, consider the house finch *Carpodacus mexicanus*, which was introduced from Southern California to New Jersey in 1940. The initial population was sedentary, but within 40 y the introduced population irrupted, colonizing the eastern half of the United States (Able and Belthoff 1998). Similarly, within several generations, sedentary Pacific Ocean red lionfish *Pterois volitans*, introduced in Florida Atlantic waters, colonized shelf waters throughout the Caribbean Sea and US Atlantic waters as far north as the New York Bight (Ruttenberg et al. 2012). Shelf populations of sardines expand rapidly in episodes of irruptive growth (chap. 4, Schooling through Food Webs). The switching of resident forms of salmon to migratory anadromous types when introduced into novel environments is well documented (Hendry et al. 2004b; O'Neal and Stanford 2011). Thus novel interactions between the norm of reaction and environmental conditions could cause populations to rapidly express new migration behaviors within a single generation, which contrasts starkly with changes that would require many generations were they due to genetic divergence and mutation alone (Pfennig et al. 2010).

Evolution of Partial Migration and Ecological Speciation

In Arctic charr and other Salmonidae (salmons, trouts, and charrs) the recurrence of migration ecotypes across so many populations presents a natural experiment on the evolution of partial migration: How do parallel ecotypes emerge and persist when opportunities exist for genetic mixing between sympatric populations that would act against polyphenism? An evolutionary legacy of colonization, phenotypic plasticity, and developmental thresholds could play a role, allowing repeated divergence of ecotypes across populations (Hughes 2012). Crispo (2008) argued that colonization success should be higher for a population exhibiting high phenotypic plasticity: a jack-of-all-trades tactic that is favored over a master-of-some tactic. A corollary is that increased variation in migration behaviors should be observed in colonizing populations. Following colonization, new selection regimes can lead to genetic accommodation of threshold traits and genetic divergence from source populations (Ghalambor et al. 2007; Klemetsen 2010; see fig. 6.6).

In threespine stickleback, parallelism is not only evident in ecotypic recurrence but also in the genetic divergence between ecotypes, where each ecotype represents its own population (Rundle et al. 2000). This small, widespread, temperate-boreal coastal species underwent postglacial radiation from an ancestral ocean ecotype to freshwater limnetic and benthic ecotypes, the latter varying in size, degree of lateral plate armature, color, and courtship and nest guarding behaviors (fig. 6.7; McKinnon and Rundle 2002; Bell et al. 2004; Shaw et al. 2007). Oceanic forms show high levels of both genotypic and phenotypic variance (Mäkinen et al. 2006). In the laboratory, oceanic sticklebacks can be induced to exhibit a much broader range of courtship and guarding behaviors than freshwater ecotypes (Shaw et al. 2007). Colonizing anadromous sticklebacks will within 12 generations diverge into freshwater ecotypes (Rundle et al. 2000; Bell et al. 2004; Furin et al. 2012). Size-assortive courtship curtails successful mating between ecotypes (Nagel and Schluter 1998; McKinnon et al. 2004). As a demonstration of this effect, mating success by adults of the same ecotypes but originating from different watersheds is higher than between differing ecotypes within the same system (Rundle et al. 2000). Further, the "hybrid" offspring of ecotype crosses are intermediate in their phenotype and experience reduced growth and reproductive success (Hatfield and Schluter 1999).

The threespine stickleback represents a unique model species for understanding sympatric population divergence and speciation, also known as ecological speciation. But the story line's arc follows that of partial migration —first, high ancestral phenotypic plasticity allows for novel spatial behaviors in new or changing environments, then selection reinforces new ecotypes, and, finally, genetic divergence follows phenotypic divergence through assortive mating and other mechanisms. In a review of ecological speciation, Hendry (2009) questioned a tendency in the literature to assume that ecological speciation requires complete reproductive isolation, highlighting examples in threespine sticklebacks and other species, where phenotypic and genotypic divergence between ecotypes occurred with incomplete reproductive isolation. This view is consistent with the threshold model, where ecotypic forms associated with partial migration need not be the result of genetic divergence (West-Eberhard 2005). Because genotypic change follows the expression of ecotypes, multiple scenarios of genetic divergence can be envisioned specific to selection regimes (environmental and social interactions) and genetic recombination and mutation. This expectation conforms to reports of differing levels of reproductive isolation between anadromous and freshwater sticklebacks where they co-occur in the same watershed (McKinnon et al. 2004; Kitamura et al. 2006; Jones et al. 2008; Hendry 2009).

An interesting phenomenon that applies to the threshold model is heterochrony, or changes in developmental timing that can result in rapid evolutionary change (Gould 1977). Developmental rates for multiple traits are coupled by a single gene's action (pleiotropy), which is regulated at certain ontogenetic switchpoints (Pigliucci 2001; West-Eberhard 2005). Heterochrony can lead to exaggerated ecotypes such as slow-developing giants and rapidly developing dwarfs (neotony and progenesis; Gould 1977). In a scenario that might apply to the evolution of polyphenism in Arctic charr, McPhee et al. (2012) suggested that, in a novel environment, charr would encounter a broad range of thermal conditions resulting in a full expression of discontinuously distributed developmental rates, which in turn would lead to modes of reproductive and migration behaviors such as the dwarf residents and large mi-

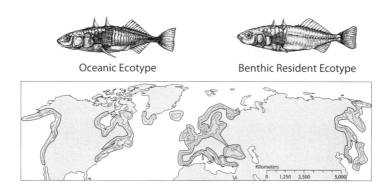

Figure 6.7. Global distribution of threespine stickleback. Throughout its range the species occurs as oceanic (anadromous) and freshwater (limnetic, benthic) ecotypes. Oceanic populations are characterized by a higher number of lateral plates than freshwater limnetic and benthic ecotypes (Bell et al. 2004). These ecotypes recur in thousands of coastal watersheds, indicative of strong latency for partial migration. Distribution map from McKinnon and Rundle (2002) and stickleback drawings reprinted from Bell 1976, with permission.

gratory ecotypes. For anadromous sticklebacks (the presumed ancestral type), early stages of development resemble limnetic (juvenile) and benthic (adult) forms of the same species, leading West-Eberhard (2005) to speculate that heterochrony could perpetuate or truncate the juvenile period, leading to the emergence of multiple ecotypes.

The threshold model implies a central role for phenotypic plasticity in the evolution of migration. The model also emphasizes carryover effects where early expression in a liability trait leads to multiple developmental pathways and associated lifetime behaviors. Because much of the genotypic variation associated with migration is latent, selection for migration behaviors can only occur across the range of the phenotype that is responsive to a population's environment (Pfennig et al. 2010). In new environments, a novel phenotypic response can cause a different portion of the underlying reaction norm to be exposed to selection, which may or may not be adaptive but permits rapid divergence (Ghalambor et al. 2007). This is clearly not a simple system of Darwinian selection, where selection operates through genetic recombination and random mutations

(West-Eberhard 2005). Indeed, the threshold model could be considered as ecological inheritance (chap. 1, Migration: A Trait, Syndrome, or Complex System?) because new environments and novel phenotypic response drive subsequent selection.

Is Partial Migration the Solution to the "Eel Problem"?

Temperate eels of the family Anguillidae—American, European, and Japanese eels—all show partial migration where juveniles dispersing from oceanic natal areas remain in marine waters, invade brackish waters, or continue to push up into freshwater habitats of all sorts (rivers, streams, creeks, lakes, ponds, etc.). Similar to charr and sticklebacks, migratory contingents recur throughout the ranges of the three species. But eels show no obvious morphological differences between contingents. Through otolith tracer approaches, Tsukamoto et al. (1998) and others (Morrison et al. 2003; Daverat et al. 2006; Arai and Chino 2012) were able to uncover the generality of partial migration in eel species that were previously thought to exhibit only one spatial behavior: juvenile migration into and residency within freshwater habitats. These findings are clearly at variance with the historical moniker for this group of "freshwater eels."

Anguilla species in the North Atlantic use the same oceanic spawning area and their larvae overlap, leading to a century-long debate on sympatric speciation—the so-called "eel problem." During the long juvenile period, American and European eels occur as spatially separated species in North American and European continental waters, but, following their emigration from these habitats and spawning in the Sargasso Sea, early stage larvae of the two species broadly overlap in time and location. The same Gulf Stream Current then transports larvae of both eel species (fig. 6.8; Schmidt 1922; Kettle and Haines 2006). Further, in contrast to sticklebacks or charr, where recurring habitat differences lead to recurring ecotype variants, there is no apparent habitat heterogeneity in the open ocean environments that larvae (leptocephali) occupy en route to the shelf waters of North America or Europe. Indeed, broad-scale synchrony in juvenile abundances between the two species during the past several decades is indicative of shared oceanic habitats (Bonhommeau et al. 2008; Fenske et al. 2009). Should we not expect their dispersal outcomes to overlap? Or, as Schmidt (1922) posed the question, "how do masses of larvae in the Western Atlantic sort themselves out?"

The eel problem provoked early controversy centering on (1) the assumption of sympatric speciation and (2) mechanisms that might lead to niche partitioning between species. Tucker (1959) challenged early accounts that the two eels were separate species, evidence for which depended on vertebral counts. Tucker suggested that these taxonomic differences were in fact phenotypic in nature, a response to differing temperatures experienced early in life. Tucker argued that there was only one panmictic eel species, whose offspring dispersed throughout the entire North Atlantic. Further, he believed that most European juveniles did not have the energetic and navigational wherewithal to make the long return journey to the Sargasso Sea. This part of the species range constituted

an ecological trap, and fishery yields in Europe depended on reproduction by adults produced from North American waters. The principal contesting viewpoint posited larval duration differences between species, where European eels had evolved longer larval durations than American eels, so that the former, while exposed to opportunities to immigrate into North American shelf habitats, was developmentally incompetent to do so (fig. 6.8; Schmidt 1922; Deelder 1960; Harden Jones 1968). Successful transport for either species would require different time spans on the order of several months to 1 y for American eel and 1.5–2.5 y for European eel (Kettle and Haines 2006). Thus larval duration differences could be selected by niche partitioning of juvenile habitats that were discretely distributed across an ocean basin—in either North American or European watersheds.

Decades of oceanographic, genetic, and otolith research have led to the current consensus that the North Atlantic does indeed contain two species of Anguilla, and that larval duration separates the two species (McCleave et al. 1998; Arai et al. 2000; Kettle and Haines 2006). Might larval duration constitute a liability trait leading to initial niche partitioning and subsequent speciation, as do traits related to growth, condition, and development in stickleback, charr, and salmon species? Two lines of evidence support this speculation: (1) documented cases of hybrid A. rostrata × A. anguilla and (2) threshold responses associated with partial migration. The range of American and European eels is not exclusive but overlaps in Iceland, the northernmost part of their range. In this same region, larval durations are intermediate between American and European eels (Kuroki et al. 2008), and molecular markers have documented the co-occurrence of both species and their hybrid (Avise et al. 1990; Albert et al. 2006). A chief tenet of the threshold model is that intermediate forms are not selected. Iceland represents a unique location where intermediate larval durations can lead to transport to juvenile habitats. This finding also solves the problem of why American and European eels

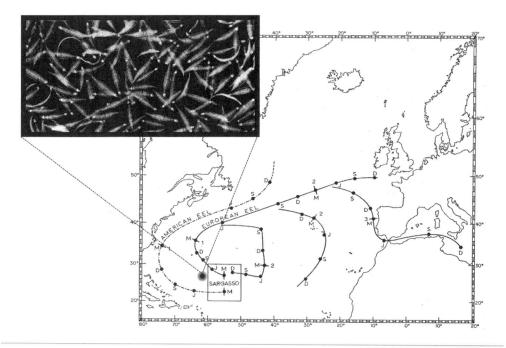

Figure 6.8. Larval durations for American and European eels estimated by Harden Jones (1968) from surface current velocities. Months and years are placed on vectors for each species on the basis of March spawning (box). Inset shows leptocephali larvae of both species (the majority are European eel) taken in a single haul (filled circle) by Schmidt and colleagues in the Sargasso Sea in June 1920. Reprinted from Harden Jones 1968; inset from Schmidt 1922, with permission of Nature Publishing Group.

do not mix—they do, only the outcome results in a minority behavior that does not affect the overall selection for differing larval durations between the two panmictic species (Als et al. 2011).

Juvenile European eels migrating into freshwater habitats respond to liability traits (condition and thyroid hormone status) in a threshold manner (Edeline et al. 2005; Edeline 2007). More active glass eels exhibit higher thyroxine levels and are more likely to respond to high flows by moving from tidal estuaries to upstream fluvial habitats. Those below threshold thyroxine levels tend to stay put. Through carryover effects, these early threshold behaviors could explain longer-term juvenile behaviors where temperate eels will persist throughout their juvenile periods (2–20+ y) in coastal, estuarine, or freshwater environments.

To conceive of how the threshold model might influence speciation in *Anguilla*, let us briefly consider zoogeography. *Anguilla* species are most concentrated in the tropical Indo-Pacific, the family's provenance for which molecular tracers indicate a 45-my history (Tsukamoto and Aoyama 1998). At the opposite end of *Anguilla*'s phylogenetic tree are the two North Atlantic temperate eels, which form their own branch with species separating some 5–10 my ago. Consider next a hypothetical ancestral tropical species, with a range of larval duration times that results in the species' spread by dominant currents throughout the tropical Indo-Pacific basin. Phenotypic plasticity causes differential outcomes, and over epochs of oceanographic change, latent variation in larval duration is generated. A shift in spawning behavior (time/location) and oceanography causes irruptive growth by this species to temperate environments, requiring longer durations and leading to phenotypic displacement of the larval duration threshold. Tropical and temperate juvenile habitats, if sufficiently distant or associated with opposing oceanographic features (Kimura et al. 1994), could result in genetic divergence should those differences persist over sufficient generations (fig. 6.9).

In their theory of temperate eel phylogeny, Tsukamoto and Aoyama (1998) proposed that variability in larval duration may have been instrumental in *Anguilla* species' radiation 30–40 my ago through the Tethys Sea, which connected the Indo-Pacific basin to the North Atlantic Ocean. The ancestral Atlantic species conveyed with them high plasticity in larval development rates associated with partial migration, which in turn contributed to their successful colonization to Western Hemisphere continen-

tal habitats. Over geological time, plate tectonics caused the juvenile habitats to become increasingly separated, leading to discrete distributions of larval durations, genetic divergence, and speciation. In summary, ecological speciation likely occurs in temperate eels as it does in stickleback and charr, the chief difference being that niche partitioning leading to speciation occurs across an ocean basin rather than between habitats within the same watershed.

Cod Ecomorphs and Biocomplexity

Fully marine versions of ecotypes are rare, but a recent example is the red and olive ecomorphs of Gulf of Maine cod (Sherwood and Grabowski 2010). The red ecomorph is the minority component, and the olive type is more frequently expressed. Red coloration in Atlantic cod is

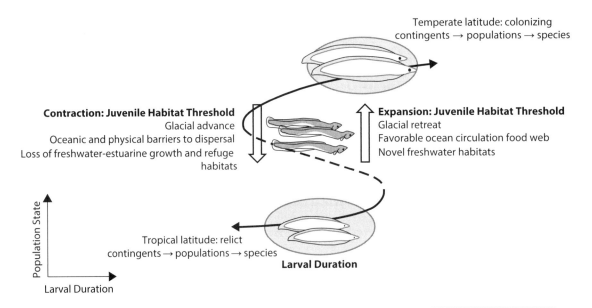

Figure 6.9. Hypothesis for radiation of temperate *Anguilla* species based on larval duration and availability of novel juvenile growth habitats. In this bifurcation diagram the distance to juvenile habitats influences the action of larval duration. Phenotypic plasticity in larval duration is promoted over epochs of oceanographic changes for a given population, some of which is assimilated in the genome. New juvenile habitats become available through shifts in ocean transport conditions and development of novel growth habitats (say, through glacial retreat). Greater phenotypic expression of larval durations causes a change in larval duration thresholds associated with metamorphosis and ingress into new juvenile habitats (see fig. 6.6). The persistence of different thresholds between relict and colonizing populations would lead to speciation.

associated with inshore foraging on demersal crustaceans (Gosse and Wroblewski 2001), but whether it is associated with seasonal/ontogenetic changes in diet or lifetime foraging and spatial behaviors remains unanswered. Toward the center of the Gulf of Maine, ~100 km offshore, is Cashes Ledge, a small distinct bank that supports a dense stand of kelp forest. The bank contains high densities of Atlantic cod, a minority of these sporting red pigmentation, far distant from where this ecomorph typically occurs. Sherwood and Grabowski (2010) compiled a compelling case for resource polymorphism and partial migration for this local group of red cod. Red cod occurred in shallower areas of the bank (<20 m) and depended upon local feeding on crabs, lobsters, and demersal fishes. In comparison to the co-occurring olive type, red cod experienced lower lifetime growth rates, deeper body shape, and did not occupy adjacent deepwater habitats.

Might partial migration also occur in populations of Atlantic cod where ecomorphs are less obvious? Local inshore populations of Atlantic cod have been noted for over a century. In Norway, for instance, they are called fjord cod. Indeed, a seminal 20th-century debate on the efficiency of marine fish hatcheries centered on whether fjord cod were self-sustaining stocks or members of larger coastal populations (Solemdal et al. 1984; Smith 1994; Secor 2004). Inshore seas, fjords, and bays are conducive to local retention of larvae, resulting in reproductively isolated populations that show genetic differences (Ruzzante et al. 2000; Salvanes et al. 2004; Bradbury et al. 2008). On the other hand, conventional tagging studies indicate that inshore juveniles and adults show both resident and migratory behaviors (Robichaud and Rose 2004). More recent application of electronic telemetry and otolith tracer studies also support these plastic behaviors (Green and Wroblewski 2000; Jamieson et al. 2004; Neat et al. 2006; Lindholm et al. 2007), although genetic studies have produced a complex picture on whether inshore cod comprise separate subpopulations or whether they represent contingents within a larger population.

A special adaptation to cold temperatures provides a test of whether genetic inshore subpopulations occur. Resident behaviors by cod confer risk associated with overwintering in frigid (<0°C) inshore habitats. Indeed, inshore cod produce a plasma antifreeze glycoprotein to contend with winter temperatures that would be otherwise lethal. Might expression of this protein be indicative of genetic divergence between resident and migratory cod? Employing mitochondrial DNA, a conservative marker of genetic lineage, Carr et al. (1995) failed to detect differences between inshore and offshore wintering cod in Newfoundland waters. But for the same aggregation, Ruzzante et al. (1996) employed more variable nuclear microsatellite DNA markers and found evidence for population structure. The two marker types depict different timescales: maternal mitochondrial DNA depends on mutation and drift in the maternal genome alone, whereas nuclear DNA is more labile owing to genetic crossover (Ruzzante et al. 1996). Further, inferences from nuclear microsatellite DNA have been shown to be sensitive to sampling design and temporal stability, and they depend on a central assumption that alleles are not undergoing selection (Wirgin et al. 2007; Pampoulie et al. 2011). At least one allele commonly used in Atlantic cod population studies showed evidence of "hitchhiking selection," where the allele is jointly expressed with other alleles that are under selection (Nielsen et al. 2006). Testing neutrality in genes, required for discriminating long-term population separation, is often difficult given assumptions related to (1) representative sampling of the gene pool, and (2) the numerous microsatellite markers are acting in a neutral manner (Nielsen et al. 2009).

An exciting trend in population genetics is the relaxation of assumptions related to neutral selection and adoption of functional molecular markers, such as some single nucleotide polymorphisms. Molecular markers associated with coding and regulatory regions of the genome can be used to uncover population structure that is responsive to ecological circumstances, such as local selection, migration, and past abundance dynamics (fig. 6.10; Hauser and Carvalho 2008; Pampoulie et al. 2008; Gaggiotti et al. 2009; Mueller et al. 2011; Williams and Oleksiak 2011). Employing such markers in populations of Atlantic cod, Bradbury et al. (2010) uncovered nonneutral allelic frequencies associated with latitudinal thermal regimes. Remarkably, these thermal clines in single nucleotide polymorphisms paralleled each other for populations in both the western and eastern North Atlantic basins—a form of natural replication in observing how certain genes were associated with regional adaptation.

Loosely termed *population biocomplexity* (mix of genetic and life history divergence within species and populations; from Ruzzante et al. 2006), breakthroughs in sequencing genomes have allowed molecular biologists to move away from typological endeavors such as strictly classifying species and populations toward a deeper understanding of how "networks of interacting genes" are expressed as complex traits that underlie population structure (Naish and Hard 2008; Nielsen et al. 2009). Populations are characterized by measuring markers of diversifying selection, responsive to ecological changes occurring over single to multiple generations (fig. 6.10; Ackerman et al. 2011; Williams and Oleksiak 2011). Not surprisingly, biocomplexity embraces collective agencies among alleles and the action of minor alleles: those that deviate from expected aggregate outcomes, including nonadditive allelic interactions such as dominance, standing genetic variance (assimilation), epistasis (emergent interactions of alleles on the same trait), and pleiotropy (single allelic control over multiple traits). Observing populations under selection yields insight into intergenerational dynamics (Stockwell et al. 2003; Naish and Hard 2008; Nielsen et al. 2009) and past adaptations to environmental gradients (Pigliucci 2001; Gaggiotti et al. 2009).

For the resident contingent of cod on Cashes Ledge, the jury is still out on whether this behavior is maintained through reproductive isolation or partial migration (Sherwood and Grabowski 2010). Reproductive isolation by

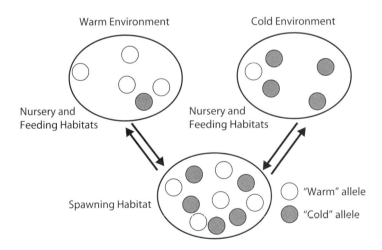

Figure 6.10. Biocomplexity underlying population structure within a single generation. In this example, alleles are responsive to thermal conditions experienced by contingents occupying differing nursery and feeding habitats. Such functional molecular markers can serve as natural tags of migration behavior, but they also serve to link population structure to seminal environmental regimes, responses shaped by short-term diversifying selection. Redrawn from Nielsen et al. 2009, with permission.

each ecomorph might occur through differences in spawning time, spawning behavior, and embryo and larval dispersal. Such locally adapted populations could persist even with moderate invasion (5%–10%) by adjacent subpopulations (Hauser and Carvalho 2008; Poulsen et al. 2011). Alternatively, conditional migration behaviors might exist under a threshold framework, where some liability trait such as early condition or growth is influenced by time and location of initial juvenile settlement, forage conditions, and competition on this isolated kelp-laden bank. A biocomplexity argument might combine elements of both explanations by employing, for instance, a particular candidate locus, the expression of which is associated with one or the other migratory behavior (Nielsen et al. 2009). As a recent example, Gaggiotti et al. (2009) observed that a single locus in Atlantic herring was responsive to salinity and discriminated Baltic Sea (salinity 6–8) and the North Sea (salinity 34–36) spawning aggregations, implicating a functional genetic marker associated with population structure.

Biocomplexity studies on locally adapted groups of cod and other marine fishes harken back to subtle phenotypic differences in groups of migrating fish, which initiated quantitative fisheries science (Smith 1994; chap. 5, Bounded Fisheries). In an early multivariate statistical analysis, Heincke (1898) detected subtle morphological differences among schools of Atlantic herring collected from bank habitats in Germany and the Netherlands. His inference of local populations (a.k.a. the "modern migration hypothesis") held sway for decades until overshadowed by Hjort's year-class phenomenon, which involved larger population entities (e.g., North Sea and Norwegian stocks of herring; Secor 2004). This is not to say that Hjort failed to recognize biocomplexity. Indeed, he was the first to coin the term *contingent*, used here and elsewhere to designate, in an agnostic way, groups of fish and birds that exhibit similar migration behaviors over major portions of their lifetimes (Alerstam and Hedenström 1998; Secor 1999). Genetic structure associated with contingents (biocomplexity) represents short-term selective forces, not necessarily leading to long-term population structure. Still, in stressed systems (overfishing, climate change), loss of this variance could convey risk to long-term resilience (Ruzzante et al. 2006; Hauser and Carvalho 2008; Petitgas et al. 2010). For instance, rapid expansion of sardines in the North Sea depended on a small resident contingent that escaped the attention of industrial fisheries for decades (Petitgas et al. 2012). Loss of this resident behavior—its genetic and behavioral repertoire—would have caused full collapse of this population. In the future, emerging metapopulation with biocomplexity perspectives will result in the next generation of transformative discoveries, where rich data sets of the molecular attributes and movements of individuals will inform scientists on how migration responds to population dynamics and ecosystem change.

Partial Migration Writ Large

The threshold model provides intriguing insight into how genetics and ecology jointly govern migration, but such explanations do not fully capture the diversity of fish migrations. A broader framework concedes that there are many interacting propensities to "stay or go" that lead to diverse sets of partial migration behaviors (table 6.2; fig. 6.11). Further, although the decision to disembark is traditionally ascribed to the individual (Dingle 1996), individual volition is impossible to separate from collective agencies such as reaction norms, population dynamics, social transmission, and food web interactions. It is these agencies that cause partial migration to be adjusted, altered, interrupted, amplified, and stabilized in countless ways as groups of fish respond conditionally to ecological and social circumstances (Bonte and de la Peña 2009; B. B. Chapman et al. 2011b).

Across animal taxa, three dominant types of partial migration have been classified: (1) non-natal divergence, (2) natal divergence, and (3) differential migration (B. B. Chapman et al. 2011b; Shaw and Levin 2011). These classifications are supplemented with other variants

Table 6.2. Names, pattern, and specific examples for types of partial migration

Type of partial migration (PM)	Pattern	Examples
Nonnatal divergence	One contingent stays put in the natal habitat, another contingent disembarks but returns for spawning. This is the classic type of PM for which the threshold model applies. Termed nonbreeding PM by Shaw and Levin (2011).	Freshwater resident and migratory eco-morphs of sticklebacks (Rundle et al. 2000) and Arctic charr (Nordeng 1983), which share the same spawning habitat Freshwater, estuarine, and coastal contingents of temperate eels (Tsukamoto et al. 1998) and striped bass (Secor 1999), which occur absent separate spawning populations Shelf- and oceanic-associated components of blue marlin (Kraus et al. 2011), albacore (Childers et al. 2011), and white shark populations (Jorgensen et al. 2012a,b) Multiple summer feeding aggregations in North Sea plaice (Hunter et al. 2004)
Natal divergence	Resident and migratory contingents comingle during nonbreeding seasons but originate from separate natal habitats (populations). More generally, natal divergence includes stock mixing between the migratory contingent of one population and the resident contingent of another. Termed breeding PM by Shaw and Levin (2011).	Atlantic herring aggregations where a migratory contingent comingles with a resident contingent on feeding and wintering grounds (McQuinn 1997b; Corten 2001; Ruzzante et al. 2006) Seasonally shared shelf habitats for inshore and coastal spawning cod stocks (Robichaud and Rose 2004) Multiple reef and wetland nursery systems contribute to the same reef adult snapper aggregation in the Red Sea (McMahon et al. 2012)
Differential Migration	Migration within a population conditionally dependent on size, age, sex, or other attribute. Size- and sex-dependent interactions can lead to other types of PM such as natal divergence, skipped spawning, and straying. Midlife abrupt shifts in migration behaviors—"switching"—constitute a type of differential migration.	Range extension with size for black drum (Jones and Wells 1998) Decreased range with size for Pacific bluefin tuna (Itoh et al. 2003) and tiger sharks (Meyer et al. 2009) Sexual segregation by mako sharks and dogfish to avoid sexual harassment (Mucientes et al. 2009) Increased foraging range by female striped bass (Secor and Piccoli 1996) and scalloped hammerhead shark (Klimley 1987) to meet large size and reproductive costs Differential estuarine ingress by juvenile cohorts of Atlantic bluefish (Hare and Cowen 1993) Patterns of residency by jack male salmon (Gross 1985)

Type of partial migration (PM)	Pattern	Examples
		Resident behavior on nesting territory by cardinal fish (Fukumori et al. 2008) Increased straying rates by male Pacific salmon (Hendry et al. 2004a)
Skipped Spawning	All fish disembark from the natal habitat, but only a portion of adults undertakes natal migrations each year.	Well known for sturgeons from hard-part analysis and aquaculture. Nonannual and delayed spawning are now believed to be common in marine fishes, particularly in longer-lived species (Rideout et al. 2005; Secor 2008)
Straying	A portion of the migratory contingent reproduces in nonnatal habitats. Considered an amended type of nonnatal divergence PM, straying propensity often varies by size and sex (differential PM).	Advective transport of English Channel–spawned European plaice to multiple nursery habitats (Bolle et al. 2009) Transoceanic displacements of spiny dogfish (McFarlane and King 2003) Invasion of Beaufort Sea watersheds by pink salmon (Babaluk et al. 2000)
Irruptive Migration	Wholesale shifts in distribution caused by self-reinforcement of the migratory contingent's behavior. Evacuation is considered as a type of irruptive migration when it occurs as a result of catastrophic environmental change.	Rapid range expansion of Japanese sardines into the central North Pacific Ocean during the 1980s (Fréon et al. 2005) Temporary evacuation by white sturgeon in the lower Columbia River as a result of Mount St. Helen's eruption (DeVore et al. 1999) Evacuation of juvenile blacktip sharks in advance of a storm (Heupel et al. 2003) Invasion of the northwest Atlantic Ocean by Pacific lionfish (Ruttenberg et al. 2012)
Vertical Migration	Only a fraction of the population exhibits diel or other recurrent vertical movements. This class can be considered a commuting or ranging behavior (table 1.2), but in some instances these movements define contingents over longer periods of time (seasons and years).	Vertical migrations in Japanese eel leptocephali, which become increasingly pronounced with age, influencing transport fates (Otake et al. 1998) Movement distances are associated with vertical migrations in Pacific halibut (Seitz et al. 2011) Juvenile plaice separate into migratory and nonmigratory individuals on the basis of their vertical behaviors (Berghahn 1987)

Note: Types are modified and amended from B. B. Chapman et al. (2011b) and Shaw and Levin (2011).

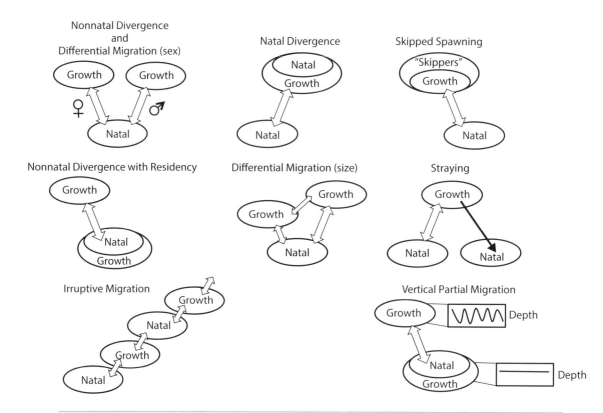

Figure 6.11. Classes of partial migration. Ellipses represent natal and nonnatal habitats (the latter generalized as growth habitats), and double open arrows represent within-generation migrations. Straying is indicated by the solid arrow. See table 6.2 and text for more explanation on each type of partial migration.

highlighted in this and preceding chapters (table 6.2). This scope for partial migration is broader than previous definitions (e.g., Lack 1944; Berthold 1996) and more fully captures the diversity of life cycles exhibited by marine fishes. These categories are not mutually exclusive nor are they likely complete. For Atlantic herring, for instance, examples can be found for both nonnatal divergence (Brophy et al. 2006) and natal divergence (McQuinn 1997b), differential migration through size-dependent entrainment (Huse et al. 2002), skipped spawning (Engelhard and Heino 2005), straying (Ruzzante et al. 2006), irruptive range expansion and contraction (Corten 2001), and diversity in diel schooling behaviors and vertical distribu-

tions (Fernö et al. 1998). The list of categories remains incomplete, anticipating the discovery of new behaviors and controls. For now, a set of partial migration classes is introduced, with special attention given to how individual and group propensities give rise to partial migration. Readers should also refer to the taxonomic reviews of partial migration in fishes by Jonsson and Jonsson (1993), Kerr et al. (2009), and B. B. Chapman et al. (2012).

Nonnatal and Natal Divergence: The Food Availability Hypothesis

Recall that in aquatic food webs, a juvenile migrates and occupies nonnatal habitats to accommodate its need for larger-sized forage (chap. 4, Ontogenetic Habitat Shifts). Extending this argument to partial migration, we might expect that the propensity to migrate will increase with diminishing resources in the occupied habitat. Invoking resource availability leads to theories already covered—density-dependent habitat selection, making the best of a bad situation, and changed liability thresholds. Regardless of the specific explanation, there is a strong expectation that migration is favored when destination habitats are more productive than occupied ones (Baker 1978), a concept that Gross et al. (1988) termed the food availability hypothesis. It follows that partial migration occurs where the relative rank in food availability between natal and nonnatal habitats is not predictable or varies in some ordered way, say, with size, condition, dominance, sex, population density, or climate oscillations.

Food availability could account for year-round resident behaviors by contingents for species that are otherwise known to be highly migratory. Food availability likely underlies the resident behavior by Cashes Ledge red cod, where dense kelp stands harbor sufficient prey to support a local resident contingent. Resident behaviors of "kelpy" snapper *Pagrus auratus* in embayments of Australia and New Zealand might also be driven by seasonal food availability (Hamer et al. 2006; Parsons et al. 2011). Large bluefish persist year round in an isolated estuary in western South Africa, whereas the species is highly migratory off eastern North America where estuaries and bays are more numerous (fig. 4.5; Shepherd et al. 2006; Grothues and Able 2007; Hedger et al. 2010). Similarly, tagging and otolith chemistry evidence supports residency by blue marlin in the Gulf of Mexico (fig. 2.6) and by albacore tuna off Baja California, panmictic species that elsewhere are known to undertake transoceanic migrations (Wells et al. 2010; Childers et al. 2011; Kraus et al. 2011).

Boreal and temperate shelf fishes move from shallow inshore waters to deeper waters owing to seasonal changes in prey availability and temperature, but partial migration intrudes on this general pattern. In the mid-Atlantic United States, superoptimal temperatures cause winter flounder *Pseudopleuronectes americanus* to depart shallow coastal waters in summer (Able and Grothues 2007). In a generalized life cycle, adults spawn in bays and estuaries during winter months and then emigrate during summer as temperature warms. Recent telemetry studies show behaviors that run contrary to this expectation. Showing nonnatal divergence, a contingent of adults remains resident during summer in a shallow Long Island (New York, USA) embayment despite superoptimal temperatures (Sagarese and Frisk 2011). In contrast, this same species in the Gulf of Maine exhibited natal divergence, where not all prespawning fish remained in the small embayment where they were tagged. Rather, a contingent departed during winter and presumably spawned elsewhere in shelf waters (DeCelles and Cadrin 2011), notwithstanding the possibility of skipped spawning. Interestingly, partial migration behaviors by winter flounder were traditionally known but subsequently discounted by fishery scientists (Frisk et al. 2013). Another flatfish, Pacific halibut *Hippoglossus stenolepis*, was presumed to be highly migratory throughout shelf waters of the northeast Pacific, but telemetry uncovered a surprising mix of restricted resident movements together with longer-distance movements over a several-month period (Seitz et al. 2011), suggestive of nonnatal divergence (fig. 6.12).

Populations arrayed across productivity gradients provide circumstantial support for the food availability hypothesis. Resident ecotypes of Norwegian Arctic charr are more prevalent in southern productive watersheds or watersheds with large lake systems, whereas migratory ecotypes predominate in less productive boreal watersheds (Kristoffersen et al. 1994; Klemetsen et al. 2003). Other trends in partial migration occur for populations that span a large latitudinal range. Along the Sea of Japan across a latitudinal distance of 1,000 km, Morita and Nagasawa (2010) documented

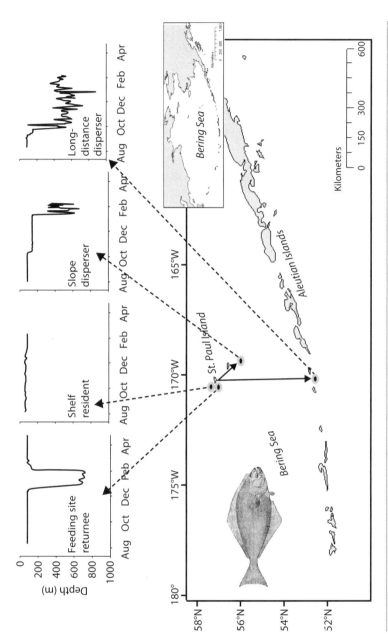

Figure 6.12. Vertical movements associated with migratory and resident behaviors by Pacific halibut, uncovered through satellite telemetry (Seitz et al. 2011). Tag recovery (pop-up) locations are indicated by black ellipses. Net movements are indicated by solid arrows. Mean fork length for nine halibut released at St. Paul Island was 121 cm. Redrawn from Seitz et al. 2011, with permission. Halibut illustration reprinted courtesy of the Alaska Fish and Game Department (Riley Woodford).

increased resident behaviors by masu salmon *Ocorhynchus masou* males in the southern range, presumably associated with increased prey resources. Temperate eel species (*A. anguilla, A. rostrata, A. japonica*) exhibit a gradient in partial migration with higher incidence of coastal and estuarine contingents at higher latitudes (Jessop et al. 2004; Daverat et al. 2006; Vélez-Espino and Koops 2010), although this pattern could be biased owing to less research effort at low latitudes.

Increased migration by invasive species is also driven by the interplay of partial migration and habitat productivity. Hendry et al. (2004a) postulated a positive feedback, where colonization into productive habitats leads to high juvenile production, which in turn causes increased density-dependent emigration and straying over subsequent generations. This type of feedback could explain the rapid development of sea-run salmon and trout in South America. Following its introduction in the 1930s into the Rio Grande River (Argentina), brown trout *Salmo trutta* rapidly expanded. Migratory and resident ecomorphs occur for this species just as in Arctic charr (smaller, pigmented, fusiform residents in comparison to migrants; Klemetsen et al. 2003). The resident form was introduced and persisted for several generations, but by the mid-1950s, anglers began reporting sea-run brown trout. During this period, resident trout depleted their forage base, which may have stimulated the expression of the migratory ecotype. In surveys of watersheds in the Rio Grande, resident trout only occurred where invertebrate biomass remained high (O'Neal and Stanford 2011). This shift toward increased migration represents a positive feedback between invasion and partial migration, where multigenerational change in migration propensity resulted from the introduced species' ecological impact—an example of ecological inheritance.

Generalizing further, freshwater and marine primary production levels vary in a coarse way with latitude, with higher relative ocean production in temperate environments and higher relative freshwater productivity in tropical ones. Might this difference in primary production and presumed forage availability result in directional selection for migration between freshwater and ocean habitats? Gross et al. (1988) noted the predominance of (1) anadromous species at high latitudes versus (2) catadromous species at low latitudes (fig. 6.13) and proposed that seaward (anadromy) and landward (catadromy) migrations resulted from past selection to exploit increased forage resources across the sea–freshwater transition. Conforming to this hypothesis, we have already noted that Arctic charr in boreal versus temperate watersheds migrate at a higher rate (Klemetsen et al. 2003). On the other hand, watershed attributes themselves may be more important (Näslund et al. 1993). For instance, emigration is curtailed in systems with less suitable growth habitats, where charr cannot attain sufficient condition or size to seasonally emigrate, or where distances and elevations between watersheds and marine environments are too great (Kristoffersen et al. 1994). Additionally, freshwater and marine transitions are not abrupt, and we find instances where large estuaries are much more productive than watershed habitats regardless of latitude (Houde and Rutherford 1993; Dodson et al. 2013). Finally, anadromy and catadromy are imperfect taxonomies, particularly catadromy, which specifies marine spawning and freshwater migration for feeding. For this class of migration, do estuaries comprise freshwater habitats? How long must an individual remain in freshwater habitats to be defined as catadromous? Need this be an obligate behavior for the entire population? McDowall (2008) proffered a well-reasoned alternative hypothesis for latitudinal trends in anadromy: that high-latitude colonization following glacial retreat by salmon and other anadromous species favored population divergence through niche expansion into boreal freshwater watersheds. Migratory behaviors were key to this divergence, not because of relative food availability but because boreal watersheds can become ecological traps

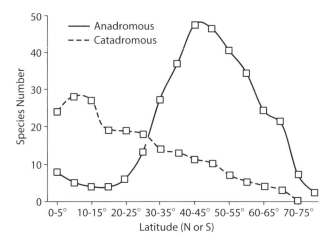

Figure 6.13. Latitudinal trends in species richness for fishes classified as anadromous or catadromous (Gross et al. 1988). Species with ranges >5° were repeatedly counted for each latitude interval where they occurred. Redrawn from Gross et al. 2008, with permission of the American Association for the Advancement of Science.

to year-round residents. In these watersheds, migration and partial migration are viewed as a type of seasonal evacuation (see below).

The provenance of anadromous species—freshwater or marine—has garnered substantial interest because it addresses the central question of whether migratory or nonmigratory behaviors came first. McDowall (2001) assembled phylogenetic evidence for anadromous salmon and, finding no compelling evidence for either origin, suggested that anadromy (migratory behavior) was too deeply buried in the taxa's ancestry to resolve the question (see also Quinn and Myers 2004). Borrowing from avian migration theory, the question of which came first entails a false dichotomy (Berthold 1996). Partial migration is likely ancestral to both resident and migratory behaviors in salmon and other fishes. Indeed, McDowall (2001) emphasized flexibility and reversibility as central attributes in the lineage of salmon and other anadromous taxa.

As an alternative example of the evolution of migratory behaviors within closely related taxa, consider temperate basses (Lateolabracidae and Moronidae), where partial migration

is likely related to divergence in nursery habitat. Juvenile fish are particularly variable in their patterns of nursery habitat use because of the linked requirements for forage and predation refuge (Beck et al. 2001). For many coastal species, juveniles utilize both estuarine and shallow ocean nurseries (Able 2005; Ray 2005). Japanese sea bass show attributes of partial migration, where shelf-spawned larvae use both coastal and estuarine nurseries (Islam et al. 2011). In the largest of Japan's estuaries, Ariake Sea, 50% of the recruits to the adult population used freshwater tidal nurseries as juveniles, a behavioral feat considering the retention mechanisms required to act against a 5-m tide. Adults in this system also exhibit resident behaviors (Secor et al. 1998). Elsewhere, Japanese sea bass complete their entire life cycles in marine water (Islam et al. 2011). Although not within the same taxonomic family, the European sea bass exhibits similar features of partial migration, including juvenile ingress and subsequent resident behaviors by juveniles and adults in England's Thames Estuary (Pawson et al. 1987). This species occurs within the same family (Moronidae) as anadromous and freshwater temperate basses of North America. Secor (2002b) proposed a scenario of radiation within this family, mediated through partial migration where ancestral use of coastal nurseries shifted toward estuaries (*D. labrax*), that led to estuarine spawning in the large and relatively stable estuaries of North America (*Morone saxatilis*, *M. americana*) followed by colonization into fully freshwater ecosystems (*M. mississippiensis* and *M. chrysops*). The molecular phylogeny of Moronidae supports this scenario of serial radiation via partial migration (Williams et al. 2012).

Differential Migration

The strong influence of body size on reproduction, physiology, and behavior can result in differential migration where the propensity to migrate varies with size and sex (Berthold 1984; Klimley 1987; Hendry et al. 2004b). Size-dependent migration behaviors result in the same individual adopting different spatial behaviors over its life span rather than adopting a single behavior, as would occur through the threshold model for partial migration. In marine fishes, swimming performance, foraging success, and predation risk change most rapidly during the larval and juvenile periods. Thus differential migration is common: cohorts of fish leave natal food webs and enter new ones in a size-dependent manner. Similarly, for older juveniles and adults, increased size will reduce the cost of movement (fig. 2.1), and larger fish often range more widely. This is seen in many shelf fishes—cods, flatfishes, snappers, drums (Rose 1993; Jones and Wells 2001; Russell and McDougall 2005; Dicken et al. 2007; Hattori et al. 2010; Jorgensen et al. 2010). Coastal sharks such as lemon sharks, bull sharks *Carcharhinus leucas*, and angel sharks *Squatina dumeril* also show increased ranges with size (reviewed by Speed et al. 2010). Still, in other instances, juveniles undertake the longest migrations. Transoceanic migrations by Atlantic and Pacific bluefin tuna occur more frequently for juveniles than adults (fig. 6.14A; Itoh et al. 2003; Rooker et al. 2008). This behavior could be akin to

that proposed for birds, where larger individuals are better able to exploit a narrower set of conditions or endure seasonal hardship (B. B. Chapman et al. 2011b). Large bluefin tuna can exploit colder waters, where forage resources are more concentrated, whereas smaller individuals range more widely in lower latitudes in search of more dispersed resources (fig. 6.14B; Polovina 1996; Fromentin and Powers 2005). Tiger *Galeocerdo cuvier* and white sharks similarly range more widely as juveniles than adults, undertaking exploratory foraging forays and avoiding cannibalism (Goldman and Anderson 1999; Meyer et al. 2009).

Sex and size interact in the propensity to migrate. As reviewed in chapter 3 (Sperm Competition), sexual dimorphism is common in fishes but varies according to mating systems. In broadcast spawning fish, larger females enjoy increased fecundity. Sexual dimorphism is also seen in females that devote large amounts of energy to individual offspring, such as live-bearing sharks. In fishes that control reproductive access and provide parental care, large size by one or the other sex favors territorial defense. Differential migration often attends these dimorphisms. In comparison to males, female striped bass mature later, attain larger sizes, and show an increased tendency to undertake extensive coastal migrations (Secor and Piccoli 2007). From the standpoint of reproductive success, striped bass females migrate to support larger adult sizes and greater fecundity, whereas fertilization success by resident and migratory males is less dependent on adult size. Similarly, immature female scalloped hammerhead sharks *Sphyrna lewini* migrate farther offshore in pursuit of energy-rich pelagic prey to grow faster and reach larger sizes than males. Following maturation they migrate to support gestation of large offspring (Klimley 1987). Opposed sex-biased migrations occur in connected marine and freshwater habitats in Australia by barramundi and tupong *Pseudophritis urvillii* (demersal predators related to Antarctic icefish). In barramundi, males differentially invade freshwater habitats (McCulloch et al.

2005), whereas tupong females occur seasonal-ly in rivers (Crook et al. 2010). In both species, spawning takes place in coastal waters, indi-cating nonnatal partial migration.

Courtship in sharks is not without hazard, often resulting in bite wounds to females. So-called sexual harassment can lead to differen-tial migration and segregation of sexes as re-ported for mako *Isurus oxyrinchus* and white sharks, catshark *Scyliorhinus canicula*, and spiny dogfish (Mucientes et al. 2009; Jorgensen et al. 2012b; Wearmouth et al. 2012). In sharks that give live birth, fidelity to certain pupping habitats causes females to show more limited ranges than males, resulting in sex-biased mi-grations (Wearmouth and Sims 2008; Vianna et al. 2013).

Sex-biased migrations can arise from re-source competition and territoriality (Johnson and Gaines 1990). Such behaviors are common in birds, where males remain resident and over-winter on nesting grounds more frequently than females (Lack 1944; Belthoff and Gauthreaux 1991). In a fish named for a bird, the cardinal fish *Apogon notatus*, females control nearshore boul-der areas as spawning habitat. Offshore winter migration requires reacquisition of spawning territories each year, but recent research de-tected a resident contingent that never leaves breeding territories (Fukumori et al. 2008). In a similar vein, small resident males in many salmon species remain proximate to spawning tributaries (chap. 3, Five Mating Systems). Dif-ferential migration is seen in male plainfin mid-shipman *Porichthys notatus* (toadfishes, family Batrachoididae), which hold nest territories in a size-dependent fashion (DeMartini 1988). Body size also plays a role in the amount of time a fish will remain resident, guarding offspring prior to moving on and seeking new mating opportu-nities (Gross 2005).

Explanations for differential migration in-clude (1) the arrival time hypothesis, where males remain resident or take shorter mi-grations so they are first to arrive and hold territories during the next breeding season; (2) the dominance hypothesis, where compe-tition drives seasonal migrations by smaller

individuals; and (3) the endurance hypothesis (a.k.a. the body size hypothesis), where larger individuals can endure year-round exposure to seasonally suboptimal habitats. Because these hypotheses are not mutually exclusive, it has been difficult to garner support for a leading theory (Bell 2005).

Skipped spawning has been reviewed pre-viously (chap. 3, Provisioning Eggs) but con-stitutes a type of differential migration be-cause only a portion of adults will undertake a spawning migration in a given year. Examples include year-round residency in feeding habi-tats for a portion of orange roughy *Hoplostethus atlanticus* that do not spawn in a given year, and the biennial spawning migrations of white and dusky sharks *Carcharinus obscurus* (Bell et al. 1992; Bonfil et al. 2005; see Secor 2008 for other examples).

Straying also varies by size and sex within populations. In theory, increased dispersal and straying by males are favored in polygamous mating systems, common in marine fishes, as a means to increase fertilization opportunities (Johnson and Gaines 1990). Thus males often predominate among strays in salmon and oth-er coastal species (Hutchings and Gerber 2002; Cano et al. 2008). Here again, classifications of partial migration overlap: differential migra-tion with natal divergence.

A broader definition of differential migra-tion considers not only seasonal migration pat-terns but also variation in timing and distances of those migrations (Brodersen et al. 2011). An illuminating example is US Atlantic bluefish, which each year produce early and late season-al cohorts of juveniles, the former preferential-ly migrating from shelf waters into estuarine nurseries (see chap. 4, fig. 4.5). Munch and Con-over (2000) showed that the relative propor-tion of the two cohorts forecasted subsequent population dynamics and fishery yields.

Changed Minds: Switching

In moderately long-lived species including eels, Arctic charr, striped bass, white perch, and barramundi, otolith profile analysis reveals that individuals can make midlife "correc-

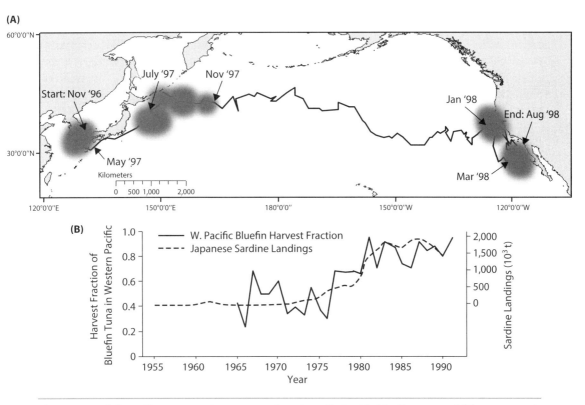

Figure 6.14. (A) Rapid transoceanic movement by a juvenile Pacific bluefin tuna and (B) transoceanic oscillations in Pacific bluefin tuna commercial harvests driven by prey availability. (A) A bluefin tuna was released at 55-cm fork length (age 1) off Kyushu, Japan, in November 1996 and then at age 2 (November 1997) undertook a 2-month transoceanic journey. Regions of sustained incidence are shown in gray. (B) The fraction of the entire Pacific Ocean harvest of bluefin tuna taken from the western Pacific is shown in comparison to Japanese sardine landings. Note that the spatial shift in Pacific bluefin tuna toward the western Pacific in the 1970s was concurrent with the period of irruptive growth by Japanese sardines (chap. 4, Schooling through Food Webs). (A) redrawn from Itoh et al. 2003; (B) redrawn from Polovina 1996, with permission.

tions," opting out of one migratory lifestyle and adopting another (Radtke et al. 1996; Secor 1999; Tsukamoto and Arai 2001; Zlokovitz et al. 2003; McCulloch et al. 2005; Kerr et al. 2009). Such abrupt changes in individuals, absent group behaviors such as evacuations (see below), often defy easy explanation but are clear demonstrations of the conditional nature of movement propensities. Examples might include a long-lived slow-growing resident charr that finally attains a size that permits migration; schools of adult herring that are overwhelmed by numerically dominant schools of juveniles, causing

them to adjust their spatial behavior; or an energetically exhausted sturgeon after spawning adjusting its seasonal migration pattern over a several-year period. Longevity provides greater opportunity for switching, but switched behaviors can occur for juveniles as well. From freshwater natal habitats, some Hudson River striped bass juveniles invade nearly full-strength marine waters during their first year of life and then reinvade freshwater, residing there for the remainder of their life spans (Secor and Piccoli 1996). Similar to differential migration, switching causes the same individual to sample environments differently over its life cycle. Hudson River striped bass that switched behaviors midlife showed intermediate levels of polychlorinated biphenyl contamination in comparison to resident (high contamination) and migratory (low contamination) contingents (fig. 6.15; Zlokovitz et al. 2003).

Switching refers to longer-term unusual behaviors, which have lifetime consequences distinguishing them from more frequent ranging behaviors. In some instances, otolith chemistry has indicated hypervariable spatial behaviors, perhaps the result of frequent lifetime switches. For instance, in Japanese eel, Tsukamoto and Arai (2001) observed >20 lifetime switching behaviors between marine, brackish, and freshwater habitats. Still, small changes in salinity can result in nonlinear increases in otolith strontium (Kraus and Secor 2004a), lead-

ing to differences that could be due to ranging or dynamic habitat conditions rather than changes in lifetime movement patterns. As an example, consider the Gironde Estuary, where otolith strontium profile analysis showed that European eels, thin-lipped mullet *Liza ramada*, and European flounder frequently switched occupancy between freshwater and estuarine waters (Daverat et al. 2011). Depending on sampling location, such apparent switching could be observed for a fish with a limited home range, through exposure to daily and weekly excursions of the tidal salt wedge.

Evacuations, Irruptions, and Straying

Evacuees depart as a result of degradation or catastrophic loss of habitats, or through rapid increases in population density (Kaitala et al. 1993). As Thorpe (1994) notes, "When the scale of

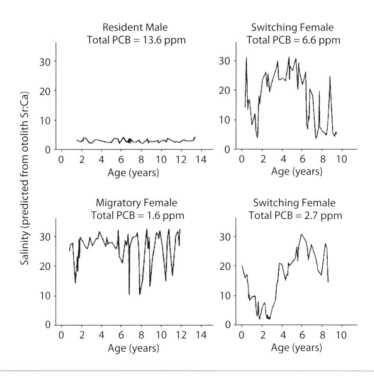

Figure 6.15. Influence of partial migration on levels of polychlorinated biphenyl (PCB) body burdens in Hudson River striped bass. The Hudson River has a legacy freshwater source of PCBs. Individuals exhibiting switching behaviors showed intermediate levels of total PCB in comparison to resident and migratory individuals. Redrawn from Zlokovitz et al. 2003, with permission.

the environmental change exceeds the animal's capacity to respond in situ, the general biological response to adversity—migration—comes into play." Evacuation is considered as a collective behavior: if the local conditions change abruptly, we might predict that a partially migratory population would respond through (1) increased representation of the migratory contingent or (2) complete evacuation. In the extreme, an individual has no choice. In the year of the Mount St. Helens volcanic eruption, for example, large numbers of returning salmon were deterred from natal tributaries and strayed into neighboring unaffected streams (Quinn and Fresh 1984; Leider 1989). Density-dependent dispersal can be considered as another type of evacuation (chap. 5, Homogenized Life Cycles: The Basin Model).

In marine fishes, seasonal evacuations are prevalent. Colder temperatures and diminished food availability in winter are associated with widespread evacuation of rivers, estuaries, and shallow shelf waters for many boreal and temperate shelf species (Able and Fahay 1998, 2010). But in many populations these departures are incomplete, and a substantial fraction of the population remains behind to endure suboptimal conditions. In other instances, evacuation occurs during summer months as a result of high temperatures or other stressful conditions such as hypoxia. Again, some individuals remain behind to endure these conditions (Wingate et al. 2009; Sagarese and Frisk 2011).

Just as partial migration leads to irruptive range expansion in a colonizing species, so too can it permit escape from ecological catastrophe. The Mount St. Helens eruption of 1980 expelled megatons of ash, silt, and debris into the Columbia River watershed, suffocating entire spawning tributaries of salmon and disrupting biotic and abiotic conditions in the lower Columbia River. Salmon nevertheless persisted (Leider 1989), buffered against the single year's catastrophe by partial migration (incidence of an exploratory contingent) and the storage effect (multiple cohorts not all engaged in the same spawning run; Thorpe 1994; Secor 2007).

Similarly, white sturgeon *Acipenser transmontanus*, which are typically resident to the lower Columbia River, were observed evacuating into shelf waters subsequent to the 1980 eruption (DeVore et al. 1999). Large storm events have also been attributed to evacuation. Black rockfish typically show strong site fidelity (<1 km^2), but a minority (10%–40%) departed home ranges to areas tens to hundreds of kilometers distant, perhaps in response to winter storms (Green and Starr 2011). Coastal sharks such as blacktip *Carcharhinus limbatus* might also depart in response to imminent storms (Heupel et al. 2003).

Massive shifts in the distribution of shelf species conceivably represent evacuations, although the specific stimulus is often a mystery. As exemplified by the decadal displacement of winter skate from US to Canadian shelf waters (reviewed in chap. 5, Bounded Fisheries), evacuation versus diminished abundance as causative agents are hard to disentangle. Atlantic mackerel *Scomber scombrus* exhibit extreme oscillations in abundance at the northernmost portion of their range in Icelandic shelf waters, with low or nil incidences during periods of cold ocean temperatures (Astthorsson et al. 2012). More subtle shifts in range coincide with recent warming in the northwest Atlantic (Overholtz et al. 2011), suggesting that the Icelandic stock fluctuations may be driven largely by temperature-dependent range dynamics. Large regional displacements have been noted for skipjack tuna related to El Niño (Lehodey et al. 1997), which in a manner similar to winter skate could be misinterpreted as diminished abundance, depending on regional management units (Myers and Worm 2003; Sibert and Hampton 2003). Such relocations are likely reinforced through social interactions as a result of schooling (chap. 2, Schooling and Flocking), possibly overriding regional changes in habitat suitability.

Rare but spectacular long-range movements in species not known to be highly migratory represent a type of evacuation. Spiny dogfish is a small (<100 cm), long-lived shark (80 y) that supports important fisheries throughout its

Northern Hemisphere distribution. In a 10-y study, 71,000 fish were tagged from British Columbia (Canada) fisheries, and recaptures of ~3,000 individuals confirmed strong homing (>95%, uncorrected for effort and mortality) to shelf regions where they were tagged (McFarlane and King 2003). On the other hand, 1% of the recaptures (N = 30) occurred in Japanese waters, >6,000 km distant. Rare transatlantic migrations by spiny dogfish have also been noted. Similarly, juvenile goliath groupers *Epinephelus itajara* typically make short migrations from mangrove nurseries to adult reef habitats <20 km distant, but a small fraction undertake migrations >200 km (Koenig et al. 2011). We have previously reviewed vagrants and navigation errors (chap. 2, Losing Their Way; chap. 5, Roamers) but here consider the fate of these neighborhood quitters: over many generations the potential for genetic interchange and metapopulation persistence supports the view that evacuation and colonization are different sides of the same coin.

Not all individuals who leave for greener pastures find them. Large black drum that spawn within the extreme northern part of their range do so in low numbers, and offspring are unlikely to survive (Cowan et al. 1992; Jones and Wells 2001). The sudden disappearance of contingents of large bluefin tuna in extreme parts of the species range is well documented (Crockford 1997; Fromentin and Powers 2005) and is representative of the risk of far-flung migrations.

When evacuation routes are cut off, ecological traps can result. Barramundi and temperate eels that ascend into inland habitats may become trapped by dewatering or by their inability to navigate a return course (Haro et al. 2000; Walther et al. 2011). Dams have led to self-sustaining resident "landlocked" populations of charr, salmon, river herring *Alosa* spp., striped bass, and sturgeon. Persistence by these local populations in landlocked freshwater ecosystems may have occurred owing to rapid divergence of latent ecotypes, similar to that described above for threespine stickleback.

In New England (USA), colonial European settlers erected thousands of milldams (Waldman 2013), resulting in numerous landlocked populations of alewife. Assimilated genetic variance in these populations likely allowed subsequent divergence of resident populations from ancestral anadromous ones (Palkovacs et al. 2008). Similarly, landlocked white-spotted charr *Salvelinus leucomaenis* adopted fully resident life cycles following dam construction on a Hokkaido river (Morita et al. 2000). In this way, partial migration in alewife and charr—reflecting an environmental legacy of natural impoundment due to glaciation, barrier island migration, and other landform changes—preconditioned the species to be resilient to dams. Still, these isolated resident subpopulations are vulnerable to extirpation, exemplified by landlocked Kootenai white sturgeon (Paragamian and Hansen 2008) and Formosa salmon *Oncorhynchus masou formosanus* (Healey et al. 2001). Another example of resilience of landlocked ecotypes includes the tropical bigmouth sleeper *Gobiomorus dormitor*, a demersal predator distributed in freshwater, estuarine, and coastal environments that is thought to reproduce principally in marine waters. Discovery of reproduction in impounded reservoirs in Puerto Rico, where they became landlocked, and crater lakes in Nicaragua, where they were introduced, suggests that partial migration preconditioned the species to exploit novel environmental settings through adoption of a fully freshwater life cycle (Bedarf et al. 2001; Bacheler et al. 2004).

Shelf species sometimes disperse or migrate to regions from which they cannot escape. Tropical and subtropical species can disperse and migrate to temperate waters during summer via major coastal currents (e.g., the Gulf Stream, Kuroshio Current, and East Australia Current) and then lack the physiological capacity to return (McBride and Able 1998). For instance, the Gulf Stream sweeps juvenile butterflyfishes *Chaetodon* sp. into temperate northern estuaries in New Jersey and New York, where they show high site fidelity through fall months as temperatures drop to sublethal temperatures (see also chap. 5, Fish of a Different Feather). Through a complement of field and laboratory experiments, McBride and Able (1998) concluded that such individuals constituted doomed expatriates, unable to return. Other such tropical species that produce juveniles common to temperate waters include filefish *Monocanthus hispidus*, Florida pompano *Trachinotus carolinus*, and crevalle jacks *Caranx hippos*.

Amphidromous Evacuations

Evacuation during the larval period is epitomized by a diverse collection of small-bodied fishes that with every generation depart and reinvade the small streams of volcanic islands. McDowall (2007) reserved the term *amphidromous* for this group, which includes taxa (60 species across a half dozen families, including Retropinnidae, Galaxiidae, Cottidae, Gobiidae, Eleotridae, and Pinguipedidae) that reproduce predominantly in island watersheds; disperse into marine habitats as larvae, often persisting there for many months; and then reinvade island streams as late stage larvae or juveniles. In New Zealand, millions of juveniles school upstream from coastal waters each spring attracting "white bait" artisanal fisheries that fetch high prices (>\$100 kg^{-1}; McDowall 2007). Other upstream migrations cause amphidromous gobies to buck up against extreme flows and elevations, climbing cliff faces behind island waterfalls hundreds of meters high. Like pelagic-spawning reef fishes, the evacuation of amphidromous larvae into pelagic ocean environments leads to varying outcomes in terms of population connectivity. But with typically longer larval durations, amphidromous species trend toward more open populations than do many reef species, although circulation features surrounding islands can sometimes favor retention and self-recruitment (Swearer et al. 1999; Murphy and Cowan 2007).

Island watersheds are ephemeral owing to dewatering, volcanism, and high-flow events —conditions that are hypothesized to drive wholesale evacuations during the larval period (McDowall 2010). Still, fully freshwater species

often occur sympatrically with amphidromous relatives, and, similar to the freshwater radiation suggested above for Moronidae (Williams et al. 2012), McDowall (2004) favored the view that freshwater species on volcanic islands radiated from amphidromous beginnings. Partial migration with fully resident freshwater contingents has been noted for this group, particularly for larger watersheds in New Zealand and Japan. In freshwater gobies *Rhinogobius* sp. distributed throughout Japan, Tsunagawa and Arai (2009) detected facultative amphidromy, where a fraction of larvae remained in freshwater. In some but not all instances, downstream barriers artificially induced this behavior. Partial migration may be less commonly expressed in small and more ephemeral island watersheds, those that do not favor larval retention (Radtke and Kinzie 1996).

Vertical Migration and Other "Ranging" Behaviors

Individual exploratory behaviors when frequently recurrent and related to short-term ends are called *ranging behaviors*. Constancy in such movements can be termed commuting (Dingle 1996) as is well known for reef fishes, which occupy different habitats on a daily basis for feeding and refuge (Annese and Kingsford 2005; Meyer and Holland 2005; Parker et al. 2008; Green and Starr 2011). Diel vertical movements, common in schooling pelagic fishes, are variable and likely constitute ranging behaviors. For zooplanktivorous sharks such as basking shark and megamouth shark *Megachasma pelagios*, diel vertical migrations can be quite rhythmic (Nelson et al. 1997; Sims et al. 2005). As reviewed previously for fish larvae, vertical behaviors can be critical to longer-range movements and migrations. Sharks are sensitive to seabed geomagnetic fields that vary with depth, leading Klimley (1993) to suggest that some migratory sharks undergo recurrent diving behaviors to navigate. Similarly, recurrent ascents could play a role in celestial navigation in salmon, tuna, and other fishes (Friedland et al. 2001; Willis et al. 2009). During their ocean migrations to the Sargasso Sea, European eels undertake vertical migrations, avoiding predators during the daytime by diving to deeper darker waters and then returning at night to warmer euphotic zone waters, where temperatures improve muscle efficiencies (Righton et al. 2012). Tunas and sharks in open ocean environments are known to undertake frequent deep diving behaviors (>200 m), the distances of which would exceed horizontal movements on a seasonal or yearly basis.

Vertical migration behaviors, persisting over seasons and years or changing with ontogeny, can make important contributions to longer-term contingent behaviors within populations (fig. 6.12; Mehner and Kasprzak 2011; B. B. Chapman et al. 2012). A good example is vertical migrations that transport fish via tidal currents over large horizontal distances. For Japanese sea bass, differential vertical positioning over a tidal cycle causes some juveniles to move up the Chikugo Estuary and others to be retained in coastal areas (Islam et al. 2011). A similar role for partial vertical migration causes the juveniles of Japanese flounder (Tanaka et al. 1989) and American eels (McCleave and Kleckner 1982) to either ascend estuaries or retain their position in coastal habitats. During spawning migrations, adult plaice make use of transporting tides, spending up to one-third of their time in the water column, entering midcolumn water during slack tide and maintaining their position riding the downstream tide; during tides opposed to directional migrations, plaice remain on the seabed (Rijnsdorp et al. 1985; Hunter et al. 2004; Metcalfe 2006). Nonmigratory plaice orient to the seabed throughout the tidal cycle and do not exhibit rhythmic vertical behaviors.

Might resident and migratory groups exhibit persistent differences in vertical migration behaviors? As reviewed previously, diel diving behaviors of whale sharks and other species may represent a form of station keeping (chap. 2, Gliding Fishes; Parker et al. 2008), but these behaviors also play a role in navigation and orientation during migrations. For North Sea plaice,

Hunter et al. (2004) deployed a large number of electronic tags and discovered nonnatal partial migration: three separate summer feeding aggregations in broad regions of the North Sea, all originating from a single southern spawning ground. Interestingly, only two summer aggregations occurred along strong tidal streams. The third northern aggregation occurred in stratified waters that lacked strong-amplitude currents for use in physical transport and guidance, and must have used other movement behaviors in migrating to southern spawning grounds (Metcalfe 2006). It seems feasible that partial migration was underlain by separate vertical behaviors, two of three contingents exploiting tidal stream transport. Whether vertical migration is a type of partial migration, as suggested by B. B. Chapman et al. (2012), remains an open question. Early evaluation of this question for marine fishes indicates that because vertical behaviors lead to multiple ecological outcomes, their contribution to seasonal and lifetime movements will be difficult to uncover (e.g., Hobson et al. 2009; Walli et al. 2009; Jorgensen et al. 2012a). Still, vertical ranging (and commuting) behavior(s) represents an important topic of research, particularly in examining the role of individual agencies (navigation, foraging, predation risk, transport) in contingent behaviors.

Summary

Partial migration, the simultaneous occurrence of resident and migratory members within a population, is a central means by which fishes respond to the exigencies of surviving, growing, and reproducing in nonstationary and patchy marine ecosystems. Harden Jones's (1968) "loose-end" of what happens to individual or groups of fish that break away from the migration triangle (chap. 1) is resolved when we consider that individuals within populations are members of multiple migration triangles (life cycles), some of which are open to emigration. Partial migration starts with an individual's propensity to stay put or move, and ends with collective outcomes: homing, range expansion, colonization, population and meta-population structure, and speciation.

Partial migration is ubiquitous among plants, insects, birds, and mammals but until recently has been emphasized in only a few marine fishes. Partial migration likely plays a central role in the evolution of migration and complex life cycles among most fishes. Through focused treatments on Arctic charr, Atlantic salmon, threespine stickleback, European eel, white perch, and Atlantic cod, we saw modes of phenotypic responses leading individuals to be resident or migratory over major phases of their life. These species exhibited migration and life history types that recurred time and again across numerous populations and ecosystems, implicating a latent genetic architecture for migratory behaviors.

The threshold model for partial migration, borrowed from the avian literature, postulates that early life thresholds in a liability trait (e.g., growth, condition, social status) carries over to lifetime migration behaviors. Because the threshold response is bimodal (or sometimes multimodal), large amounts of genetic variance are expressed only as a single outcome, resulting in genetic assimilation. Genetic assimilation in turn preserves the capacity to adopt migratory or resident behaviors should environments change. The threshold itself can change through selection on the norm of reaction (a.k.a. genetic accommodation).

Important ecological consequences of partial migration include niche partitioning (Arctic charr and Atlantic cod ecomorphs), compensating for a poor start (slow-growing white perch larvae and Arctic charr juveniles), capitalizing on a good start (early migrating Atlantic salmon and river-ascending ayu), reproductive success among male ecotypes (hooknose and jack coho salmon), and colonization success (recurring colonization by threespine stickleback ecomorphs and brown trout invasion into the South Atlantic Ocean). Further, partial migration likely underlies speciation in some marine fishes, as exemplified in the molecular lineage of temperate eel and sea bass families. Functional molecular markers that differentiate populations and contingents on the basis of selection regimes (e.g., exposure of Atlantic cod to <0°C) hold great promise in further linking the ecological outcomes of partial migration with systems of genetic control. Biocomplexity, the mix of genetic and life history divergence within populations, will likely result in the next generation of transformative discoveries in migration ecology.

Partial migration "writ large" concedes that there is unlikely a single cause, genetic or otherwise, that captures the diversity of fish migrations. Because we remain in an era of discovery, caution should be exercised in classifying types of partial migration and associated causes. Still, borrowing from avian ecology, we can accommodate a large range of contingent behaviors into classes of partial migration, including nonnatal divergence (classic resident and migratory contingents), natal divergence (mixing between populations in certain habitats), differential migration (principally age- and sex-dependent migrations), switching, skipped spawning, straying, irruptions, and vertical migration (each contingent exhibits different vertical behaviors). These categories of behavior have been linked to myriad causes, including food availability, competition, differences in growth between males and females, mating opportunities, sexual harassment, increased swimming performance and roaming with size, increased tolerance of suboptimal environmental conditions with size, changed reproductive investments with size, exposure to novel food webs and ecosystems, catastrophic ecosystem changes, threshold changes in food webs and ecosystems, physical evacuation of larvae (amphidromy), land-locking, and navigation and physical transport related to vertical behaviors. These explanations overlap broadly, causing partial migration to be adjusted, altered, interrupted, amplified, and stabilized in countless ways as groups of fish respond conditionally to their ecological and social circumstances.

Segue

In the early 20th century, Hjort working on Norwegian herring and Gilbert on sockeye salmon simultaneously arrived on the term *contingents* to describe groups of fish within populations that exhibited similar seasonal migration patterns (Secor 1999). Although its intended usage was likely martial (i.e., a group of soldiers), the term coincidentally applies to the contingencies of marine life. Seascapes are anything but constant, resulting in no optimum life history type, no single migration pattern, no population that is exclusively open or closed. Responses at the collective scale, even under relatively stable conditions, are diverse with evidence of hyperdispersion, modality, and error. Consider, for instance, (1) the oversupply of larvae, (2) the production of minority resident behaviors during generations where migrants experience higher growth and replacement rates, and (3) the full evacuation of amphidromous offspring into marine dispersive environments. These collective outcomes may seem manifestly wasteful and maladaptive, but they represent contingency plans that contribute to population growth, stability, and yield over generations.

Recapitulation: Scaling Collective Movements from Fertilization to Speciation

Migration Is Not Just about Abundance

Natural historians have long viewed migration as a response to abundance, a means to escape the constraints of local resources and seek them elsewhere. The birds and fish of two worlds migrate seasonally because temperate latitudes fail to sustain abundances year round. Marine fishes often expand their ranges during periods of high abundance. In *Fish Migration*, Harden Jones (1968) generalized migration as an adaptation to abundance. But he went further—rather than demonstrating how distributions simply responded to density, he showed how migrations were collectively organized in dispersive environments as closed life cycles. The dispersal of young, the active selection of nursery and foraging habitats, and the return migrations by adults to spawn in particular locations were agencies by which population numbers were distributed across marine food webs. During the 47 y since *Fish Migration* was published, improved measures of fish movements have resulted in an appreciable increase in types of collective agencies governing migration. The generalization that migration is an adaptation to abundance is appealing in its simplicity but too broad.

Migration Closes the Deal

Population thinking dominates views on the ecology of marine fishes, arguably more so than other vertebrates. This perspective has been driven by the faithful itineraries of salmon; the evidence of spawning fidelity in cod, herring, eels, and North Sea plaice; the genetic markers of philopatry; the oceanographic retention and transport of larvae; and finally the age of telemetry, which continues to show evidence for closed populations. Migration by adults offsets the dispersal of gametes and larvae and the different habitat needs of juveniles, causing spawners to reset the life cycle generation after generation by inextricably linking mating systems with migration. For sharks and other marine fishes with nondispersive young, reproduction depends on migrations to specific regions for courtship and nurseries, just as they do for birds. In his *Essays on Marine Populations*, Sinclair (1988) advances the theory that marine ecosystems are not simply dispersive environments but instead contain regions of closed circulation, which serve to retain larvae and to structure mating systems and life cycles of marine organisms. Migration and spawning behaviors by adults thus close the deal by causing otherwise dispersive young to arrive into these retention features. Retention features by their frequency and extent define the number and size of populations. But as we look more carefully at larval dispersal and juvenile and adult migrations, we find they often do not follow the straight and narrow pathway required for life cycle closure. The marine environment is highly dispersive, exacting large losses to closed populations owing to dispersive young, lost juveniles, and roaming adults. Such losses could be conceived of ecological waste, but another view is that this is ecological overhead, resulting in exploration, colonization, and insurance as emergent features, which are occasionally adaptive in nonstationary marine ecosystems.

Individual Propensities

The individual movements by marine fishes are constrained by vertebrate design, development and size, energetics and physiology, mode of personality, and navigation capabilities. In Nathan et al.'s (2008) "A Movement Ecology Paradigm," movements result from external drivers acting upon the internal state of the organism. In his *Migration: The Biology of Life on the Move*, Dingle (1996) proposed that seasonal movements were toggled on and off according to environmental and internal drivers. These ideas align well with the recent focus on individual-based ecology, driven by new empirical approaches and modeling capacity. A focus on the individual is intuitively attractive, observing the manifest purposefulness of directed migrations in so many marine fishes, birds, and other vertebrates. Fish migrate long distances each year between habitats that are often separated by equally suitable habitats, yet efforts to divert them—say, through physical displacement—most often fail.

Although the movement ecology paradigm is a tenable causal framework linking external and individual states to observed movements, it is insufficient once we move beyond the individual, where collective agencies quickly take hold. As gametes, embryos, and small larvae are buffeted by simple diffusion kinetics, the behavior of a single offspring is noninformative. We must track thousands and millions of individual larvae as they are both individually and collectively influenced by density, turbulence, and geostrophic flow in their marine settings. So-called individual-based models of larval dispersal fate are thus designed to discover collective rather than individual outcomes. Reductio ad absurdum, fertilization is a collective outcome depending on high spermiation, alternate mating styles, and diverse outcomes for individual spermatozoa. How individual fish move respective to each other in shoals and schools depends on individual proximity rules, but these rules can only be understood by their aggregate interactions and outcome. When diversity is imposed on individual behavior—through personality modes, for example—nonlinear outcomes result, such as changes in school conformation, directed migrations, enhanced navigation, adopted migration, and straying. In many instances, understanding individual internal states and capacities will be an important starting point, but individual movement propensities are arguably inseparable from collective agencies that propagate into migration and life cycle outcomes.

What Are Collective Agencies?

If individual movements are not uniform, then it logically follows that we must consider collective agencies. As we sum individual behaviors, statistical properties quickly emerge that translate the diversity of individual properties into group propensities—skewness, stochasticity, bias, error, hyperdispersion (oversampling), and modality.

When individuals interact in certain ways, other statistical properties emerge—positive and negative feedbacks; threshold responses and path dependency; redundancy, synchrony, and stability. Understandably, some may be discomfited by explanations that are statistical in nature, rather than ones reliant on a specific set of processes. But regardless of whether classified as inductive or deductive science, empirical and modeling studies have both arrived at the same conclusion: fish movements are not uniform within schools, contingents, populations, and metapopulations. Rather, movement behaviors are complex and contribute to important ecological outcomes that cannot be understood by scaling up from an average individual. Of course, this same argument for emergent properties has occurred in other disciplines, such as molecular ecology, physiology, evolution, and human ecology (Gould 2002; Sumpter 2006).

Because statistical attributes are often viewed in the context of underlying processes, the idea that these attributes themselves reflect agency might strike one as odd at first blush. But agency stipulates operation (propensities), not cause. Collective agency refers to the multiple contributions of many individuals at some aggregate level. Structural agency references repeatable patterns in aggregate properties. Collective and structural agency need not be divorced from process-oriented studies. In studies on larval dispersal, schooling behaviors, and the threshold model for conditional migrations, we see excellent examples of scientists embracing multiple levels of inference from laboratory experiments, field studies, and simulation modeling. In this book, ecological outcomes that were associated with collective agency included mating systems, embryo and larval dispersal, schooling, patchiness, hyperdispersion, competition, conservatism, straying, redundancy, carryover, and error. Higher-order emergent properties that derive from migration as a collective behavior include population dynamics, partial migration, population structure, the evolution of complex life cycles, range dynamics, ecological overhead, and stability and resilience (chap. 1, fig. 1.4).

How audacious is it that we attempt to manage free-ranging, wild-living species over such broad spatiotemporal scales?
—Link (2010)

Problems come into view when considering and ecology of possibilities versus the ecology of optima.
—Cury (1994)

Resilience Theory

Resilience theory, which specifies sudden shifts between alternative regimes, applies to marine fish populations that undergo threshold changes in their abundance. The concept of sustainable yield, the mainstay of traditional fisheries management, depends on yield levels that are alternatively controlled by positive feedback and negative feedback according to the logistic growth model (fig. 7.1A; Beisner et al. 2003; see also chap. 4, Life Cycle Schedules). At low population size, birth rates exceed death rates, and the population experiences exponential growth (positive feedback). At some intermediate population size, density-dependent forces (negative feedback) come to bear, and production declines as the population approaches carrying capacity. The production curve exhibits two equilibria where yield is nil: when the population is at carrying capacity (stable equilibrium) and when the population is extirpated (unstable equilibrium; fig. 7.1B). The goal of fisheries management is to hold the stock at an intermediate equilibrium population state where yield is highest—the maximum sustainable yield (MSY). In this example, population states (abundance levels) respond continuously to harvest regulations, with important limiting assumptions related to constant carrying capacity and the form of density dependence.

Threshold shifts in abundance and landings are common in marine fisheries, particularly apparent for stocks that have undergone a cycle of fisheries development and overexploitation, such as the western stock Atlantic bluefin tuna (fig. 7.2). Still, the ability of assessment scientists to parameterize the logistic model is often confounded by an incomplete range of low and high abundance population states, which are limited by socioeconomic constraints (management, markets, etc.). As Hilborn and Walters (1992) warn, accurate estimation of yield functions would require overexploitation. Further, the yield curve does not necessarily pass through zero, indicating that some minimum abundance is required for population persistence. Assessment science has therefore advanced precautionary harvest levels, employing reference points other than maximum yield, but these too rely on the assumption of enhanced production at an intermediate population state. To the uninitiated it may seem surprising that this simple stewardship principle works. Yet in many instances it does. With improved controls in fisheries management, the number of sustainably harvested marine species has increased in some regions (Mace 2004; Worm et al. 2009; Neubauer et al. 2013), representing a noteworthy success. On the other hand, some humility is called for because (1) there are few controlled experiments on

209

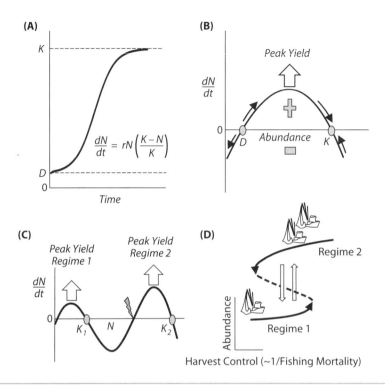

Figure 7.1. Threshold changes and feedbacks relevant to fisheries yield. (A) The logistic population growth model (dN/dt) is a function of density-independent rate of growth r and abundance at carrying capacity K, and D is the depensation abundance threshold. (B) The production (yield) curve is shaped by stable and unstable attractors associated with carrying capacity and depensation. Maximum sustainable yield is represented as peak yield, which occurs at some intermediate population state. (C) Alternate population states influence yield functions. The lightning bolt is the tipping point from Regime 2 to Regime 1. Note that the range of yield overlaps between the two different population states. (D) Bifurcation plot of a regime shift as a result of habitat expansion for western stock Atlantic bluefin tuna where harvest control is the forcing variable and abundance the state variable (see text for details).

the effect of harvest on yield for marine fishes (Smith 1994), (2) there are a substantial number of instances where marine fisheries are not responding to harvest controls in intended ways (Hutchings and Reynolds 2004), and (3) assumptions related to MSY are unrealistic for marine ecosystems and unlikely to hold over generations (Link 2010). The latter two issues direct our attention to how nonstationary conditions (e.g., climate change) influence population states through regime shifts (Perry et al. 2010; Frank et al. 2011).

Managers, scientists, and stakeholders are often frustrated by long-term collapses in ma-

rine fish stocks despite the implementation of best-available science and stringent harvest controls. Examples include northwest Atlantic populations of Atlantic cod, western stock Atlantic bluefin tuna, North Sea herring, Japanese sardines, Atlantic menhaden, and protected species such as American shad and green sturgeon *A. medirostris*. Conversely, management entities are sometimes caught flat-footed as stocks suddenly increase beyond forecasted levels. Recent examples include the irruptive growth of fisheries for dogfish and Atlantic herring in the northwest Atlantic (Rago et al. 1998; Overholtz and Friedland 2002), and Atlantic mackerel and Atlantic monkfish *Lophius piscatorius* in Iceland (Solmundsson et al. 2010; Astthorsson et al. 2012). Perhaps the assumption of continuous response to harvest controls needs reexamination. Might the response of some stocks to fishing and other forcing variables exhibit discontinuous alternative state dynamics?

Resilience is defined by the capacity to recover after some disturbance, or the propensity of systems to converge toward a particular state, represented in dynamic systems theory by an attractor (Scheffer 2009). For the logistic growth model, a single attractor exists—the tendency for the population to move toward carrying capacity. By imposing an external force, fishing, the population is moved from this attractor toward a lower biomass but a higher yield. So long as that additional mortality is applied, a new population state can persist over time, but once fishing is removed, the population is predicted to gravitate back toward carrying capacity. Multiple attractors can result in discontinuous transitions between stable states (fig. 7.1C), often represented by a branching bifurcation plot, which tracks

a state variable against a forcing variable between two attractors (fig. 7.1D; Scheffer and Carpenter 2003; Collie et al. 2004). Note that for the same intermediate range in forcing variable levels, two alternative population states can exist. Such plots have been introduced elsewhere in case studies of curtailed and expanded spatial range for American eel (habitat accessibility was the forcing variable; chap. 6, Is Partial Migration the Solution to the "Eel Problem"?) as well as depressed and irruptive population states of Japanese sardines (predation rate was the forcing variable; chap. 4, Schooling through Food Webs).

Discontinuous alternative states would prove vexing in traditional fisheries management. Early in a developing fishery the lightly exploited stock is assumed to be near carrying capacity, and the manager might wish to move toward an intermediate state where yield is believed to be highest, doing so by increasing fishing rates. As fishing is increased, the population state changes smoothly but only to a critical transition point, after which it descends to a markedly lower population state (fig. 7.3). The manager is now pressed to recover the population

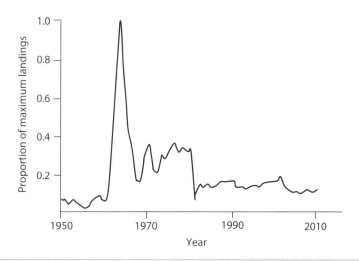

Figure 7.2. Historical landings for western stock Atlantic bluefin tuna. Landings data are from International Commission for the Conservation of Atlantic Tunas (Standing Committee on Research and Statistics 2013).

by imposing strict harvest controls but must now work through a wider range of population states to transition back to the initial regime. Note also that the order of population states differs as one moves forward or backward across regimes, exhibiting path dependency (fig. 7.3). Collie et al. (2004) provided compelling evidence for alternative states in the influence of fishing on Georges Bank haddock during the period 1931–2000. A threshold change from high to low adult biomass occurred in the early 1960s. Over similar levels of exploitation, yield was approximately threefold less in comparison to those observed before the population collapsed.

The discontinuous nature of change between population states implicates another factor beyond the forcing variable, a factor that catalyzes the rapid change between alternative states

(fig. 7.4; Beisner et al. 2003; Collie et al. 2004; Scheffer 2009). These are elements that structure populations—metaschools of sardines colonizing new food webs, lost spawning contingents of collapsed North Sea herring, an expanding subpopulation of irruptive North Sea anchovy, collapsed age structure in diminished northern cod, and retained age structure in recovered Chesapeake striped bass. Such factors modify attractors within population states and can cause them to coalesce by enhancing stability.

The existence of stable states is controversial given that most evidence (albeit some quite compelling) is epiphenomenal, coming from ecological case studies (Scheffer and Carpenter 2003; Folke et al. 2004). Further, complex systems are not stationary. Still, their dynamic nature can be conveyed by the propensity to

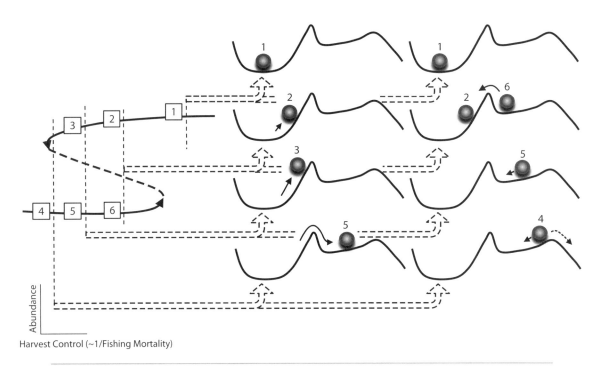

Figure 7.3. Representation of stable population states associated with forward and backward shifts across a tipping point. Population states are given as boxed numbers in the left-hand bifurcation plot and as balls in the right-hand stability diagrams. Solid arrows in the right-hand diagrams indicate changes in exploitation as a forcing variable. Note that population state 4 is also vulnerable to depensation as it approaches an unstable equilibrium.

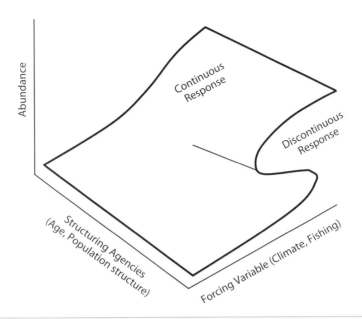

Figure 7.4. Coalescence of two regimes (population states) to a continuous regime through the action of a structuring variable. Redrawn from Collie et al. 2004.

fluctuate around an attractor or equilibrium state (Collie et al. 2004). In marine systems, catastrophic food web and ecosystem changes have been highlighted, such as the loss of coral reef function through (1) overexploitation of fish that regulate reef browsers or (2) threshold responses to temperature and pH (Hughes et al. 2003). Similarly, the loss of large predators has been implicated in cascading food web effects, resulting in dominance by jellyfish in coastal seas (Scheffer and Carpenter 2003). Regime shifts often involve wholesale changes in abundance and distribution (Perry et al. 2010). Below are possible examples of threshold shifts in population states resulting from fishing and other forcing variables for Atlantic bluefin tuna, northern cod, and American eel.

Western Stock Atlantic Bluefin Tuna: One or Two Regimes?

For western stock Atlantic bluefin tuna, commercial fisheries initially spun up in the 1960s (purse seine fisheries for canned pet food) and later in the 1970s (long-line fisheries for the Japanese sashimi market; fig. 7.2). Estimated juvenile production was high during the 1960s and early 1970s, but then the population plummeted, evident in depressed landings and a three- and fivefold decline in adult and juvenile abundances, respectively (fig. 7.5A; Standing Committee on Research and Statistics 2013). Harvest controls were applied in the 1980s through the actions of the International Commission for the Conservation of Atlantic Tunas (ICCAT), but to the frustration of managers, assessment scientists, and fishermen the western stock has failed to recover during the past three decades (fig. 7.5A).

Following imposition of harvest reductions, lack of recovery has caused some to believe that harvest regulations were insufficiently stringent—that bluefin tuna had life history characteristics that made them vulnerable to overexploitation. Indeed, some have called for a complete harvest ban to recover the population (Convention on International Trade in Endangered Species 2010). Such a view was in keeping with the standard logistic model. We should expect a recovery in recruitment (juvenile production) if we were sufficiently protective of

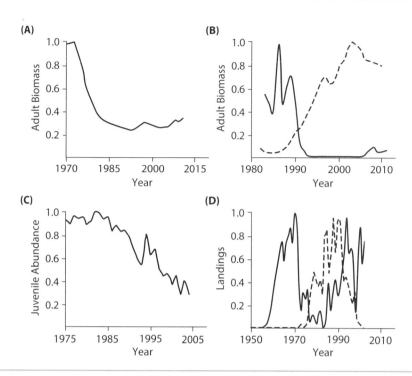

Figure 7.5. Patterns of abundance and landings for (A) western stock Atlantic bluefin tuna, (B) northern Atlantic cod (solid line) and Atlantic striped bass (dashed line), (C) American eel, and (D) Peruvian-Chilean anchovy (solid line) and sardine (dashed line). Abundance expressed as proportion of maximum. Abundance data for bluefin tuna, cod, striped bass, and American eel are from stock assessments (Atlantic States Marine Fisheries Commission 2011, 2012; Fisheries and Oceans Canada 2012; Standing Committee on Research and Statistics 2013). Landings data for sardines and anchovy are from Tarazona et al. 2003.

spawners. Another view held that the western stock was mixing with the eastern stock of Atlantic bluefin tuna, centered in European and North African waters, in a manner that hindered their recovery. Higher exploitation rates in the Mediterranean Sea and northeastern Atlantic would curtail recovery of the western stock depending on the level of connectivity between the two stocks (National Research Council 1994).

With the sustained period of >20 y of poor recruitments, scientists, managers, and stakeholders began to speculate that it was no longer possible to recover to historical recruitment levels, perhaps because of a regime shift, which led ICCAT commissioners to require management advice dependent on two premises—a high recruitment scenario (single regime model) and a low-recruitment scenario (alternative state model; Standing Committee on Research and Statistics 2013). Controversy has ensued because under the single-regime model the western stock is overfished and greater harvest controls are called for, but under the two-regime model the stock is not overfished because only lower yield is expected under the current regime. Some argue that the two-regime premise is a contrivance to maintain status quo, but others point out that the sharp decline of the western stock emulates a critical transition between alternative regimes. Still others note that the decline of Atlantic bluefin tuna mirrors that of other large pelagic fishes (Myers and Worm 2003), the result of sustained

overharvest and "shifting baseline expectations" (Pauly 1995) for how such populations
should perform.

In advance of the western stock's decline, a
large shift occurred in its spatial distribution
(Fromentin et al. 2014). The current distribution
of Atlantic bluefin tuna is centered in temperate North Atlantic waters, but during the 1960s
a large contingent also occurred in equatorial
waters off Brazil, targeted by Japanese long-line
vessels but somewhat of a mystery given bluefin tuna's thermal preferences for cooler waters. Coincident with the decline in the western
stock, this contingent suddenly disappeared
and has not since occurred despite heavy fishing effort in that region. Seasonal distributions
of catch suggested that the contingent was connected to aggregations of the western stock,
centered in the Caribbean and Gulf of Mexico
(Fromentin et al. 2014). Still, questions remain.
How did this migration pathway open and then

suddenly close? Was the disappearance of this
contingent related to diminished recruitment
by the western stock? In an analysis of oceanographic conditions at that time, Fromentin et
al. (2014) found evidence for a transient "ecological bridge" of tolerable thermal conditions
between the North and South Atlantic Oceans
(fig. 7.6), which opened and closed during the
period of the long-line fishery off Brazil. The
bridge was actually a tunnel—stratified water
conditions that enabled Atlantic bluefin tuna
of the western stock to transit through deeper
cooler waters to equatorial and South Atlantic
waters. Given the seasonal nature of the equatorial fishery that exploited bluefin tuna, Fromentin et al. (2014) surmised that the contingent represented a unique life cycle involving
seasonal migrations between North and South
Atlantic waters.

We now have a premise for a regime shift—
a change in spatial behavior allowed exploita-

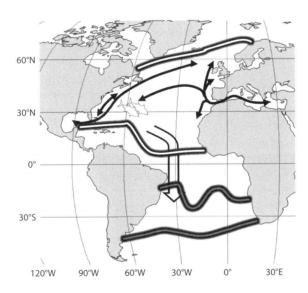

Figure 7.6. Latitudinal ranges of potential habitat for Atlantic bluefin tuna, estimated from a habitat
suitability model (temperature, salinity, and bathymetry), shown as shadowed lines for the North (white)
and South (gray) Atlantic Oceans. During the 1960s, an ecological bridge opened as a result of changed
oceanographic conditions (open arrow), when a major fishery occurred in Brazilian waters. Current major migration routes are shown as solid arrows. Redrawn from Fromentin et al. 2014, with permission.

tion of new food webs in the equatorial and
South Atlantic basins, increasing the overall
carrying capacity for the western stock of Atlantic bluefin tuna. Collapse of the corridor
caused a threshold shift to a lower carrying
capacity. Alternatively, overexploitation of the
contingent could have caused loss of the behavior before the bridge collapsed. Evidence is
only circumstantial, so the two potential causes cannot be distinguished. Another important
part of the story is what precipitated the initial
invasion of this bridge. Fromentin et al. (2014)
suggest that initial exploratory behaviors by
some small contingent were conveyed to a larger segment of the population through adopted
migration (chap. 5, Recalcitrant Nature: Adopted Migration and Tradition). As with Japanese
sardine, when the expanded range was no longer sustainable, behavioral components were
lost, resulting in an overall population decline
beyond what would have been expected by exploitation alone. Interestingly, Atlantic bluefin
tuna continue to be incidentally captured in
the South Atlantic. Are these roamers, or might
these would-be contingent leaders be testing
the waters for future expansions into more distant waters?

In the coupled changes in abundance, oceanography, and spatial behaviors, western Atlantic bluefin tuna are a plausible case study of
how resilience theory accommodates the positive and negative feedbacks that pervade ecological systems, those highlighted in this book
as collective agencies. With changed oceanographic conditions or fishing levels, the same
collective agencies that are stabilizing can become destabilizing. For instance, schooling and
association rules can contribute to seasonal
homing behaviors that tend to stabilize populations within a given oceanographic regime, but
those same propensities can cause threshold
shifts in population states when oceanographic
features change abruptly.

The Atlantic bluefin case study is built on circumstantial, correlative evidence. Other feasible explanations of a population shift could be
erected, such as collapsed age structure and
changes in spawning habitat suitability. Lest

the evidence for threshold shifts in Atlantic bluefin tuna seem too speculative, there is
strong evidence for 120-y cycles in historical
(1600–1960) Mediterranean harvests (Ravier
and Fromentin 2001). These oscillations suggest an influence of climate, yet a careful examination indicated that age structure acted
to regulate these rhythms in population states.
Termed *age resonance* (Bjørnstad et al. 2004),
strong and weak year-classes (stochastic recruitments) are reflected onto the population
over generations, causing oscillations (emulating regime shifts) that do not directly correspond to changes in climate, exploitation,
or other forcing variables. Density-dependent
regulation (e.g., cannibalism) at high-abun-
dance states was a suggested mechanism that
entrains the cycle (Bjørnstad et al. 2004; Royer
and Fromentin 2006). Recovery from low-pop-
ulation states may depend on rebuilding age
structure, but this may lead to oscillations of
similar wavelength to generation time (Brunel
and Piet 2013).

A Tale of Two Species: Northern Atlantic Cod and Chesapeake Striped Bass

Northern cod off Labrador and Newfoundland
and Chesapeake Bay striped bass both experienced a period of intense overfishing, stimulat-
ing fishing moratoria to promote their recovery. That striped bass showed a dramatic recovery and northern cod did not is instructive of a
central resilience mechanism—the storage effect (chap. 4, The Storage Effect). The northern
cod population (chap. 3, Five Mating Systems)
collapsed in the early 1990s, but when viewed
over a longer period, the crash was a spectacular termination of an industrial fishery with a
peak yield of >800,000 t, which supported tens
of thousands of jobs in the Canadian Maritimes
(Hutchings and Myers 1994b). Just prior to the
fishing moratorium in 1992, landings were still
moderately high at >200,000 t, but assessments
indicated that the population was dangerously
overexploited (fig. 7.5B). Indeed, total adult biomass in the mid-1990s was estimated at ~2,000
t (Fisheries and Oceans Canada 2012). This
~99% hyperdepletion is extreme among world

fisheries (Hutchings and Reynolds 2004; Wilberg et al. 2011). Chesapeake Bay striped bass declined tenfold in the 1970s, but recovery was incredibly rapid—the stock was deemed fully recovered in 1995 (Richards and Rago 1999), with record years of adult biomass and recruitment observed during the 2000s (fig. 7.5B).

As reviewed in chapter 5, northern stock cod exhibited a positive feedback—increasing their pattern of aggregation with decreased abundance, which led to increased exposure to fisheries (Hutchings 1996) and exploitation rates that exceeded 75% per year. Chesapeake striped bass experienced similar rates of exploitation prior to their crash (Secor 2000a). Larger juveniles and subadults were targeted prior to their coastal migrations, captured in gauntlets of gill nets and intense sport fisheries in the more restricted waters of the Chesapeake Bay. Yet when moratoria were imposed, one stock recovered and the other did not—why?

Age structure served as a catalyst for Chesapeake Bay striped bass recovery that did not similarly exist for the northern cod stock. Atlantic cod and striped bass share moderate longevity. Both species can live for over 20 y, are highly fecund, mature moderately late (age ~6–8 y for females), and spawn frequently throughout their reproductive life span, epitomizing periodic strategists (Hutchings and Myers 1993; Secor 2000a). Fishery preferences and age-specific migration behaviors by striped bass caused older age-classes (fishes >100 cm), mostly distributed in coastal waters, to be lightly exploited in comparison to younger age-classes. These fish, 10–35 y in age, persisted during the period of stock depletion and contributed to the formation of the next generation—a clear example of the storage effect (Secor 2000a). Age structure allowed increased expression of differential migration, where larger, older fish migrated out of reach of the more intensive fisheries within the Chesapeake Bay. For the northern cod stock, age structure was severely truncated through overfishing on the relative sedentary aggregations of cod, resulting in adults that were mostly composed of first-time spawners (Myers et al. 1996). Wholesale loss of

age structure may have also contributed to lost spawning and migration behaviors (Hutchings and Myers 1993; Rose 1993). A historical spring spawning behavior is now completely absent (Miller et al. 1995; Ciannelli et al. 2013). The loss of generational age structure preconditioned this stock for collapse and arguably shifted it to a different production state, where it persists today despite efforts to rebuild biomass, age structure, and spatial diversity.

The loss of massive cod stocks from the northwest Atlantic Ocean carried over to food web changes that also exhibited threshold dynamics (Frank et al. 2011). Forage fishes that were prey to cod—sand lance *Ammodytes americanus*, capelin, and Atlantic herring—released from predation expanded ninefold in abundance. The forage fish outgrew their zooplankton prey resources and within 10 y overshot the ecosystem's carrying capacity. But on the return trip toward more typical abundance levels, the forage fish populations oscillated at a resonance approximating their average generation time. Increased abundance of pelagic fishes may have further dampened cod's capacity to recover through increased predation on cod larvae by these planktivores (Köster and Möllmann 2000; Perry et al. 2010). Still, in their recent assessment of reversibility in northern cod, Frank et al. (2011) were optimistic, noting moderate recovery in age structure and increasing abundance in other demersal species (haddock).

American Eels: When Does Diversity Matter?

Temperate eels span nearly all types of aquatic environments from open ocean to subterranean streams, leading to a conundrum common to many coastal fishes: which environments are most critical for population persistence and growth? Should we prioritize the stream, the lake, the river, the tidal creek, or all of the above? In one scheme proposed by Beck et al. (2001), adults are surveyed using life cycle tracers to ascertain past nursery habitats. Habitats producing the majority of recruits are ranked highest, and the majority wins. Controversy ensued determining whether to score nursery habitats on the basis of area or availability (Dahlgren et al. 2006), but a more fundamental issue is whether the exercise of ranking majority habitats was sufficiently protective. In particular, do minority habitats also play a role in population growth, persistence, stability, and resilience (Kraus and Secor 2005)?

American eel as a panmictic population species has been in free fall for the last two decades (fig. 7.5C). The period of decline is synchronous throughout the species range, suggesting a common species-wide influence (Atlantic States Marine Fisheries Commission 2012), perhaps because of forcing variables such as climate or oceanographic conditions. Regional stresses during their continental phases of life history would seem less likely to produce this global response. In 2004, the US Fish and Wildlife Service (USFWS) was petitioned to provide federal protections of American eel under the Endangered Species Act (USFWS 2007). The petition arose from concerns about the decline of American eel due to historical loss and degradation of their freshwater habitats. Large hydroelectric and navigation dams are an important legacy issue for diadromous species like eels, but an emerging issue has been the geometric growth of low-head dams associated with increased residential development in suburban and rural (exurban) regions. Dams substantially impede eel movements, even those as low as 2 m (Machut et al. 2007), meaning that this proliferation of low-head dams in recent decades likely curtailed eel access to watersheds as much as 80% (Atlantic States Marine Fisheries Commission 2006).

On the other hand, eel abundance, growth, and biomass are much higher in estuaries and coastal areas than in freshwater habitats (Jessop et al. 2002; Morrison and Secor 2003). On an area or available habitat basis, freshwater habitats are likely to rank lower than coastal ones (Cairns et al. 2009). Further, otolith tracer findings indicate that most eels found in coastal waters persist there throughout their juvenile period (Jessop et al. 2002; Morrison et al. 2003). The previous presumption of an obligate

freshwater eel phase had been proven wrong (chap. 6, Is Partial Migration the Solution to the "Eel Problem"?), which served as the principal justification for a negative finding by USFWS (2007) regarding American eels: "The status of the American eel and the effects of freshwater loss must be examined in light of the American eel's habitation in fresh, estuarine, and marine habitats." If estuarine and marine eel contingents are the most productive, aren't these sufficient to compensate for lost freshwater eel production?

A central conservation issue for American eel is why diversity should matter in a panmictic species. An important type of diversity is response diversity, which relates to redundancy and ecological overhead wherein a set of similar entities samples important system attributes, reducing variability in the aggregate outcomes (Elmqvist et al. 2003). For American eels, differential migration occurs where female distributions are biased toward freshwater habitats and males tend to occur in coastal waters. Also, the oldest individuals occur in slow-growth upstream habitats. The vast historical distribution of American eels in US and Canadian Atlantic watersheds means that eels were sampling large gradients in habitat quality, contributing to response diversity. From the perspective of alternative stable states, an ocean forcing variable may indeed be the underlying cause of decline for American eels, but the severe amount of lost freshwater habitat likely preconditioned a threshold response, one that will be difficult to overcome when ocean conditions again become favorable.

By emphasizing majority habitats in its decision, the USFWS was following recent consensus guidance (Beck et al. 2001), but that guidance was flawed in not considering the importance of maintaining diverse classes of habitats. With respect to managing around regime change, Folke et al. (2004) warned against conserving only the "hot spots" of diversity and production without due attention on how other habitats contributed to functional and redundant diversity in dynamic ecosystems.

Resilience to Fishing and Climate Change: Uncharted Territory

Regardless of whether one accepts the premise that marine fishes undergo continuous or discontinuous changes in their abundance and distributions, there can be little dispute that marine fishes are responding to novel ecosystems resulting from climate change. As one example, a unique mix of decadal climate oscillations and long-term global warming now influences the production of major fished stocks in the North and Baltic Seas—cod, plaice, and herring (Perry et al. 2005; Brunel and Boucher 2007). The trajectory of climate change will often thwart traditional restoration goals, and it will not be possible to return to a previous ecosystem or population state (Perry et al. 2010). Differential sensitivities to warming among species will result in unique assemblages and altered food webs and fisheries (Murawski 1993; Perry et al. 2005). Hobbs et al. (2009) suggested that we are now faced with "hybrid" ecosystems comprising both historical and novel attributes. By recognizing that climate trajectories and associated ecosystem change cannot be overcome, this perspective still allows for conservation aims centered on stabilizing features related to desired ecosystem functions and services, albeit at future states that are likely different than past ones. Hobbs et al. (2009) note, "Retaining the somewhat static view of ecosystems as particular assemblages in particular places will become increasingly unrealistic and is likely to shackle conservation and restoration efforts to ever more unrealistic expectations and objectives." Finding ourselves in the midst of hybrid ecosystems emphasizes the need for adaptive scientific and management frameworks that can be responsive to ecosystem trajectories (Folke et al. 2002).

Climate variability acts on fish populations through multiyear oscillations in temperature, currents, upwelling, and mixing, which are modified as they propagate through food webs and marine populations (Collie et al. 2004; Drinkwater et al. 2010). At-

mospheric forcing—the effects of wind and temperature—is decelerated as energy is transferred into ocean waters, but sustained forcing will alter vertical and horizontal mixing and temperature (Cushing 1962b). Such bottom-up influences are conveyed to the size spectra of food webs (e.g., phytoplankton and zooplankton), which influences early survival and growth of marine fishes (chap. 4, Gleaning Production). Changed ocean circulation and temperature can also influence transport success of young (chap. 3, Alignment of Larval Dispersal with Mating Systems) and migration circuits and ranges of adults (Planque et al. 2010; Pinsky et al. 2013; Fromentin et al. 2014).

Climate variability is typically indexed by multiyear changes in regional atmospheric and oceanographic circulation patterns. The well-known El Niño–Southern Oscillation (ENSO) is indexed by the eastern extension of equatorial Pacific waters centered in the western Pacific Ocean, yet ENSO is manifest in both the Pacific and Atlantic Oceans in various atmospheric, ocean circulation, and food web outcomes (Ottersen et al. 2010). Similarly, the North Atlantic Oscillation (NAO) references atmospheric pressure differences between boreal central Atlantic (Iceland) and subtropical eastern Atlantic (Azores) basins, yet it is manifest in diverse regional differences in winds, currents, warming or cooling, and precipitation (Alheit and Bakun 2010). Intriguing differential outcomes can occur during the same climatic phase. Cycles of sardine and anchovy populations in the Humboldt upwelling system (Peru and Chile) are opposed to one another despite similar feeding ecologies (fig. 7.5D). Anchovies are sensitive to warming that is associated with positive phases of ENSO because of their restricted distribution; sardines are not. As a separate example, during a recent positive phase of NAO, three Atlantic cod populations exhibited unique population responses (Lehodey et al. 2006). North Sea cod recruitment was low during this phase as a result of increased ocean circulation, warming, and changes in the timing and production of zooplankton for first-feeding cod (Beaugrand et al. 2003; chap. 4, Match-Mismatch). NAO-associated warming for the Norwegian cod population had an opposing effect, with increased recruitment. Over the northwest Atlantic Ocean, positive phases of NAO resulted in cooler conditions, which, as reviewed in chapter 3 (Shelf Spawning: Atlantic Cod), caused more southerly shelf spawning and advective loss of eggs and larvae (Rose et al. 2000).

Although it is convenient to associate climate forcing with the familiar indices such as ENSO and NAO, these are coarse multivariate signals ("blunt instruments"; Bakun 2010), and these usual culprits do not necessarily catalyze threshold population changes. Through a detailed narrative on the collapse of Peruvian-Chilean anchovy, Alheit and Bakun (2010) point out that the 1.3-million-t fishery collapsed in 1971, ahead rather than on the heels of the 1972–1973 El Niño event to which the collapse is traditionally ascribed. Still, coastal conditions presaging the collapse were similar (relatively warm) to those associated with positive phases of ENSO (Alheit and Bakun 2010).

Climate oscillation signals have been associated with region-specific food web and population responses that may be synchronous with these signals or "flipped" (antiphasic) (Lehodey et al. 2006; Alheit and Bakun 2010). Opposed regional responses can dampen aggregate outcomes. In the case of northern Atlantic cod, for instance, inshore resident cod can offset poor recruitments of offshore spawning cod during positive phases of NAO. On the other hand, a population may show an amplified response, as when positive phases of ENSO curtail suitable thermal habitats and alter the size spectrum of prey for Peruvian-Chilean anchovy. In the shadow of these decadal regime shifts are much longer-term changes, such as the Pacific Decadal and Atlantic Multidecadal Oscillations. Regardless of the span of climate oscillations, these changes do not comprise regimes per se, but they can serve to catalyze, reinforce, sustain, and amplify changed population and ecosystem states (Collie et al. 2004; Alheit and Bakun 2010).

The relative influence of climate versus fisheries on marine fishes has been an enduring question in fisheries science (Hjort 1914; Cushing 1975; Rothschild 1986), often resulting in inaction on overfished stocks in the face of uncertainty about their relative importance (Smith 1994). The issue is of course a false dichotomy (Planque et al. 2010). Perry et al. (2010) usefully suggest that we should evaluate how fishing preconditions populations to be increasingly sensitive to climate influences and vice versa. For instance, a population diminished by climate forcing might exhibit a more aggregated response to exploitation. In general, exploitation can make populations more sensitive to climate variability owing to changes in abundance, age structure, loss of spatial structure, alteration in life history traits, and changes in food web structure (Perry et al. 2010; Planque et al. 2010). Species that are characterized by life cycle closure, utilize narrow sets of habitats, and exhibit strong seasonal fidelity in their migrations will be more vulnerable to linked fishing-climate influences (Harden Jones 1968; Murawski 1993; Yates et al. 2012; Petitgas et al. 2013).

Prediction of Tipping Points

In ecosystem-based fisheries management, a guiding principle is to increase stability or resilience in fisheries and the ecosystems upon which they rely (Link 2010). This principle would imply that critical transitions should be avoided, although salutatory changes might conceivably bring about societal benefits rather than harm. For instance, a central justification for even small improvements to dissolved oxygen in the bottom waters in the Chesapeake Bay is that the benthic food web is near a critical threshold in how it processes nutrients (Kemp et al. 2009). As another example, one could conceive that it would be unwise and impractical to curtail the periodic outbreaks of sardines and anchovies, even though these irruptions potentially destabilize food webs and fisheries through lags in predator populations and overcapitalization in fisheries. Whether we wish to avoid thresholds or not, from a practical point of view, can we determine our position on the resilience landscape and forecast imminent tipping points?

In keeping with the role of redundancy as an intermediate agency, Scheffer et al. (2009) suggest that complex systems approaching a tipping point exhibit greater synchrony or autocorrelation in their response to forcing variables. Greater synchrony would also be expected in overexploited stocks with truncated age distributions, which tend to be more coupled to environmental forcing. Flickering or greater stochastic variation is a second warning sign, where positive feedbacks amplify variation before a critical transition (Scheffer et al. 2009). Further, if sufficiently high, stochasticity becomes its own forcing variable, leading to a forward or backward shift. Elevated variance in population states was observed for exploited versus unexploited stocks in the California Current ecosystem (Hsieh et al. 2006). As demographic and spatial structure is lost, heightened sensitivity and exaggerated response by populations to environmental forcing are expected (Perry et al. 2010). Shepherd and Cushing (1990) proposed that increased variance in recruitment at diminished stock size was common for marine fishes. In their view, this increased stochasticity led to persistence (diminished propensity for extirpation). An opposing view is that increased flickering in recruitment states could lead to a regime shift or depensation toward extirpation.

Although there have been no instances of species extinctions for marine fishes during the past century, extirpation of individual populations has occurred in species with more vulnerable life cycles, such as anadromous fishes, long-lived rockfish *Sebastes* spp., and inshore habitat specialists (Musick et al. 2000). Collapsed stocks are typically defined as those occurring at <80% of their historical abundance (Hutchings and Reynolds 2004). This population state may not be desirable from an ecological or economic point of view; however, under the assumptions of the logistic growth model it can be conceived of as stable. The issue is what happens as low-abundance states continue to decline toward zero. Is there a point of no return? Two potential thresholds come to bear. Depensation describes a downward spiral after some threshold where diminishing abundances cause accelerated decline (Frank and Brickman 2000). For instance, predation and fishing mortality on a per capita basis increases in a population that becomes increasingly aggregated at low abundance. So-called Allee effects are subtly different than depensation, consisting of an abundance threshold below which the population can no longer replace itself.

From a practical point, however, it will be difficult to know enough about a system (or even a single fish population) to predict critical transitions in a manner timely for management purposes. Still, reconstructions of regime shifts and the behavior of fisheries and marine ecosystems before and after threshold changes are invaluable. Such exercises help us build conceptual models in adaptive management as well as propose and evaluate performance metrics that relate to stability and resilience (Bakun 2010). On the other hand, retrospective assessments are typically biased in that we rarely have the complete picture on population states and forcing variables; and structural elements that serve as catalysts for thresholds may be cryptic or difficult to assess. The inability to practically forecast regime shifts should not condemn us to inaction. Rather, by recognizing that stocks interact with their food webs, fishing, and climate and oceanographic changes through positive and negative feedbacks, we can advance metrics that reduce the propensity for transitions that relate to yield, stability, resilience, and persistence. These reference points emphasize ways to build biodiversity within fish populations.

Collective Agencies and Biodiversity

Diversity contributes to ecological function according to the aggregate outcome of many discrete entities: the so-called averaging effect (Doak et al. 1998). Recall quorum decisions in flocking birds and schooling fish, overdispersion of numerous small larvae seeking patchy plankton, or the repeated spawning bouts of a periodic strategist. These behaviors all generate statistical (collective) outcomes that relate to stability and resilience under the organizing concept of the central limit theorem. With increased sampling by multiple entities, variance about some central attractor is diminished. Thus, in a nonstationary ecosystem, individual movements, foraging, and reproduction will not always be successful, but at some larger level of structural organization, increased sampling will cause an ecological function to persist. This averaging effect relies on two broad collective agencies: (1) sampling units (alleles, individuals, schools), which experience different outcomes under the same background environmental conditions, and (2) self-organizing structures (schools, contingents, subpopulations) that lead to ecological (population) consequences.

Whether surfing through food webs as larvae or migrating across ocean basins as adults, marine fishes throughout their life cycle interact with multiple food webs across a range of scales unique among vertebrate taxa (chap. 4, Marine and Terrestrial Food Webs: How Different?). As they do so, marine fishes perform different ecological functions. Fish larvae serve as predator and prey, linking primary consumers to higher trophic levels. Through seasonal migrations, predators such as sharks and tunas exploit production at the highest trophic level and exert periodic top-down control in the local food webs they visit. Abundant schooling fishes cause large energy inputs and removals as they migrate seasonally among marine ecosystems. Peterson et al. (1998) proposed that such actors contributed to cross-scale resilience where ecological function (e.g., trophic role) was repeated across multiple spatiotemporal scales, contributing to overall resilience in that function. For many marine fishes this role can be played by a single population as its members grow and transit through food webs. Cross-scale resilience owing to seasonal migrations has also been described for a population of roach *Rutilis rutilis*, which regulates lake ecosystem states through its migrations (Brönmark et al. 2010).

Leaving aside more theoretical and controversial aspects of the diversity–ecological stability debate (May 1973; Peterson et al. 1998; McCann 2000), let us consider how we might practically build stability and resilience into marine fish populations. The rivet model (Ehrlich and Ehrlich 1981) considers equivalent contributions by species to ecological function. In aviation, the more rivets there are fastened to the wings of planes, the more reliably they fly (Naeem 1998). Rivets may be lost, but their function persists—up to a point. Similarly, species overlap sufficiently in their niches such that removal of one or a few species is compensated by remaining species until some minimum threshold is reached. This type of resilience is known as functional redundancy (Naeem 1998; Peterson et al. 1998). Functional redundancy was observed in the persistence of principal trophic guilds of Georges Bank fishes, which persisted despite large changes in species composition during 30 y of heavy exploitation (Garrison and Link 2000). Alternatively, large sharks have not been replaced by functional equivalents in many marine ecosystems, resulting in cascading effects to lower trophic levels (Myers and Worm 2003; Ferretti et al. 2010).

A more realistic assumption for functional redundancy is that some species are of greater importance in their contributions than others. The drivers and passengers model gives greater weight to certain species that play keystone roles in the food web (Walker 1992). In an analysis of over 100 experimental studies representing both terrestrial and aquatic food webs, Cardinale et al. (2006) found support for the drivers and passengers model, suggesting also that in more diverse food webs a sampling effect (i.e., increased number of species) ensured the inclusion of functions that dominant species imparted.

Functional redundancy relates to how ecosystems, food webs, and populations maintain integral functions in the face of structural (species) loss. A second class of diversity, response diversity (see American Eels: When Does Diversity Matter? above), addresses how complex systems adapt to temporal changes through their collective responses (Elmqvist et al. 2003; Nyström 2006). In transient nonstationary systems, an increasing number of entities provides a greater number of potential fates. Some set of these fates is likely to be suited to large changes in the ecosystem, such as regime shifts. Stability is enhanced through oversampling in time, resulting in a so-called insurance effect (Loreau et al. 2001; Elmqvist et al. 2003). Structurally, the two classes of diversity rely on similar elements. For instance, functionally redundant entities can become increasingly independent over time, resulting in greater system asynchrony and response diversity.

Building Resilience into Populations: The Storage and Portfolio Effects

Are there ways to build resilience and stability into marine fisheries? What does the fisheries scientist or manager have at their disposal to develop reference points that relate to resilience, stability, and persistence? From a fisheries perspective, what do resilience, stability, and persistence mean, and how do we define and implement these terms as fishery reference points? As indicated above, resilience is defined by the duration or speed of return to some stable state following a disturbance. Fisheries' recovery plans specify levels of expected resilience, where harvest controls and other measures are designed to recover a population over a period that is typically some multiple of their generation time (e.g., 10 y or 10 y + 1 generation time; Restrepo et al. 1998). Resilient populations then show rapid recovery, returning to their original state. Marine populations that are stable or resistant against fishing pressure are those less likely to depart from their original state in the first place. Stability is measured easily enough through variance, and resistance is estimated by how quickly a changed state occurs following a disturbance. Persistence simply relates to continued presence, the duration of existence for a population. All of these measures depend on the level of forcing variable (stress) applied and the duration over which we observe pop-

ulations. Simulation models in which multiple realizations over repeated generations can be generated have been useful to uncover how resilience reference points would likely perform in the real world.

Broadening the scope of yield-based fisheries management will center on reference points that can effectively translate goals, science, and assessment into a decision framework (Link 2005). In broad terms, reference points establish how close we are to a threshold condition we wish to avoid, or a target condition we wish to attain. These threshold or target benchmarks are informed by stewardship goals. In the recent history of fisheries management these goals include (1) maintaining a stock at MSY or other level where replenishment rate exceeds removals, (2) recovering a depleted stock, and (3) considering other ecological services such as a stock's role in the food web (e.g., multispecies management and ecosystem-based fisheries management). Once a stewardship goal is established (often through law or consensus of managers and stakeholders), a range of models can be brought to bear to develop a quantitative representation of that goal against which a stock or population is assessed. Two central reference points are biomass and fishing rate at MSY. These benchmarks rely on our best estimates of stock production (B_{MSY}) and the manner in which the fishery interacts with the stock (F_{MSY}), where indicators of current stock performance are stock biomass (B) and fishing rate (F). In the most elementary case, when B/B_{MSY} is lower than 1.0 (or some fraction of 1.0 when precaution and other considerations are incorporated), then the stock is overfished and harvest control rules are implemented. When F/F_{MSY} exceeds 1.0 (or some fraction of unity), then overfishing is occurring and management is responsive to this departure.

Reference points and the three-tiered management system (goal/reference point → assessment → decision/action) sounds logical enough, but in fact they represented a profound change in fisheries management during the past two decades. Reference points are the lynchpin in a cycle of adaptive management initiated by goal setting, establishing benchmarks, and evaluating stock status against those benchmarks. Frequent management interventions are responsive to stock performance, and the cycle is renewed with periodic reevaluations, sometimes termed *benchmark assessments*, which examine anew central assumptions implicit in the goals and models that guide reference points and how stocks are assessed. Increasingly, precautionary buffers against uncertainty in model assumptions, assessment outputs, and management implementation (e.g., harvest reporting, enforcement) are being incorporated at all stages of the adaptive management cycle.

Link (2010) argues that stewardship of fisheries arose from traditions of cultivation in agriculture and forestry, leading to the "audacious" belief that we could manage wild, fluctuating, and migratory marine fishes in a similar manner to the harvests of wheat, trees, pigs, and chickens. We can see evidence for these traditions in the reliance on simplifying assumptions of closed populations with delimited stock boundaries (chap. 5, Bounded Fisheries). There is considerable debate on the effectiveness of reference point frameworks in single-species fisheries management (reviewed in Link 2010). The common goal of maximizing yield has been widely criticized as too generic. Depending on societal aims and ecological constraints, other goals such as long-term sustainability, profit, employment, or various ecological roles may be of greater importance. Many species are not well assessed owing to poor information on landings and fishing effort or lack of information on abundance, recruitment, and age and size composition. The management system critically depends on effective external or self-enforcement of harvest limits (King and Sutinen 2010). Despite these continuing struggles, managers and policymakers now frame their decisions and communicate to stakeholders in the vernacular of reference points. To manage fisheries in nonstationary systems or implement ecosystem-based fisheries management, the challenge is to expand the vocabulary of reference points, demonstrating their relevance and applicability to broader socioecological aims.

Below we review two classes of reference points pertaining to temporal and spatial aspects of stability and resilience—diversity in age and spatial structure within populations. These are just a subset of many others that are now being considered under the rubric of ecosystem-based management (Rochet and Trenkel 2003). Although it might be precautionary to protect natural diversity wherever we see it, such protection will be impractical and difficult to justify to managers and the public. Criteria for useful diversity-based reference points can be determined by asking a few questions. (1) How does the reference point relate to theory or predictions on the significance of diversity on population resilience? (2) Can the performance of the stock be assessed against the reference point? (3) Is there a means for managers to be responsive to the reference point? A critical issue for resilience is whether a population's store of phenotypic and behavioral plasticity can keep pace with ecosystem change.

An important value in developing diversity-based reference points is to better articulate to ourselves, managers, and stakeholders when and where principal classes of diversity matter. Simulation modeling and management strategy evaluations are invaluable tools for exploring likely resilience outcomes for populations characterized by varying levels of diversity.

Diversity in Age Structure: The Storage Effect

Age structure builds functional redundancy, response diversity, and cross-scale resilience into populations. Production (yield) becomes more stable through the redundant contributions of multiple age-classes (year-classes) to egg production. The storage effect—defined by repeated spawning into varying conditions over a reproductive lifespan—is an example of response diversity. Offspring survival and recruitment vary each year, but by "oversampling" across years, a population buffers against this variability on a generational basis (Longhurst 2002; Secor 2007). Cross-scale stability occurs when reproductive behaviors or attributes change over a fish's life, such as when older females spawn earlier or produce larger offspring than smaller females (Lambert 1987; Hutchings and Myers 1993; Rijsndorp 1994; Marteinsdottir and Thorarinsson 1998; Berkeley et al. 2004). It is not necessarily important whether specific spawning times, egg attributes, or other maternal effects confer a consistent survival advantage to offspring (e.g., Marshall et al. 2010), but rather that these attributes cause a range of outcomes over a generation's lifespan. Additionally, variations in spawning behaviors will carry over to larval dispersal and partial migration, which will confer stability through spatial behaviors (Secor 2007).

Fishing often truncates age structure either by selecting larger fish directly because they are more valuable or accessible, or indirectly by increasing overall mortality on the population (Law 2000; Planque et al. 2010). By removing cohorts of older lower-turnover individuals, density-dependent regulation is relaxed, favoring increased yield. The remaining cohorts exhibit faster life histories and are more strongly coupled to climate oscillations and other forcing variables (Perry et al. 2005). During the past 60 y, the mean age of adult Norwegian cod has declined from 11 to 7 y, associated with sustained fishing. This faster life history has resulted in population dynamics that are increasingly controlled by warm-cool cycles in the Barents Sea (Ottersen et al. 2006). Thus, with age truncation, buffering associated with the storage effect diminishes, and populations become more tightly coupled to the environment (Brander 2005; Planque et al. 2010).

Reference points for age structure have been proposed that capture all three resilience functions. Functional redundancy relies on equivalent contributions between age-classes (rivet model) and is indexed by the number of age-classes (Lambert 1987) or through the uniformity of their distribution (e.g., Shannon diversity index estimated from age-class abundances; Marteinsdottir and Thorarinsson 1998; Marshall et al. 2003). By weighting age-classes by their expected contributions to fecundity (drivers and passengers model), Secor (2000a)

developed a reference point that indexed current age structure against expected age structure in an unfished population (fig. 7.7). Response diversity relates to generation time and can be indexed by mean age at first spawning, mean age of adults, or older age-class abundances (Hutchings and Baum 2005; Brunel and Piet 2013). All of the aforementioned indices and reference points represent aspects of cross-scale resilience, but the key will be conservation of longevity. Secor (2000b) showed that despite a collapsed age structure (low age diversity index, low number of age-classes), Chesapeake Bay striped bass showed resilience arising from the persistence of older fish. Building reference points from these indices will require some background information on how different age-classes function in their contributions to resilience. Secor (2000a) sought to relate age diversity metrics to embryo dispersion, for instance, which related to variation in spawning times.

A central problem with age structure reference points is that they are sensitive to strong year-classes. A dominant year-class can result in increased yield and resilience but will collapse the age structure and even destabilize populations by affecting migration or food webs. Brunel and Piet (2013) evaluated the response of simulated populations of cod, herring, and plaice to management measures designed to conserve age structure. Following a period of overfishing, resistance was assessed as the rate of biomass (egg production) loss. Resilience was assessed as the rate of biomass recovery after fishing pressure was reduced. Their exercise showed that age structure indices could be effectively implemented through length-based harvest controls. On the other hand, increased age structure (presence of older age-classes) showed clear trade-offs in management goals. Decreased selectivity on older age-classes resulted in increased stability (decreased variance in adult biomass) and increased resistance to fishing and other forcing variables. On the other hand, conservation of older age-classes translated into increased selectivity on young age-classes, which resulted in decreased yield

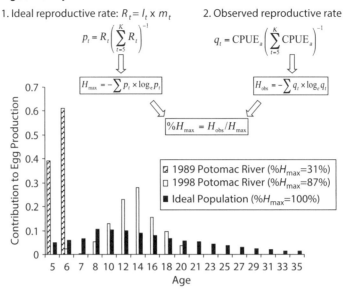

Figure 7.7. Age structure reference point estimated from expected age-specific reproductive rates (R) in an unexploited striped bass population. The steps for calculating the reference point are presented. Here l_t, survival to age t; m_t, fecundity at age t; $CPUE_a$, catch per unit effort of adult biomass at age $t \times m_t$; K, maximum age. Reprinted from Secor 2007, with permission.

and decreased resilience. The management objective of persistence and long-term stability in yield traded off against higher productivity and capacity to recover following a stress. As we will see below, similar trade-offs occur for reference points that relate to spatial structure.

Diversity in Spatial Structure: The Portfolio Effect

The overall response of a metapopulation or a population to environmental forcing will depend upon the dynamics and pattern of connectivity among constituent subpopulations or contingents. Contingents and subpopulations sample different portions of the same broad-scale environmental variability (Manderson 2008). These and other population structures are often cryptic or require special empirical approaches to uncover them, but these same structures can lead to misperceptions on stock productivity and resilience (Hutchings and Reynolds 2004). Contingents, subpopulations,

and other structures are associated with functional and redundant components of diversity. Functional diversity will relate to demographic exchange among contingents (contingents used here in generic way for spatially discrete groups). Response diversity is associated with how contingents respond asynchronously to the same environmental conditions. Asynchronous contingent responses should result in reduced variance at the aggregate (population) level, which can be measured as the portfolio effect (Doak et al. 1998; Tilman et al. 1998).

In the context of marine fishes, Secor et al. (2009) initially defined the portfolio effect (PE) as the degree to which variance in population responses is reduced owing to asynchronous contingent responses to climate change and other environmental forcing. A reference point for PE can be constructed through examining how variances at the contingent level contribute to the overall population stability. Coefficients of variation (or CV, the arithmetic mean/

standard deviation) for each contingent (C) are compared to those observed at the population (P, or aggregate) level (CV_P; Secor et al. 2009). Then, by weighting the individual CV_C by their respective adult biomasses (S_C), the population CV_P^* is estimated as if the two contingents were responding in complete synchrony to environmental forcing,

$$CV_P^* = \sum_{C=1}^{C=k} \left(\frac{S_C}{S_P} CV_C \right),$$

where k is the number of contingents. A comparison of CV_P with CV_P^* provides an estimate of the degree to which variance is dampened owing to independence between components in their contribution to an aggregate population biomass, or the PE:

$$PE = 1 - (CV_P / CV_P^*).$$

PE thus indexes how contingent independence contributes to overall stability. In a two-contingent system with no correlation between components ($r = 0$), the PE would be 30% according to the modified formula from Doak et al. (1998):

$$CV_P = CV_C \frac{[1 + r (k - 1]^{0.5}}{k^{0.5}}.$$

Imposing synchrony at $r = 0.5$ and 0.8 would reduce the PE to 13% and 5%, respectively. Conversely, negative covariance substantially increases the PE ($r = -0.5$ results in 50% PE). As abundance becomes less equitable between contingents, the PE is diminished (fig. 7.8; Doak et al. 1998; Tilman et al. 1998). Variances and mean abundances may differ between contingents and species, resulting in unpredictable PEs across series of years and systems. Here we are following the drivers and passengers model by weighting variances according to numerical abundance.

In an application on Bristol Bay sockeye salmon populations, Schindler et al. (2010) showed cross-scale resilience owing to the portfolio effect. An earlier case study (Hilborn et al. 2003) had suggested patterns of independence between climate oscillations (Pacific Decadal Oscillation) and structural components (populations, subpopulations, spawning contingents), which in aggregate stabilized abundance and fisheries yield for this salmon metapopulation. Through careful examination of a long-term data set (50 y), Schindler and colleagues compiled CV statistics across population structures. They were further able to incorporate variances specific to age structure (see above). An overall PE of 50% was observed as a result of the independent contributions of stream and river population components and age-classes. The greatest degree of dampening was attributed to stream components, likely caused by their high number (averaging effect) in comparison to number of rivers and age-classes. Genetic, ecomorphic, life history, and run time differences attend the diverse stream and lake spawning components that contributed to the PE. Conserving these characteristics is feasible through control of seasonal fisheries that would otherwise diminish certain runs disproportionately, and through measures that protect spawning and nursery habitats. Hatchery stocking programs need to be carefully managed to conserve genetic and other sources of diversity associated with the stream components (Schindler et al. 2010).

Partial migration is an important source of diversity that can be evaluated through the PE. For eight subpopulations of estuarine white perch in the Chesapeake Bay, resident and migratory juvenile populations varied in their dominance and independence (Kerr and Secor 2012), yielding a range of portfolio effects associated with partial migration (fig. 7.9). In a simulation model, Kerr et al. (2010a) evaluated stability (PE), resilience (rate of recovery after loss due to an environmental event), and yield in white perch populations structured by partial migration. Relative abundance and levels of covariance were simulated in resident and migratory contingents, the latter driven by climate oscillations (Kraus and Secor 2004a). They found that greater contingent synchrony and dominance by the migratory contingent favored yield and resilience. Increased con-

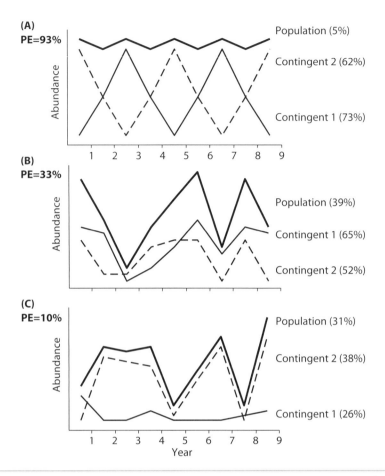

Figure 7.8. Scenarios of asynchronous abundances leading to varying portfolio effects (PEs). The first two scenarios are for contingents with equivalent abundances but where covariance between the contingents is (A) negative or (B) independent. The third scenario is for (C) one contingent that is substantially less abundant, resulting in a lower portfolio effect.

tingent asynchrony and more equitable abundances favored stability—a result similar to that described above for simulations on the effect of fishing stresses on age structure diversity (Brunel and Piet 2013).

The portfolio effect is relevant to the concept of essential fish habitat, where a diverse set of habitats might offset risk from habitat loss, catastrophe, or overexploitation. Kraus and Secor (2005) initially called attention to risks associated with only conserving the most productive fish habitats (Beck et al. 2001) and proposed maintenance of a habitat mosaic that supported

important population structures (e.g., resident and migratory contingents). The aforementioned eel example, in which the freshwater (minority) contingent was discounted in its significance, represents such a risk-prone management tactic. On the other hand, for many sharks, intrapopulation diversity in habitat use is apparently limited, suggesting a curtailed capacity for population structure to function in a stabilizing role (Yates et al. 2012), although this perspective could change with recent evidence for partial migration in elasmobranchs (e.g., Jorgensen et al. 2012b). Catastrophic loss

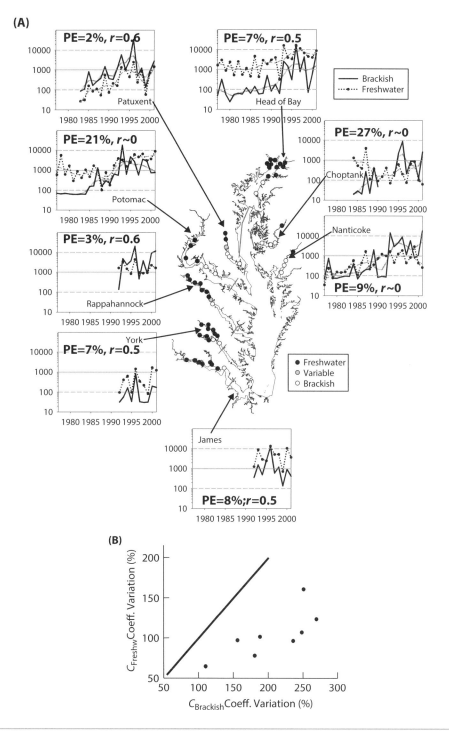

Figure 7.9. Estimates of the portfolio effect (PE) for eight populations of Chesapeake Bay white perch structured by resident (freshwater) and migratory (brackish) contingents. (A) Time series represent surveyed juvenile abundance used to index contingent structure. Note that PE varies according to expectations of contingent covariance (*r*) and relative abundance. (B) The migratory contingent (C_{Brackish}) is more variable than the resident contingent (C_{Freshw}), suggesting differing roles in population resilience. From Kraus and Secor 2005, with permission.

is a relevant risk for sharks and other species exhibiting high habitat fidelities in lieu of partial migration and other diversifying collective agencies (Baum et al. 2003). The risk of catastrophic loss can be elevated when fishing or other stresses cause increased (1) homogenization of spatial behaviors and (2) increased aggregation in populations. We have already reviewed increased coalescence of Atlantic cod under heavy exploitation. Another scenario explored by McGilliard et al. (2011) is when marine protected areas result in increased aggregation (decreased dispersion) by marine fishes owing to high exploitation rates outside of the protected area. In simulation models of both local and global catastrophes (modeled as short periods of high mortality), increased spatial concentration of the population increased risk of population collapse over that which would have occurred absent the marine protected area. These results are highlighted not to discount the value of spatial management tactics such as marine protected areas, but rather to emphasize the important role of population structure in controlling resilience and persistence. As we have seen elsewhere, when we introduce functional aspects to population structure (straying, entrainment, irruptions), both stabilizing and destabilizing influences can emerge (Kell et al. 2009; Secor et al. 2009; Kerr et al. 2010b; Petitgas et al. 2010; Fromentin et al. 2014).

Summary

Traditional fisheries management depends on population states that are alternatively controlled by positive and negative feedback. The objective is to control mortality as a forcing variable in a manner to sustain a population at maximum yield. The stewardship premise of single-species management is that marine fish populations exist in stationary systems and smoothly transition from one population state to another through density-dependent regulation. In many instances, however, fishing and other forcing variables such as climate change cause populations to change abruptly in a manner unexpected. Resilience theory applies to populations that exhibit discontinuous threshold changes from one state to another, moving rapidly through a tipping point. An important consequence is that threshold shifts reconfigure how populations respond to fishing and other forcing variables, such that the same level of fishing can attend large and unexpected differences in yield and other population attributes.

During much of the past century, many marine fish stocks have been overexploited, resulting in large oscillations in population states. Some of these population changes—particularly in instances of collapse—have properties that emulate regime shifts. Northern Atlantic cod and western stock Atlantic bluefin tuna experienced a cycle of overfishing, decline, and stringent fishing controls, yet over a 30-y period they have remained at a low abundance state. Following a population collapse, the same level of fishing effort on Georges Bank haddock generated only one-third of historical lower harvests. In other instances, climate, oceanographic, and habitat changes catalyze threshold changes in population states. Examples highlighted in this chapter included the 80% loss of historical freshwater habitats for American eel and the sudden emergence and closure of an ecological bridge for Atlantic bluefin tuna.

Also challenging the premise of stationarity in single-species management are cycles of abundance in marine fishes that occur in phase with multiyear climate oscillations such as ENSO or NAO, indicative of regime shifts controlled by oceanographic and food web changes. Regime shifts associated with these global climate indices may be realized through less obvious forcing variables such as regional warming (Peruvian-Chilean anchovy) and life history feedbacks (Atlantic bluefin tuna). Overshadowing these regime changes is the overall vector of climate change and the prospect for novel hybrid ecosystems. Stewardship goals must broaden to accommodate nonstationary ecosystems through adaptive management frameworks and expanded reference points, those that relate to population resilience, stability, and persistence.

The impact of regime shifts can be moderated through the internal structure of populations. In community ecology, stability is enhanced through the redundant action of many species (averaging effect). Response diversity relates to how species through their collective responses buffer against ecosystem transience and change. Cross-scale diversity describes how different species perform redundant ecosystem functions by acting at multiple spatiotemporal scales. These same stabilizing sources of diversity are observed in how marine fish populations are internally structured. Progress is being made on reference points asso-

ciated with population structure (the storage and portfolio effects), which complement those that relate to ecosystem-based fisheries management. Whether applicable to population or ecosystem states, reference points are the lynchpin in the three-tiered system of adaptive management (goal/reference point → assessment → decision/action). Reference points that relate to resilience and stability should be vetted for their responsiveness to forcing variables and population state changes through simulation models. Their implementation will require engaging managers and stakeholders on the relevance and feasibility of stability-related reference points as an important complement to existing yield-based reference points.

Bibliography

Abecassis, M., H. Dewar, D. Hawn, and J. Polovina. 2012. Modeling swordfish daytime vertical habitat in the North Pacific Ocean from pop-up archival tags. Marine Ecology Progress Series 452:219–236.

Able, K. P., and J. R. Belthoff. 1998. Rapid "evolution" of migratory behaviour in the introduced house finch of eastern North America. Proceedings of the Royal Society B: Biological Sciences 265(1410):2063–2071.

Able, K. W. 2005. A re-examination of fish estuarine dependence: Evidence for connectivity between estuarine and ocean habitats. Estuarine Coastal and Shelf Science 64(1):5–17.

Able, K. W., and M. P. Fahay. 1998. The First Year in the Life of Estuarine Fishes in the Middle Atlantic Bight. Rutgers University Press, New Brunswick, NJ.

———. 2010. Ecology of Temperate Waters of the Western North Atlantic Estuarine Fishes. Johns Hopkins University Press, Baltimore.

Able, K. W., and T. M. Grothues. 2007. An approach to understanding habitat dynamics of flatfishes: Advantages of biotelemetry. Journal of Sea Research 58(1):1–7.

Able, K. W., T. M. Grothues, J. T. Turnure, D. M. Byrne, and P. Clerkin. 2012. Distribution, movements, and habitat use of small striped bass (*Morone saxatilis*) across multiple spatial scales. Fishery Bulletin 110(2):176–192.

Ackerman, M. W., C. Habicht, and L. W. Seeb. 2011. Single-nucleotide polymorphisms (SNPs) under diversifying selection provide increased accuracy and precision in mixed-stock analyses of sockeye salmon from the Copper River, Alaska. Transactions of the American Fisheries Society 140(3):865–881.

Adams, C. E. 1999. Does the underlying nature of polymorphism in the Arctic charr differ across the species? International Society of Arctic Char Fanatics Information Series 7:61–67.

Adams, C. E., and F. A. Huntingford. 2004. Incipient speciation driven by phenotypic plasticity? Evidence from sympatric populations of Arctic charr. Biological Journal of the Linnean Society 81(4):611–618.

Adams, C. E., A. J. Wilson, and M. M. Ferguson. 2008. Parallel divergence of sympatric genetic and body size forms of Arctic charr, *Salvelinus alpinus*, from two Scottish lakes. Biological Journal of the Linnean Society 95(4):748–757.

Åkesson, S., and A. Hedenström. 2007. How migrants get there: Migratory performance and orientation. Bioscience 57(2):123–133.

Albert, V., B. Jonsson, and L. Bernatchez. 2006. Natural hybrids in Atlantic eels (*Anguilla anguilla*, *A. rostrata*): Evidence for successful reproduction and fluctuating abundance in space and time. Molecular Ecology 15(7):1903–1916.

Alerstam, T. 1991. Bird flight and optimal migration. Trends in Ecology and Evolution 6(7):210–215.

———. 2006. Conflicting evidence about long-distance animal navigation. Science 313(5788):791–794.

Alerstam, T., and A. Hedenström. 1998. The development of bird migration theory. Journal of Avian Biology 29(4):343–369.

Alerstam, T., M. Rosén, J. Bäckman, P. G. P. Ericson, and O. Hellgren. 2007. Flight speeds among bird species: Allometric and phylogenetic effects. PLoS Biology 5(8):1656–1662.

Alheit, J., and A. Bakun. 2010. Population synchronies within and between ocean basins: Apparent teleconnections and implications as to physical-biological linkage mechanisms. Journal of Marine Systems 79(3–4):267–285.

Almany, G. R., M. L. Berumen, S. R. Thorrold, S. Planes, and G. P. Jones. 2007. Local replenishment of coral reef fish populations in a marine reserve. Science 316(5825):742–744.

Almany, G. R., and 8 coauthors. 2009. Connectivity, biodiversity conservation and the design of marine reserve networks for coral reefs. Coral Reefs 28(2):339–351.

Als, T. D., and 9 coauthors. 2011. All roads lead to home: Panmixia of European eel in the Sargasso Sea. Molecular Ecology 20(7):1333–1346.

Amores, A., and 13 coauthors. 1998. Zebrafish hox

clusters and vertebrate genome evolution. Science 282(5394):1711–1714.

Andersen, K. H., J. E. Beyer, M. Pedersen, N. G. Andersen, and H. Gislason. 2008. Life-history constraints on the success of the many small eggs reproductive strategy. Theoretical Population Biology 73(4):490–497.

Anderson, J. T., and B. deYoung. 1995. Application of a one-dimensional model to vertical distributions of cod eggs on the northeastern Newfoundland shelf. Canadian Journal of Fisheries and Aquatic Sciences 52(9):1978–1989.

Anderson, R. C., M. S. Adam, and J. I. Goes. 2011. From monsoons to mantas: Seasonal distribution of *Manta alfredi* in the Maldives. Fisheries Oceanography 20:104–113.

Annese, D. M., and M. J. Kingsford. 2005. Distribution, movements and diet of nocturnal fishes on temperate reefs. Environmental Biology of Fishes 72(2):161–174.

Arai, T., and N. Chino. 2012. Diverse migration strategy between freshwater and seawater habitats in the freshwater eel genus *Anguilla*. Journal of Fish Biology 81(2):442–455.

Arai, T., T. Otake, and K. Tsukamoto. 1997. Drastic changes in otolith microstructure and microchemistry accompanying the onset of metamorphosis in the Japanese eel *Anguilla japonica*. Marine Ecology Progress Series 161:17–22.

———. 2000. Timing of metamorphosis and larval segregation of the Atlantic eels *Anguilla rostrata* and *A. anguilla*, as revealed by otolith microstructure and microchemistry. Marine Biology 137(1):39–45.

Argos. 1996. User's manual. CLS/Service Argos, Toulouse. http://www.argos-system.org/manual/.

Astthorsson, O. S., H. Valdimarsson, A. Gudmundsdottir, and G. J. Óskarsson. 2012. Climate-related variations in the occurrence and distribution of mackerel (*Scomber scombrus*) in Icelandic waters. ICES Journal of Marine Science 69(7):1289–1297.

Atlantic States Marine Fisheries Commission. 2006. Terms of Reference and Advisory Report to the American Eel Assessment Peer Review. Stock Assessment Report 6-01. ASMFC, Arlington, VA.

———. 2011. Striped Bass Stock Assessment Update 2011. ASMFC, Arlington, VA.

———. 2012. American Eel Benchmark Stock Assessment. Stock Assessment Report 12-01. ASMFC, Arlington, VA.

Auditore, P. J., R. G. Lough, and A. Broughton. 1994. A review of the comparative development of Atlantic cod (*Gadus morhua* L.) and haddock (*Melanogrammus aeglefinus* L.) based on an illustrated series of larvae and juveniles from Georges Bank. NAFO Scientific Council Studies 20:7–18.

Avise, J. C., W. S. Nelson, J. Arnold, R. K. Koehn, G. C. Williams, and V. Thorsteinsson. 1990. The evolutionary genetic status of Icelandic eels. Evolution 44(5):1254–1262.

Axelsen, B. E., L. Nottestad, A. Fernö, A. Johannessen, and O. A. Misund. 2000. "Await" in the pelagic: Dynamic trade-off between reproduction and survival within a herring school splitting vertically during spawning. Marine Ecology Progress Series 205:259–269.

Ayling, T., and G. Cox. 1982. Fishes of New Zealand. William Collins, Auckland, New Zealand.

Babaluk, J. A., N. M. Halden, J. D. Reist, A. H. Kristofferson, J. L. Campbell, and W. J. K. Teasdale. 1997. Evidence for non-anadromous behavior of arctic charr (*Salvinus alpinus*) from Lake Hazen, Ellesmere Island, Northwest Territories, Canada, based on scanning probe microprobe analysis of otolith strontium distribution. Arctic 50:224–233.

Babaluk, J. A., J. D. Reist, J. D. Johnson, and L. Johnson. 2000. First records of sockeye (*Oncorhynchus nerka*) and pink salmon (*O. gorbuscha*) from Banks Island and other records of Pacific salmon in Northwest Territories, Canada. Arctic 53(2):161–164.

Bacheler, N. M., J. W. Neal, and R. L. Noble. 2004. Reproduction of a landlocked diadromous fish population: Bigmouth sleepers *Gobiomorus dormitor* in a reservoir in Puerto Rico. Caribbean Journal of Science 40(2):223–231.

Baker, R. R. 1978. The Evolutionary Ecology of Animal Migration. William Clowes and Sons, London.

Bakun, A. 1996. Patterns in the Ocean: Ocean Processes and Marine Population Dynamics. California Sea Grant, San Diego.

———. 2001. "School-mix feedback": A different way to think about low frequency variability in large mobile fish populations. Progress in Oceanography 49(1–4):485–511.

———. 2006. Wasp-waist populations and marine ecosystem dynamics: Navigating the "predator pit" topographies. Progress in Oceanography 68 (2–4):271–288.

———. 2010. Linking climate to population variability in marine ecosystems characterized by non-simple dynamics: Conceptual templates and schematic constructs. Journal of Marine Systems 79(3–4):361–373.

Bakun, A., and P. Cury. 1999. The "school trap": A mechanism promoting large-amplitude out-of-phase population oscillations of small pelagic fish species. Ecology Letters 2(6):349–351.

Bakun, A., E. A. Babcock, S. E. Lluch-Cota, C. Santora, and C. J. Salvadeo. 2010. Issues of ecosystem-based management of forage fisheries in "open" non-stationary ecosystems: The example of the sardine fishery in the Gulf of California. Reviews in Fish Biology and Fisheries 20(1):9–29.

Ballerini, M., and 10 coauthors. 2008. Empirical investigation of starling flocks: A benchmark study in collective animal behaviour. Animal Behaviour 76:201–215.

Balon, E. K. 1984. Reflections on some decisive events in the early life history of fishes. Transactions of the American Fisheries Society 113:178–185.

———. 2002. Epigenetic processes, when natura non facit saltum becomes a myth, and alternative ontogenies a mechanism of evolution. Environmental Biology of Fishes 65(1):1–35.

Barange, M., and 7 coauthors. 2009. Habitat expansion and contraction in anchovy and sardine populations. Progress in Oceanography 83(1–4):251–260.

Barbaro, A., and 7 coauthors. 2009. Modelling and simulations of the migration of pelagic fish. ICES Journal of Marine Science 66(5):826–838.

Barraclough, W. E., and D. G. Robinson. 1971. Anomalous occurrence of carp (Cyprinus carpio) in marine environment. Journal of the Fisheries Research Board of Canada 28(9):1345–1347.

Bartsch, J., K. Brander, M. Heath, P. Munk, K. Richardson, and E. Svendsen. 1989. Modeling the advection of herring larvae in the North Sea. Nature 340(6235):632–636.

Bartumeus, F. 2007. Lévy processes in animal movement: An evolutionary hypothesis. Fractals: Complex Geometry Patterns and Scaling in Nature and Society 15(2):151–162.

———. 2009. Behavioral intermittence, Lévy patterns, and randomness in animal movement. Oikos 118(4):488–494.

Batty, R. S., J. H. S. Blaxter, and K. Fretwell. 1993. Effect of temperature on the escape responses of larval herring, Clupea harengus. Marine Biology 115(4):523–528.

Baum, J. K., R. A. Myers, D. G. Kehler, B. Worm, S. J. Harley, and P. Doherty. 2003. Collapse and conservation of shark populations in the northwest Atlantic. Science 299(5605):389–392.

Baumgartner, T. R., A. Soutar, and V. Ferreira-

Batrina. 1992. Reconstruction of the history of Pacific sardine and northern anchovy populations over the past two millennia from sediments of the Santa Barbara Basin, California. California Cooperative Oceanic Fisheries Investigations Reports 33:24–40.

Beacham, T. D., and 9 coauthors. 2004. Stock identification of Fraser River sockeye salmon using microsatellites and major histocompatibility complex variation. Transactions of the American Fisheries Society 133(5):1117–1137.

Beamish, R. J., and G. A. McFarlane. 1983. The forgotten requirement for age validation in fisheries biology. Transactions of the American Fisheries Society 112:735–743.

Beaugrand, G., K. M. Brander, J. A. Lindley, S. Soulssi, and P. C. Reid. 2003. Plankton effect on cod recruitment in the North Sea. Nature 426:661–664.

Beck, M. W., and 12 coauthors. 2001. The identification, conservation, and management of estuarine and marine nurseries for fish and invertebrates. Bioscience 51:633–641.

Bedarf, A. T., K. R. McKaye, E. P. Van Den Berghe, L. J. L. Perez, and D. H. Secor. 2001. Initial six-year expansion of an introduced piscivorous fish in a tropical Central American lake. Biological Invasions 3:391–404.

Begg, G. A., K. D. Friedland, and J. B. Pearce. 1999. Stock identification and its role in stock assessment and fisheries management: An overview. Fisheries Research 43:1–8.

Begon, M., C. R. Townsend, and L. Harper. 1990. Ecology from Individuals to Ecosystems. Blackwell, Malden, MA.

Beisner, B. E., D. T. Haydon, and K. Cuddington. 2003. Alternative stable states in ecology. Frontiers in Ecology and the Environment 1(7):376–382.

Bekkevold, D., M. M. Hansen, and V. Loeschcke. 2002. Male reproductive competition in spawning aggregations of cod (Gadus morhua, L.). Molecular Ecology 11(1):91–102.

Bell, C. P. 2005. The origin and development of bird migration: Comments on Rappole and Jones, and an alternative evolutionary model. Ardea 93(1):115–123.

Bell, J. D., J. M. Lyle, C. M. Bulman, K. J. Graham, G. M. Newton, and D. C. Smith. 1992. Spatial variation in reproduction, and occurrence of nonreproductive adults, in orange roughy, Hoplostethus atlanticus Collett (Trachichthyidae), from southeastern Australia. Journal of Fish Biology 40(1): 107–122.

Bell, M. A. 1976. Evolution of phenotypic diversity in *Gasterosteus aculeatus* superspecies on the Pacific coast of North America. Systematic Zoology 25(3):211–227.

Bell, M. A., W. E. Aguirre, and N. J. Buck. 2004. Twelve years of contemporary armor evolution in a threespine stickleback population. Evolution 58(4):814–824.

Belthoff, J. R., and S. A. Gauthreaux. 1991. Partial migration and differential winter distribution of house finches in the eastern United States. Condor 93(2):374–382.

Bennett, W. A., W. J. Kimmerer, and J. R. Burau. 2002. Plasticity in vertical migration by native and exotic estuarine fishes in a dynamic low-salinity zone. Limnology and Oceanography 47(5):1496–1507.

Bergey, L. L., R. A. Rulifson, M. L. Gallagher, and A. S. Overton. 2003. Variability of Atlantic coast striped bass egg characteristics. North American Journal of Fisheries Management 23(2):558–572.

Berghahn, R. 1987. Effects of tidal migration on growth of 0-group plaice (*Pleuronectes platessa* l) in the north Frisian Wadden Sea. Meeresforschung: Reports on Marine Research 31(3–4): 209–226.

Berkeley, S. A., M. A. Hixon, R. J. Larson, and M. S. Love. 2004. Fisheries sustainability via protection of age structure and spatial distribution of fish populations. Fisheries 29(8):23–32.

Berthold, P. 1984. The endogenous control of bird migration: A survey of experimental evidence. Bird Study 31(Mar):19–27.

———. 1996. Control of Bird Migration. Chapman and Hall, London.

———. 1999. A comprehensive theory for the evolution, control and adaptability of avian migration. Ostrich 70(1):1–11.

———. 2001. Bird Migration: A General Survey. 2nd ed. Oxford Ornithology Series. Oxford University Press, Oxford.

Berthold, P., and F. Pulido. 1994. Heritability of migratory activity in a natural bird population. Proceedings of the Royal Society B: Biological Sciences 257(1350):311–315.

Berthold, P., G. Mohr, and U. Querner. 1990. Control and evolutionary potential of obligate partial migration: Results of a 2-way selective breeding experiment with the blackcap (*Sylvia atricapilla*). Journal für Ornithologie 131(1):33–45.

Beyer, J. E. 1989. Recruitment stability and survival: Simple size-specific theory with examples from the early life dynamics of marine fish. Dana 7:45–147.

Bigelow, H. B., and W. C. Schroeder. 1953. Fishes of Gulf of Maine. Fishery Bulletin 74. US Fish and Wildlife Service, Washington, DC.

Biro, P. A., and J. R. Post. 2008. Rapid depletion of genotypes with fast growth and bold personality traits from harvested fish populations. Proceedings of the National Academy of Sciences of the United States of America 105:2919–2922.

Biro, P. A., and J. A. Stamps. 2008. Are animal personality traits linked to life-history productivity? Trends in Ecology and Evolution 23(7):361–368.

Bjorkstedt, E. P., L. K. Rosenfeld, B. A. Grantham, Y. Shkedy, and J. Roughgarden. 2002. Distributions of larval rockfishes *Sebastes* spp. across nearshore fronts in a coastal upwelling region. Marine Ecology Progress Series 242:215–228.

Bjørnstad, O. N., R. M. Nisbet, and J. M. Fromentin. 2004. Trends and cohort resonant effects in age-structured populations. Journal of Animal Ecology 73(6):1157–1167.

Blaxter, J. H. S. 1986. Development of sense-organs and behavior of teleost larvae with special reference to feeding and predator avoidance. Transactions of the American Fisheries Society 115(1):98–114.

Block, B. A., and 8 coauthors. 2005. Electronic tagging and population structure of Atlantic bluefin tuna. Nature 434:1121–1127.

Block, B. A., and 20 coauthors. 2011. Tracking apex marine predator movements in a dynamic ocean. Nature 475(7354):86–90.

Bobko, S. J., and S. A. Berkeley. 2004. Maturity, ovarian cycle, fecundity, and age-specific parturition of black rockfish (*Sebastes melanops*). Fishery Bulletin 102(3):418–429.

Boehlert, G. W., and M. M. Yoklavich. 1984. Reproduction, embryonic energetics, and the maternal fetal relationship in the viviparous genus *Sebastes* (Pisces, Scorpaenidae). Biological Bulletin 167(2):354–370.

Bolle, L. J., and 6 coauthors. 2009. Variability in transport of fish eggs and larvae. III. Effects of hydrodynamics and larval behaviour on recruitment in plaice. Marine Ecology Progress Series 390:195–211.

Bonadonna, F., and 6 coauthors. 2005. Orientation in the wandering albatross: Interfering with magnetic perception does not affect orientation performance. Proceedings of the Royal Society B: Biological Sciences 272(1562):489–495.

Bone, Q., and N. B. Marshall. 1982. Biology of Fishes: Tertiary Level Biology. Blackie, Glasgow.

Bonfil, R., and 8 coauthors. 2005. Transoceanic migration, spatial dynamics, and population linkages of white sharks. Science 310(5745):100–103.

Bonhommeau, S., E. Chassot, B. Planque, E. Rivot, A. H. Knap, and O. Le Pape. 2008. Impact of climate on eel populations of the Northern Hemisphere. Marine Ecology Progress Series 373:71–80.

Bonte, D., and E. de la Peña. 2009. Evolution of body condition-dependent dispersal in metapopulations. Journal of Evolutionary Biology 22(6): 1242–1251.

Booke, H. E. 1981. The conundrum of the stock concept: Are nature and nurture definable in fishery science? Canadian Journal of Fisheries and Aquatic Science 38:1479–1480.

Boreman, J. 1997. Sensitivity of North American sturgeon and paddlefish populations to fishing mortality. Environmental Biology of Fishes 48:399–405.

Botsford, L. W., and 8 coauthors. 2009. Connectivity, sustainability, and yield: Bridging the gap between conventional fisheries management and marine protected areas. Reviews in Fish Biology and Fisheries 19(1):69–95.

Boudreau, P. R., and L. M. Dickie. 1992. Biomass spectra of aquatic ecosystems in relation to fisheries yield. Canadian Journal of Fisheries and Aquatic Sciences 49(8):1528–1538.

Bradbury, I. R., and P. V. R. Snelgrove. 2001. Contrasting larval transport in demersal fish and benthic invertebrates: The roles of behaviour and advective processes in determining spatial pattern. Canadian Journal of Fisheries and Aquatic Sciences 58(4):811–823.

Bradbury, I. R., P. V. R. Snelgrove, and S. Fraser. 2000. Transport and development of eggs and larvae of Atlantic cod, Gadus morhua, in relation to spawning time and location in coastal Newfoundland. Canadian Journal of Fisheries and Aquatic Sciences 57(9):1761–1772.

Bradbury, I. R., P. V. R. Snelgrove, and P. Pepin. 2003. Passive and active behavioural contributions to patchiness and spatial pattern during the early life history of marine fishes. Marine Ecology Progress Series 257:233–245.

Bradbury, I. R., and 7 coauthors. 2008. Discrete spatial dynamics in a marine broadcast spawner: Re-evaluating scales of connectivity and hab-itat associations in Atlantic cod (Gadus morhua L.) in coastal Newfoundland. Fisheries Research 91(2–3):299–309.

Bradbury, I. R., and 13 coauthors. 2010. Parallel adaptive evolution of Atlantic cod on both sides of the Atlantic Ocean in response to temperature. Proceedings of the Royal Society B: Biological Sciences 277(1701):3725–3734.

Brander, K. M. 2005. Cod recruitment is strongly affected by climate when stock biomass is low. ICES Journal of Marine Science 62(3):339–343.

Brannon, E. L., 1987. Mechanisms stabilizing salmonid fry emergence timing. Canadian Special Publication of Fisheries and Aquatic Sciences 96:120–124.

Brannon, E. L., and T. P. Quinn. 1990. Field test of the pheromone hypothesis for homing by Pacific salmon. Journal of Chemical Ecology 16(2):603–609.

Breder, C. M., and D. E. Rosen. 1966. Modes of Reproduction in Fishes. T. F. H. Publications, Neptune City, NJ.

Brochier, T., and 8 coauthors. 2009. Small pelagic fish reproductive strategies in upwelling systems: A natal homing evolutionary model to study environmental constraints. Progress in Oceanography 83(1–4):261–269.

Brodersen, J., P. Anders Nilsson, B. B. Chapman, C. Skov, L.-A. Hansson, and C. Brönmark. 2011. Variable individual consistency in timing and destination of winter migrating fish. Biology Letters 8(1):21–23.

Brönmark, C., and 6 coauthors. 2010. Regime shifts in shallow lakes: The importance of seasonal fish migration. Hydrobiology 646:91–100.

Brophy, D., and B. S. Danilowicz. 2002. Tracing populations of Atlantic herring (Clupea harengus L.) in the Irish and Celtic Seas using otolith microstructure. ICES Journal of Marine Science 59(6):1305–1313.

Brophy, D., B. S. Danilowicz, and P. A. King. 2006. Spawning season fidelity in sympatric populations of Atlantic herring (Clupea harengus). Canadian Journal of Fisheries and Aquatic Sciences 63(3):607–616.

Brown, C., and K. N. Laland. 2003. Social learning in fishes: A review. Fish and Fisheries 4(3): 280–288.

Brown, E. D., J. H. Churnside, R. L. Collins, T. Veenstra, J. J. Wilson, and K. Abnett. 2002. Remote sensing of capelin and other biological features in the North Pacific using lidar and video technology. ICES Journal of Marine Science 59(5):1120–1130.

Brown, J. H. 1984. On the relationship between abundance and distribution of species. American Naturalist 124(2):255–279.

Bruderer, B., T. Steuri, and M. Baumgartner. 1995. Short-range high-precision surveillance of nocturnal migration and tracking of single targets. Israel Journal of Zoology 41(3):207–220.

Brunel, T., and J. Boucher. 2007. Long-term trends in fish recruitment in the northeast Atlantic related to climate change. Fisheries Oceanography 16(4):336–349.

Brunel, T., and G. J. Piet. 2013. Is age structure a relevant criterion for the health of fish stocks? ICES Journal of Marine Science 70(2):270–283.

Buchheister, A., and R. J. Latour. 2010. Turnover and fractionation of carbon and nitrogen stable isotopes in tissues of a migratory coastal predator, summer flounder (Paralichthys dentatus). Canadian Journal of Fisheries and Aquatic Sciences 67(3):445–461.

Burd, A. C. 1962. Growth and recruitment in the herring of the southern North Sea. Fishery Investigations of the Ministry of Agriculture, Fish Food, London II 23:1–42.

Burgner, R. L. 1991. Life history of sockeye salmon (Oncorhynchus nerka). In Pacific Salmon Life Histories, ed. C. Groot and L. Margolis, 3–117. University of British Columbia Press, Vancouver.

Burke, J. S., M. Tanaka, and T. Seikai. 1995. Influence of light and salinity on behaviour of larval Japanese flounder (Paralichthys olivaceus) and implications for inshore migration. Netherlands Journal of Sea Research 34(1–3):59–69.

Burrows, M. T., R. N. Gibson, L. Robb, and A. Maclean. 2004. Alongshore dispersal and site fidelity of juvenile plaice from tagging and transplants. Journal of Fish Biology 65(3):620–634.

Butts, I. A. E., E. A. Trippel, and M. K. Litvak. 2009. The effect of sperm to egg ratio and gamete contact time on fertilization success in Atlantic cod Gadus morhua L. Aquaculture 286(1–2):89–94.

Butts, I. A. E., M. K. Litvak, and E. A. Trippel. 2010. Seasonal variations in seminal plasma and sperm characteristics of wild-caught and cultivated Atlantic cod, Gadus morhua. Theriogenology 73(7):873–885.

Cadrin, S. X., and D. H. Secor. 2009. Accounting for spatial population structure in stock assessment: Past, present and future. In The Future of Fishery Science in North America, ed. R. J. Beamish and B. J. Rothschild, 405–426. Springer Science, New York.

Cadrin, S. X., L. A. Kerr, and S. Mariani, eds. 2013. Stock Identification Methods: Applications in Fisheries Science. 2nd ed. Academic, London.

Cairns, D. K., D. H. Secor, W. E. Morrison, and J. A. Hallet. 2009. Salinity-linked growth in anguillid eels and the paradox of temperate-zone anadromy. Journal of Fish Biology 74:2094–2114.

Campana, S. E. 1999. Chemistry and composition of fish otoliths: Pathways, mechanisms and applications. Marine Ecology Progress Series 188:263–297.

——. 2005. Otolith science entering the 21st century. Marine and Freshwater Research 56(5):485–495.

Campana, S. E., S. J. Smith, and P. C. F. Hurley. 1989. A drift-retention dichotomy for larval haddock (Melanogrammus aeglefinus) spawned on Browns Bank. Canadian Journal of Fisheries and Aquatic Science 46:93–102.

Campana, S. E., W. Joyce, and M. Fowler. 2010. Subtropical pupping ground for a cold-water shark. Canadian Journal of Fisheries and Aquatic Sciences 67(5):769–773.

Campfield, P. A., and E. D. Houde. 2010. Ichthyoplankton community structure and comparative trophodynamics in an estuarine transition zone. Fishery Bulletin 109(1):1–19.

Candy, J. R., and T. D. Beacham. 2000. Patterns of homing and straying in southern British Columbia coded-wire tagged chinook salmon (Oncorhynchus tshawytscha) populations. Fisheries Research 47(1):41–56.

Cano, J. M., H. S. Mäkinen, and J. Merilä. 2008. Genetic evidence for male-biased dispersal in the three-spined stickleback (Gasterosteus aculeatus). Molecular Ecology 17(14):3234–3242.

Cardinale, B. J., and 6 coauthors. 2006. Effects of biodiversity on the functioning of trophic groups and ecosystems. Nature 443(7114):989–992.

Carr, S. M., A. J. Snellen, K. A. Howse, and J. S. Wroblewski. 1995. Mitochondrial DNA sequence variation and genetic stock structure of Atlantic cod (Gadus morhua) from bay and offshore locations on the Newfoundland continental shelf. Molecular Ecology 4(1):79–88.

Casey, J. M., and R. A. Myers. 1998. Near extinction of a large, widely distributed fish. Science 281(5377):690–692.

Caswell, H. 1982. Life-history theory and the equilibrium status of populations. American Naturalist 120(3):317–339.

Champalbert, G., and C. Koutsikopoulos. 1995. Behavior, transport and recruitment of Bay of Biscay sole (*Solea solea*): Laboratory and field studies. Journal of the Marine Biological Association of the United Kingdom 75(1):93–108.

Chapman, B. B., and 6 coauthors. 2011a. To boldly go: Individual differences in boldness influence migratory tendency. Ecology Letters 14(9):871–876.

Chapman, B. B., C. Brönmark, J.-A. Nilsson, and L.-A. Hansson. 2011b. The ecology and evolution of partial migration. Oikos 120(12):1764–1775.

Chapman, B. B., and 6 coauthors. 2012. Partial migration in fishes: Definitions, methodologies and taxonomic distribution. Journal of Fish Biology 81(2):479–499.

Chapman, J. W., and 8 coauthors. 2011. Animal orientation strategies for movement in flows. Current Biology 21(20):R861–R870.

Charnov, E. L., and W. M. Schaffer. 1973. Life-history consequences of natural selection: Cole's result revisited. American Naturalist 107(958):791–793.

Chernetsov, N., P. Berthold, and U. Querner. 2004. Migratory orientation of first-year white storks (*Ciconia ciconia*): Inherited information and social interactions. Journal of Experimental Biology 207(6):937–943.

Chesson, P. L., and R. R. Warner. 1981. Environmental variability promotes coexistence in lottery competitive systems. American Naturalist 117:923–943.

Childers, J., S. Snyder, and S. Kohin. 2011. Migration and behavior of juvenile North Pacific albacore (*Thunnus alalunga*). Fisheries Oceanography 20(3):157–173.

Chong, Y. C. 1814. Jasaneobo. In The History of Korea, ed. K. S. Lee, 1–332. Eul-Yoo Press, Seoul. (Translation by S. Jung, Jejung National University, Korea, 1970).

Chow, S., H. Kurogi, N. Mochioka, S. Kaji, M. Okazaki, and K. Tsukamoto. 2009. Discovery of mature freshwater eels in the open ocean. Fisheries Science 75(1):257–259.

Chow, S., and 21 coauthors. 2010. Japanese eel *Anguilla japonica* do not assimilate nutrition during the oceanic spawning migration: Evidence from stable isotope analysis. Marine Ecology Progress Series 402:233–238.

Churnside, J. H., and J. J. Wilson. 2001. Airborne lidar for fisheries applications. Optical Engineering 40(3):406–414.

Ciannelli, L., and 6 coauthors. 2013. Theory, consequences and evidence of eroding population spatial structure in harvested marine fishes: A review. Marine Ecology Progress Series 480:227–243.

Clark, J. 1968. Seasonal movements of striped bass contingents of Long Island Sound and the New York Bight. Transactions of the American Fisheries Society 97:320–343.

Clemens, B. J., T. R. Binder, M. F. Docker, M. L. Moser, and S. A. Sower. 2010. Similarities, differences, and unknowns in biology and management of three parasitic lampreys of North America. Fisheries 35(12):580–594.

Codling, E. A., J. W. Pitchford, and S. D. Simpson. 2007. Group navigation and the "many-wrongs principle" in models of animal movement. Ecology 88(7):1864–1870.

Cohen, J. E. 1994. Marine and continental food webs: Three paradoxes? Philosophical Transactions of the Royal Society B: Biological Sciences 343:57–69.

Cole, L. C. 1954. The population consequences of life history phenomena. Quarterly Review of Biology 29(2):103–137.

Colin, P. L. 1992. Reproduction of the Nassau grouper, *Epinephaelus striatus* (Pises: Serranidae) and its relationship to environmental condition. Environmental Biology of Fishes 34:357–377.

Collie, J. S., K. Richardson, and J. H. Steele. 2004. Regime shifts: Can ecological theory illuminate the mechanisms? Progress in Oceanography 60(2–4):281–302.

Colosimo, P. F., and 6 coauthors. 2004. The genetic architecture of parallel armor plate reduction in threespine sticklebacks. PLoS Biology 2(5):635–641.

Compagno, L. J. V. 1984. Sharks of the World: An Annotated and Illustrated Catalogue of Shark Species Known to Date, vols. 1 and 2. FAO Species Catalogue 4. Food and Agriculture Organization of the United Nations, Rome.

Conover, D. O., and T. M. C. Present. 1990. Countergradient variation in growth rate: Compensation for lengths of the growing season among Atlantic silversides from different latitudes. Oecologia 83:316–324.

Conover, D. O., S. B. Munch, and A. Arnott. 2009. Reversal of evolutionary downsizing caused by selective harvest of large fish. Proceedings of the Royal Society B: Biological Sciences 276:2015–2020.

Conrad, J. L., K. L. Weinersmith, T. Brodin, J. B. Saltz, and A. Sih. 2011. Behavioural syndromes in fishes: A review with implications for ecology and fisheries management. Journal of Fish Biology 78(2):395–435.

Conradt, L., and T. J. Roper. 2005. Consensus decision making in animals. Trends in Ecology and Evolution 20(8):449–456.

Convention on International Trade in Endangered Species. 2010. Draft Resolution on Atlantic Bluefin Tuna. CoP15 Doc. 52. Presented at Doha, Qatar.

Cooke, S. J., and 8 coauthors. 2004. Abnormal migration timing and high en route mortality of sockeye salmon in the Fraser River, British Columbia. Fisheries 29(2):22–33.

Cooke, S. J., and 19 coauthors. 2008. Developing a mechanistic understanding of fish migrations by linking telemetry with physiology, behavior, genomics and experimental biology: An interdisciplinary case study on adult Fraser River sockeye salmon. Fisheries 33(7):321–338.

Cooper, A. B., and M. Mangel. 1999. The dangers of ignoring metapopulation structure for the conservation of salmonids. Fishery Bulletin 97:213–226.

Cope, J. M., and A. E. Punt. 2009. Drawing the lines: Resolving fishery management units with simple fisheries data. Canadian Journal of Fisheries and Aquatic Sciences 66:1256–1273.

Cordes, J. F., and J. E. Graves. 2003. Investigation of congeneric hybridization in and stock structure of weakfish (Cynoscion regalis) inferred from analyses of nuclear and mitochondrial DNA loci. Fishery Bulletin 101(2):443–450.

Corten, A. 2001. The role of "conservatism" in herring migrations. Reviews in Fish Biology and Fisheries 11(4):339–361.

Cortes, E., 2000. Life history patterns and correlation in sharks. Reviews in Fisheries Science 8:299–344.

Cosson, J., A. L. Groison, M. Suquet, C. Fauvel, C. Dreanno, and R. Billard. 2008a. Marine fish spermatozoa: Racing ephemeral swimmers. Reproduction 136:277–294.

———. 2008b. Studying sperm motility in marine fish: An overview on the state of the art. Journal of Applied Ichthyology 24(4):460–486.

Cote, J., S. Fogarty, T. Brodin, K. Weinersmith, and A. Sih. 2010. Personality-dependent dispersal in the invasive mosquitofish: Group composition matters. Proceedings of the Royal Society B: Biological Sciences 278(1712):1670–1678.

Cote, J., S. Fogarty, and A. Sih. 2012. Individual sociability and choosiness between shoal types. Animal Behaviour 83(6):1469–1476.

Couzin, I. D., J. Krause, R. James, G. D. Ruxton, and N. R. Franks. 2002. Collective memory and spatial sorting in animal groups. Journal of Theoretical Biology 218(1):1–11.

Couzin, I. D., J. Krause, N. R. Franks, and S. A. Levin. 2005. Effective leadership and decision-making in animal groups on the move. Nature 433(7025):513–516.

Cowan, J. H., R. S. Birdson, E. D. Houde, J. S. Priest, W. C. Sharp, and G. B. Mateja. 1992. Enclosure experiments on survival and growth of black drum eggs and larvae in lower Chesapeake Bay. Estuaries 15(3):392–402.

Cowen, R. K., and S. Sponaugle. 2009. Larval dispersal and marine population connectivity. Annual Review of Marine Science 1:443–466.

Cowen, R. K., K. M. M. Lwiza, S. Sponaugle, C. B. Paris, and D. B. Olson. 2000. Connectivity of marine populations: Open or closed? Science 287:857–859.

Cowen, R. K., C. B. Paris, and A. Srinivasan. 2006. Scaling of connectivity in marine populations. Science 311(5760):522–527.

Crispo, E. 2008. Modifying effects of phenotypic plasticity on interactions among natural selection, adaptation and gene flow. Journal of Evolutionary Biology 21(6):1460–1469.

Crockford, S. J. 1997. Archeological evidence of large northern bluefin tuna, Thunnus thynnus, in coastal waters of British Columbia and northern Washington. Fishery Bulletin 95(1):11–24.

Cronin, L. E., and A. J. Mansueti. 1971. The biology of the estuary. In A Symposium on the Biological Significance of Estuaries, ed. P. Douglas and R. Stroud, 14–39. Sport Fishing Institute, Washington, DC.

Crook, D. A., and 9 coauthors. 2010. Catadromous migrations by female tupong (Pseudaphritis urvillii) in coastal streams in Victoria, Australia. Marine and Freshwater Research 61(4):474–483.

Crossin, G. T., and 6 coauthors. 2004. Energetics and morphology of sockeye salmon: Effects of upriver migratory distance and elevation. Journal of Fish Biology 65(3):788–810.

Crossin, G. T., and 8 coauthors. 2007. Behaviour and physiology of sockeye salmon homing through coastal waters to a natal river. Marine Biology 152(4):905–918.

Crowder, L. B., S. J. Lyman, W. F. Figueira, and J. Priddy. 2000. Source-sink population dynamics and the problem of siting marine reserves. Bulletin of Marine Science 66:799–820.

Cury, P. 1994. Obstinate nature: An ecology of individuals. Thoughts on reproductive behavior and biodiversity. Canadian Journal of Fisheries and Aquatic Sciences 51:1664–1673.

Cushing, D. H. 1962a. Recruitment to the North Sea herring stocks. Fisheries Investigations Series II 23(5):45–71.

———. 1962b. An alternative method of estimating the critical depth. Journal du Conseil pour l'Exploration de la Mer 27(2):131–140.

———. 1969. The regularity of spawning season of some fishes. Journal du Conseil pour l'Exploration de la Mer 33(1):81–92.

———. 1975. Marine Ecology and Fisheries. Cambridge University Press, Oxford.

———. 1990. Plankton production and year-class strength in fish populations: An update of the match mismatch hypothesis. Advances in Marine Biology 26:249–293.

———. 1995. Population Production and Regulation in the Sea: A Fisheries Perspective. Cambridge University Press, Cambridge.

Dadswell, M. J., R. J. Klauda, C. M. Moffitt, R. L. Saunders, R. A. Rulifson, and J. E. Cooper, eds. 1987. Common Strategies of Anadromous and Catadromous Fishes. American Fisheries Society Symposium 1:1–561.

Dahl, K. 1909. The problem of sea fish hatching: In Rapport sur les Travaux de la Commission A, dans la Période 1902–1907. Rapports et Procès-verbaux 10, Special Part B, No. 5, ed. J. Hjort, 1–40. Conseil International pour l'Exploration de la Mer, Copenhagen.

Dahlgren, C. P., and D. B. Eggleston. 2000. Ecological processes underlying ontogenetic habitat shifts in a coral reef fish. Ecology 81(8):2227–2240.

Dahlgren, C. P., and 8 coauthors. 2006. Marine nurseries and effective juvenile habitats: Concepts and applications. Marine Ecology Progress Series 312:291–295.

Davenport, J. 1994. How and why do flying fish fly? Reviews in Fish Biology and Fisheries 4(2):184–214.

———. 2003. Allometric constraints on stability and maximum size in flying fishes: Implications for their evolution. Journal of Fish Biology 62(2):455–463.

Daverat, F., and 8 coauthors. 2006. Phenotypic plasticity of habitat use by three temperate eel species, Anguilla anguilla, A. japonica and A. rostrata. Marine Ecology Progress Series 308:231–241.

Daverat, F., J. Martin, R. Fablet, and C. Pécheyran. 2011. Colonisation tactics of three temperate catadromous species, eel Anguilla anguilla, mul-

let Liza ramada and flounder Plathychtys flesus, revealed by Bayesian multielemental otolith microchemistry approach. Ecology of Freshwater Fish 20(1):42–51.

Davies, P. C. W. 2012. The epigenome and top-down causation. Interface Focus 2(1):42–48.

Davis, C. S. 1987. Components of the zooplankton production cycle in the temperate ocean. Journal of Marine Research 45(4):947–983.

Dean, M. J., W. S. Hoffman, and M. P. Armstrong. 2012. Disruption of an Atlantic cod spawning aggregation resulting from the opening of a directed gill-net fishery. North American Journal of Fisheries Management 32:124–134.

DeAngelis, D. L., and L. J. Gross, eds. 1992. Individual-Based Models and Approaches in Ecology: Populations, Communities and Ecosystems. Chapman and Hall, New York.

DeCelles, G. R., and S. X. Cadrin. 2010. Movement patterns of winter flounder (Pseudopleuronectes americanus) in the southern Gulf of Maine: Observations with the use of passive acoustic telemetry. Fishery Bulletin 108(4):408–419.

Deelder, C. L. 1960. The Atlantic eel problem. Nature 4713:589–591.

DeMartini, E. E. 1988. Spawning success of the male plainfin midshipman. 1. Influences of male body size and area of spawning site. Journal of Experimental Marine Biology and Ecology 121:177–192.

Denny, M. W. 1993. Air and Water: The Biology and Physics of Life's Media. Princeton University Press, Princeton, NJ.

DeVore, J. D., B. James, and R. Beamesderfer. 1999. Lower Columbia River white sturgeon: Current stock status and management implications. Report No. SS 99-08. US Department of Fish and Wildlife, Washington, DC.

deYoung, B., and G. A. Rose. 1993. On recruitment and distribution of Atlantic cod (Gadus morhua) off Newfoundland. Canadian Journal of Fisheries and Aquatic Sciences 50(12):2729–2741.

Dicken, M. L., A. J. Booth, M. J. Smale, and G. Cliff. 2007. Spatial and seasonal distribution patterns of juvenile and adult raggedtooth sharks (Carcharias taurus) tagged off the east coast of South Africa. Marine and Freshwater Research 58(1):127–134.

Dickey-Collas, M., M. Clarke, and A. Slotte. 2009. "Linking Herring": Do we really understand plasticity? ICES Journal of Marine Science 66(8):1649–1651.

Dingemanse, N. J., and 7 coauthors. 2009. Individual experience and evolutionary history of predation affect expression of heritable variation in fish personality and morphology. Proceedings of the Royal Society B: Biological Sciences 276(1660):1285-1293.

Dingle, H. 1996. Migration: The Biology of Life on the Move. Oxford University Press, Oxford.

Dittman, A. H., and T. P. Quinn. 1996. Homing in Pacific salmon: Mechanisms and ecological basis. Journal of Experimental Biology 199(1):83-91.

Doak, D. F., D. Bigger, E. K. Harding, M. A. Marvier, R. E. O'Malley, and D. Thomson. 1998. The statistical inevitability of stability-diversity relationships in community ecology. American Naturalist 151(3):264-276.

Dodson, J. J. 1988. The nature and role of learning in the orientation and migratory behavior of fishes. Environmental Biology of Fishes 23(3):161-182.

Dodson, J. J., N. Abin-Horth, V. Thériault, and D. J. Páez. 2013. The evolutionary ecology of alternative migratory tactics in salmonid fishes. Biological Reviews 88(3):602-625.

Dokter, A. M., F. Liechti, H. Stark, L. Delobbe, P. Tabary, and I. Holleman. 2011. Bird migration flight altitudes studied by a network of operational weather radars. Journal of the Royal Society Interface 8(54):30-43.

Domeier, M. L., and P. L. Colin. 1997. Tropical reef fish spawning and aggregations: Defined and reviewed. Bulletin of Marine Science 60:698-726.

Dommasnes, A., F. Rey, and I. Röttingen. 1994. Reduced oxygen concentrations in herring wintering areas. ICES Journal of Marine Science 51(1):63-69.

Doroshov, S. I. 1970. Biological features of the eggs, larvae and young of the striped bass (Roccus saxatilis (Walbaum)) in connection with the problem of its acclimation in the USSR. Ichthyology 10:235-248.

Drinkwater, K. F., and 8 coauthors. 2010. On the processes linking climate to ecosystem changes. Journal of Marine Systems 79(3-4):374-388.

Dr. Seuss. 1960. One Fish, Two Fish, Red Fish, Blue Fish. Random House, New York.

Duarte, C. M., and M. Alcaraz. 1989. To produce many small or few large eggs: A size-independent reproductive tactic of fish. Oecologia 80:401-404.

Dufour, F., H. Arrizabalaga, X. Irigoien, and J. Santiago. 2010. Climate impacts on albacore and bluefin tunas migrations phenology and spatial distribution. Progress in Oceanography 86:283-290.

Dunlop, E. S., K. Enberg, C. Jørgensen, and M. Heino. 2009. Toward Darwinian fisheries management. Evolutionary Applications 2:245-259.

Edeline, E. 2007. Adaptive phenotypic plasticity of eel diadromy. Marine Ecology Progress Series 341:229-232.

Edeline, E., A. Bardonnet, V. Bolliet, S. Dufour, and E. Pierre. 2005. Endocrine control of Anguilla anguilla glass eel dispersal: Effect of thyroid hormones on locomotor activity and rheotactic behavior. Hormones and Behavior 48(1):53-63.

Edeline, E., L. Beaulaton, R. Le Barh, and P. Elie. 2007. Dispersal in metamorphosing juvenile eel Anguilla anguilla. Marine Ecology Progress Series 344:213-218.

Eder, S. H., H. Cadiou, A. Muhamad, P. A. McNaughton, J. L. Kirshvink, and M. Winklhofer. 2012. Magnetic characterization of isolated candidate vertebrate magnetoreceptor cells. Proceedings of the National Academy of Sciences of the United States of America 109(30):12,022-12,027.

Edrén, S. M. C., and S. H. Gruber. 2005. Homing ability of young lemon sharks, Negaprion brevirostris. Environmental Biology of Fishes 72(3):267-281.

Edwards, A. M., and 10 coauthors. 2007. Revisiting Lévy flight search patterns of wandering albatrosses, bumblebees and deer. Nature 449(7165):1044-1048.

Ehrlich, P. R., and A. H. Ehrlich. 1981. Extinction: The Causes and Consequences of the Disappearance of Species. Random House, New York.

Eldridge, M. B., J. A. Whipple, M. J. Bowers, B. M. Jarvis, and J. Gold. 1991. Reproductive-performance of yellowtail rockfish, Sebastes flavidus. Environmental Biology of Fishes 30(1-2):91-102.

Eliason, E. J., and 9 coauthors. 2011. Differences in thermal tolerance among sockeye salmon populations. Science 332(6025):109-112.

Ellis, T., and R. N. Gibson. 1995. Size-selective predation of 0-group flatfishes on a Scottish coastal nursery ground. Marine Ecology Progress Series 127(1-3):27-37.

Elmqvist, T., and 6 coauthors. 2003. Response diversity, ecosystem change, and resilience. Frontiers in Ecology and the Environment 1(9):488-494.

Elsdon, T. S., and 8 coauthors. 2008. Otolith chemistry to describe movements and life-history parameters of fishes: Hypotheses, assumptions, limitations and inferences. Oceanography and Marine Biology: An Annual Review 46:297-330.

Elsdon, T. S., S. Ayvazian, K. W. McMahon, and S. R.

Thorrold. 2010. Experimental evaluation of stable isotope fractionation in fish muscle and otoliths. Marine Ecology Progress Series 408:195–205.

Engelhard, G. H., and M. Heino. 2005. Scale analysis suggests frequent skipping of the second reproductive season in Atlantic herring. Biology Letters 1(2):172–175.

Erickson, D. L., and J. E. Hightower. 2007. Oceanic distribution and behavior of green sturgeon. American Symposium 57:197–211.

Evans, J. P., and A. J. Geffen. 1998. Male condition, sperm traits, and reproductive success in winter-spawning Celtic Sea herring Clupea harengus. Marine Biology 132:179–186.

Falconer, D. S. 1981. Introduction to Quantitative Genetics. Longman, London.

Faria, J. J., and 7 coauthors. 2010. A novel method for investigating the collective behaviour of fish: Introducing "Robofish." Behavioral Ecology and Sociobiology 64(8):1211–1218.

Feldheim, K. A., S. H. Gruber, and M. V. Ashley. 2004. Reconstruction of parental microsatellite genotypes reveals female polyandry and philopatry in the lemon shark, Negaprion brevirostris. Evolution 58(10):2332–2342.

Fenske, K. H., M. J. Wilberg, D. H. Secor, and M. C. Fabrizio. 2009. An age- and sex-structured assessment model for American eels (Anguilla rostrata) in the Potomac River, Maryland. Canadian Journal of Fisheries and Aquatic Sciences 68(6):1024–1037.

Fernö, A., and 6 coauthors. 1998. The challenge of the herring in the Norwegian Sea: Making optimal collective spatial decisions. Sarsia 83(2):149–167.

Ferretti, F., B. Worm, G. L. Britten, M. R. Heithaus, and H. K. Lotze. 2010. Patterns and ecosystem consequences of shark declines in the ocean. Ecology Letters 13(8):1055–1071.

Figueira, W. F., and L. B. Crowder. 2006. Defining patch contribution in source-sink metapopulations: The importance of including dispersal and its relevance to marine systems. Population Ecology 48(3):215–224.

Fiksen, O., and A. Folkvord. 1999. Modelling growth and ingestion processes in herring Clupea harengus. Fish and Fisheries 6:50–72.

Fisher, R., S. M. Sogard, and S. A. Berkeley. 2007. Trade-offs between size and energy reserves reflect alternative strategies for optimizing larval survival potential in rockfish. Marine Ecology Progress Series 344:257–270.

Fisheries and Oceans Canada. 2012. Stock Assessment Update of Northern (2J3KL) Cod. Canadian Science Advisory Secretariat Science Response 2012/009. DFO, Ottawa.

Fitzpatrick, J. L., J. K. Desjardins, N. Milligan, R. Montgomerie, and S. Balshine. 2007. Reproductive-tactic-specific variation in sperm swimming speeds in a shell-brooding Cichlid. Biology of Reproduction 77(2):280–284.

Fleming, I. A., and M. R. Gross. 1990. Latitudinal clines: A trade-off between egg number and size in Pacific salmon. Ecology 71(1):1–11.

Fogarty, M. J., and S. A. Murawski. 1998. Large-scale disturbance and the structure of marine system: Fishery impacts on Georges Bank. Ecological Applications 8(1):S6–S22.

Folke, C., S. Carpenter, T. Elmqvist, L. Gunderson, C. S. Holling, and B. Walker. 2002. Resilience and sustainable development: Building adaptive capacity in a world of transformations. Ambio 31(5):437–440.

Folke, C., and 6 coauthors. 2004. Regime shifts, resilience, and biodiversity in ecosystem management. Annual Review of Ecology Evolution and Systematics 35:557–581.

Fonteneau, A., and P. P. Soubrier. 1996. Interactions between tuna fisheries: A global review with specific examples from the Atlantic Ocean. In Status of Pacific Tuna Fisheries in 1995: Proceedings of the Second FAO Expert Consultation in Interactions on Pacific Tuna Fisheries. FAO Tech. Paper 365, 84–123. Food and Agriculture Organization of the United Nations, Rome.

Ford, M. J. 2004. Conserving units and preserving diversity. In Evolution Illuminated: Salmon and Their Relatives, ed. A. P. Hendry and S. C. Stearns, 338–357. Oxford University Press, New York.

Formicki, K., M. Sadowski, A. Tanski, A. Korzelecka-Orkisz, and A. Winnicki. 2004. Behaviour of trout (Salmo trutta L.) larvae and fry in a constant magnetic field. Journal of Applied Ichthyology 20(4):290–294.

Forward, R. B., J. S. Burke, D. Rittschof, and J. M. Welch. 1996. Photoresponses of larval Atlantic menhaden (Brevoortia tyrannus Latrobe) in offshore and estuarine waters: Implications for transport. Journal of Experimental Marine Biology and Ecology 199(1):123–135.

Frank, K. T., and D. Brickman. 2000. Allee effects and compensatory population dynamics within a stock complex. Canadian Journal of Fisheries and Aquatic Sciences 57(3):513–517.

———. 2001. Contemporary management issues confronting fisheries science. Journal of Sea Research 45:173–187.

Frank, K. T., B. Petrie, J. A. D. Fisher, and W. C. Leggett. 2011. Transient dynamics of an altered large marine ecosystem. Nature 477(7362):86–89.

Fraser, D. F., J. F. Gilliam, M. J. Daley, A. N. Le, and G. T. Skalski. 2001. Explaining leptokurtic movement distributions: Intrapopulation variation in boldness and exploration. American Naturalist 158(2):124–135.

Freedman, J. A., and D. L. G. Noakes. 2002. Why are there no really big bony fishes? A point-of-view on maximum body size in teleosts and elasmobranchs. Reviews in Fish Biology and Fisheries 12(4):403–416.

Fréon, P., F. Gerlotto, and M. Soria. 1996. Diel variability of school structure with special reference to transition periods. ICES Journal of Marine Science 53(2):459–464.

Fréon, P., P. Cury, L. Shannon, and C. Roy. 2005. Sustainable exploitation of small pelagic fish stocks challenged by environmental and ecosystem changes: A review. Bulletin of Marine Science 76(2):385–462.

Fretwell, S. 1972. Populations in a Seasonal Environment. Princeton University Press, Princeton, NJ.

Fretwell, S. D., and H. L. Lucas. 1970. On territorial behavior and other factors influencing habitat distribution in birds. 1. Theoretical development. Acta Biotheory 19:136–156.

Friedland, K. D., and 6 coauthors. 2001. Open-ocean orientation and return migration routes of chum salmon based on temperature data from data storage tags. Marine Ecology Progress Series 216:235–252.

Frisk, M. G., T. J. Miller, and M. J. Fogarty. 2001. Estimation and analysis of biological parameters in elasmobranch fishes: A comparative life history study. Canadian Journal of Fisheries and Aquatic Sciences 58(5):969–981.

———. 2002. The population dynamics of little skate Leucoraja erinacea, winter skate Leucoraja ocellata, and barndoor skate Dipturus laevis: Predicting exploitation limits using matrix analyses. ICES Journal of Marine Science 59(3):576–586.

Frisk, M. G., T. J. Miller, S. J. D. Martell, and K. Sosebee. 2008. New hypothesis helps explain elasmobranch "OutBurst" on Georges Bank in the 1980s. Ecological Applications 18:234–245.

Frisk, M. G., A. Jordaan, and T. J. Miller. 2013. Moving beyond the current paradigm of marine population connectivity: Are adults the missing link? Fish and Fisheries 15(2):242–254.

Fromentin, J. M., and J. E. Powers. 2005. Atlantic bluefin tuna: Population dynamics, ecology, fisheries and management. Fish and Fisheries 6(4):281–306.

Fromentin, J. M., R. A. Myers, O. N. Bjornstad, N. C. Stenseth, J. Gjosaeter, and H. Christie. 2001. Effects of density-dependent and stochastic processes on the regulation of cod populations. Ecology 82(2):567–579.

Fromentin, J. M., G. Reygondeau, S. Bonhommeau, and G. Beaugrand. 2014. Oceanographic changes and exploitation drive the spatiotemporal dynamics of Atlantic bluefin tuna. Fisheries Oceanography 23(2):147–156.

Frost, B. W. 1987. Grazing control of phytoplankton stock in the open Sub-Arctic Pacific Ocean: A model assessing the role of mesozooplankton, particularly the large calanoid copepods Neocalanus spp. Marine Ecology Progress Series 39(1):49–68.

Fry, B. 2002. Conservative mixing of stable isotopes across estuarine salinity gradients: A conceptual framework for monitoring watershed influences on downstream fisheries production. Estuaries 25(2):264–271.

Fry, B., and M. M. Chumchal. 2011. Sulfur stable isotope indicators of residency in estuarine fish. Limnology and Oceanography 56(5):1563–1576.

Fudge, S. B., and G. A. Rose. 2009. Passive- and active-acoustic properties of a spawning Atlantic cod (Gadus morhua) aggregation. ICES Journal of Marine Science 66(6):1259–1263.

Fuiman, L. A. 2002. Special considerations of fish eggs and larvae. In Fisheries Science: The Unique Contributions of Early Life Stages, ed. L. A. Fuiman and R. G. Werner, 1–32. Blackwell, Malden, MA.

Fuiman, L. A., and R. S. Batty. 1997. What a drag it is getting cold: Partitioning the physical and physiological effects of temperature on fish swimming. Journal of Experimental Biology 200(12):1745–1755.

Fuiman, L. A., and J. H. Cowan. 2003. Behavior and recruitment success in fish larvae: Repeatability and covariation of survival skills. Ecology 84(1):53–67.

Fukumori, K., N. Okuda, H. Hamaoka, T. Fukumoto, D. Takahashi, and K. Omori. 2008. Stable isotopes reveal life history polymorphism in the coastal fish Apogon notatus. Marine Ecology Progress Series 362:279–289.

Furin, C. G., F. A. von Hippel, and M. A. Bell. 2012. Partial reproductive isolation of a recently derived resident-freshwater population of threespine stickleback (*Gasterosteus aculeatus*) from its putative anadromous ancestor. Evolution 66(10): 3277–3286.

Gage, M. J. G., P. Stockley, and G. A. Parker. 1995. Effects of alternative male mating strategies on characteristics of sperm production in the Atlantic salmon (*Salmo salar*): Theoretical and empirical investigations. Philosophical Transactions of the Royal Society B: Biological Sciences 350(1334):391–399.

Gaggiotti, O. E., and 6 coauthors. 2009. Disentangling the effects of evolutionary, demographic, and environmental factors influencing genetic structure of natural populations: Atlantic herring as a case study. Evolution 63(11):2939–2951.

Gaines, S. D., B. Gaylord, L. R. Gerber, A. Hastings, and B. P. Kinlan. 2007. Connecting places: The ecological consequences of dispersal in the sea. Oceanography 20(3):90–99.

Garrison, L. P., and J. S. Link. 2000. Fishing effects on spatial distribution and trophic guild structure of the fish community in the Georges Bank region. ICES Journal of Marine Science 57(3):723–730.

Gauthreaux, S. A., and C. G. Belser. 2003. Radar ornithology and biological conservation. Auk 120(2):266–277.

Gedamke, T., J. M. Hoenig, W. D. DuPaul, and J. A. Musick. 2009. Stock-recruitment dynamics and the maximum population growth rate of the barndoor skate on Georges Bank. North American Journal of Fisheries Management 29(2):512–526.

Ghalambor, C. K., J. K. McKay, S. P. Carroll, and D. N. Reznick. 2007. Adaptive versus non-adaptive phenotypic plasticity and the potential for contemporary adaptation in new environments. Functional Ecology 21(3):394–407.

Gibson, R. N. 2005. The behaviour of flatfishes. In Flatfishes: Biology and Exploitation, ed. R. N. Gibson, 213–239. Blackwell, Oxford.

Gilbert, C. H. 1915. Contributions to the Life History of the Sockeye Salmon. British Columbia Fisheries Department, Vancouver.

Gillanders, B. M. 2002. Connectivity between juvenile and adult fish populations: Do adults remain near their recruitment estuaries? Marine Ecology Progress Series 240:215–223.

Gilliam, J. F., and D. F. Fraser. 1987. Habitat selection under predation hazards: Test of a model with foraging minnows. Ecology 68:1856–1862.

Giraldeau, L. 1997. The ecology of information use. In Behavioural Ecology: An Evolutionary Approach, ed. J. R. Krebs and N. B. Davies, 42–68. Blackwell, Oxford.

Glebe, B. D., and W. C. Leggett. 1981. Latitudinal differences in energy allocation and use during the freshwater migrations of American shad (*Alosa sapidissima*) and their life history consequences. Canadian Journal of Fisheries and Aquatic Science 38:806–820.

Gleiss, A. C., B. Norman, and R. P. Wilson. 2011. Moved by that sinking feeling: Variable diving geometry underlies movement strategies in whale sharks. Functional Ecology 25(3):595–607.

Goethel, D. R., T. J. Quinn, and S. X. Cadrin. 2011. Incorporating spatial structure in stock assessment: Movement modeling in marine fish population models. Reviews in Fisheries Science 19:119–136.

Gökçe, F., and E. Şahin. 2010. The pros and cons of flocking in the long-range "migration" of mobile robot swarms. Theoretical Computer Science 411(21):2140–2154.

Goldman, K. J., and S. D. Anderson. 1999. Space utilization and swimming depth of white sharks, *Carcharodon carcharias*, at the South Farallon Islands, central California. Environmental Biology of Fishes 56(4):351–364.

Gosse, K. R., and J. S. Wroblewski. 2004. Variant colourations of Atlantic cod (*Gadus morhua*) in Newfoundland and Labrador nearshore waters. ICES Journal of Marine Science 61(5):752–759.

Gotelli, N. J. 1991. Metapopulation models: The rescue effect, the propagule rain, and the core-satellite hypothesis. American Naturalist 138(3): 768–776.

Gould, S. J. 1977. Ontogeny and Phylogeny. Harvard University Press, Cambridge, MA

———. 2002. The Structure of Evolutionary Theory. Harvard University Press, Cambridge, MA.

Gould, S. J., and R. C. Lewontin. 1979. Spandrels of San Marco and the Panglossian paradigm: A critique of the adaptationist program. Proceedings of the Royal Society B: Biological Sciences 205(1161):581–598.

Govoni, J. J., and R. B. Forward. 2008. Buoyancy. In Fish Larval Physiology, ed. R. N. Finn and B. G. Kapoor, chap. 15, 495–522. Science Publishers, Enfield, NH.

Goyette, J. L., R. W. Howe, A. T. Wolf, and W. D. Robinson. 2011. Detecting tropical nocturnal birds using automated audio recordings. Journal of Field Ornithology 82(3):279–287.

Grant, G. C., and J. E. Olney. 1991. Distribution of striped bass *Morone saxatilis* (Walbaum) eggs and larvae in major Virginia Rivers. Fishery Bulletin 89:187–193.

Grant, W. S., I. B. Spies, and M. F. Canino. 2006. Biogeographic evidence for selection on mitochondrial DNA in North Pacific walleye pollock *Theragra chalcogramma*. Journal of Heredity 97(6):571–580.

Green, J. M., and J. S. Wroblewski. 2000. Movement patterns of Atlantic cod in Gilbert Bay, Labrador: Evidence for bay residency and spawning site fidelity. Journal of the Marine Biological Association of the United Kingdom 80(6):1077–1085.

Green, K. M., and R. M. Starr. 2011. Movements of small adult black rockfish: Implications for the design of MPAs. Marine Ecology Progress Series 436:219–230.

Greenberg, R., and P. P. Marra, eds. 2005. Birds of Two Worlds: The Ecology and Evolution of Migration. Johns Hopkins University Press, Baltimore.

Greenwood, A. K., A. R. Wark, K. Yoshida, and C. L. Peichel. 2013. Genetic and neural modularity underlie evolution of schooling behavior in threespine stickleback. Current Biology 23:1884–1888.

Groot, C. 1982. Modifications on a theme: A perspective on migratory behavior of Pacific salmon. In Proceedings of the Salmon and Trout Migratory Behavior Symposium, ed. E. L. Brannon and E. O. Salo, 1–21. University of Washington Press, Seattle.

Groot, C., and T. P. Quinn. 1987. Homing migration of sockeye salmon, *Oncorhynchus nerka*, to the Fraser River. Fishery Bulletin 85(3):455–469.

Gross, M. R. 1985. Disruptive selection for alternative life histories in salmon. Nature 313:47–48.

———. 1991. Salmon breeding behavior and life history evolution in changing environments. Ecology 72:1180–1186.

———. 2005. The evolution of parental care. Quarterly Review of Biology 80(1):37–45.

Gross, M. R., and J. Repka. 1998. Stability with inheritance in the conditional strategy. Journal of Theoretical Biology 192(4):445–453.

Gross, M. R., R. M. Coleman, and R. M. McDowall. 1988. Aquatic productivity and the evolution of diadromous fish migration. Science 239:1291–1293.

Grothues, T. M., and K. W. Able. 2007. Scaling acoustic telemetry of bluefish in an estuarine observatory: Detection and habitat use patterns. Transactions of the American Fisheries Society 136(6):1511–1519.

Grothues, T. M., K. W. Able, J. McDonnell, and M. M. Sisak. 2005. An estuarine observatory for real-time telemetry of migrant macrofauna: Design, performance, and constraints. Limnology and Oceanography Methods 3:275–289.

Grothues, T. M., K. A. Able, J. Carter, and T. W. Arienti. 2009. Migration patterns of striped bass through nonnatal estuaries of the U.S. Atlantic Coast. American Fisheries Society Symposium 69:135–150.

Guttal, V., and I. D. Couzin. 2010. Social interactions, information use, and the evolution of collective migration. Proceedings of the National Academy of Sciences of the United States of America 107(37):16,172–16,177.

Guttridge, T. L., A. A. Myrberg, I. F. Porcher, D. W. Sims, and J. Krause. 2009a. The role of learning in shark behaviour. Fish and Fisheries 10(4):450–469.

Guttridge, T. L., S. H. Gruber, K. S. Gledhill, D. P. Croft, D. W. Sims, and J. Krause. 2009b. Social preferences of juvenile lemon sharks, *Negaprion brevirostris*. Animal Behaviour 78:543–548.

Gyllenberg, M., I. Hanski, and A. Hastings. 1997. Structured metapopulation models. In Metapopulation Biology: Ecology, Genetics, and Evolution, ed. I. A. Hanski and M. E. Gilpin, 93–122. Academic, New York.

Hall, C. A. S. 1988. An assessment of several of the historically most influential theoretical models used in ecology and of the data provided in their support. Ecological Modelling 43(1–2):5–31.

Halpern, B. S., and R. R. Warner. 2003. Matching marine reserve design to reserve objectives. Proceedings of the Royal Society B: Biological Sciences 270(1527):1871–1878.

Hamann, E. J., and B. P. Kennedy. 2012. Juvenile dispersal affects straying behaviors of adults in a migratory population. Ecology 93(4):733–740.

Hamer, P. A., G. P. Jenkins, and P. Coutin. 2006. Barium variation in *Pagrus auratus* (Sparidae) otoliths: A potential indicator of migration between an embayment and ocean waters in southeastern Australia. Estuarine Coastal and Shelf Science 68(3–4):686–702.

Hamlett, W. C., A. M. Eulitt, R. L. Jarrell, and M. A. Kelly. 1993. Uterogestation and placentation in elasmobranchs. Journal of Experimental Zoology 266(5):347–367.

Hansen, G. J. A., and M. L. Jones. 2008. A rapid assessment approach to prioritizing streams for control of Great Lakes sea lampreys (*Petromyzon marinus*): A case study in adaptive management.

Canadian Journal of Fisheries and Aquatic Sciences 65(11):2471–2484.

Hanski, I. 1997. Metapopulation dynamics: From concepts and observations to predictive models. In Metapopulation Biology: Ecology, Genetics, and Evolution, ed. I. A. Hanski and M. E. Gilpin, 69–91. Academic, New York.

Hanski, I., T. Pakkala, M. Kuussaari, and G. C. Lei. 1995. Metapopulation persistence of an endangered butterfly in a fragmented landscape. Oikos 72(1):21–28.

Harden Jones, F. R. 1968. Fish Migration. Edward Arnold, London.

Hare, J. A., and R. K. Cowen. 1993. Ecological and evolutionary implications of the larval transport and reproductive strategy of bluefish Pomatomus saltatrix. Marine Ecology Progress Series 98:1–16.

———. 1994. Ontogeny and otolith microstructure of bluefish Pomatomus saltatrix (Pisces: Pomatomidae). Marine Biology 118:541–550.

———. 1996. Transport mechanisms of larval and pelagic juvenile bluefish (Pomatomus saltatrix) from South Atlantic Bight spawning grounds to Middle Atlantic Bight nursery habitats. Limnology and Oceanography 41(6):1264–1280.

Haro, A., and 7 coauthors. 2000. Population decline of the American eel: Implications for research and management. Fisheries 25(9):7–16.

Haro, A., and 8 coauthors, eds. 2009. Challenges for Diadromous Fishes in a Dynamic Global Environment. American Fisheries Society Symposium 69:1–943.

Harris, P. J., and J. C. McGovern. 1997. Changes in the life history of red porgy, Pagrus pagrus, from the southeastern United States, 1972–1994. Fishery Bulletin 95(4):732–747.

Harrison, S., and A. D. Taylor. 1997. Empirical evidence for metapopulation dynamics. In Metapopulation Biology: Ecology, Genetics, and Evolution, ed. I. A. Hanski and M. E. Gilpin, 27–42. Academic, New York.

Hartvig, M., K. H. Andersen, and J. E. Beyer. 2011. Food web framework for size-structured populations. Journal of Theoretical Biology 272(1):113–122.

Hasler, A. D., and A. T. Scholz. 1983. Olfactory imprinting and homing in salmon: Investigations into the mechanism of the imprinting process. Springer-Verlag, Berlin.

Hatfield, T., and D. Schluter. 1999. Ecological speciation in sticklebacks: Environment-dependent hybrid fitness. Evolution 53(3):866–873.

Hattori, T., T. Okuda, Y. Narimatsu, and M. Ito.

2010. Distribution patterns of five pleuronectid species on the continental slope off the Pacific coast of northern Honshu, Japan. Fisheries Science 76(5):747–754.

Hauser, L., and G. R. Carvalho. 2008. Paradigm shifts in marine fisheries genetics: Ugly hypotheses slain by beautiful facts. Fish and Fisheries 9(4):333–362.

Hay, D. E. 1985. Reproductive-biology of Pacific herring (Clupea harengus pallasi). Canadian Journal of Fisheries and Aquatic Sciences 42:111–126.

———. 1990. Tidal influence on spawning time of Pacific herring (Clupea harengus pallasi). Canadian Journal of Fisheries and Aquatic Sciences 47(12):2390–2401.

Hay, D. E., and J. R. Brett. 1988. Maturation and fecundity of Pacific herring (Clupea harengus pallasi): An experimental study with comparisons to natural populations. Canadian Journal of Fisheries and Aquatic Science 45:399–406.

Hay, D. E., P. B. McCarter, and K. S. Daniel. 2001. Tagging of Pacific herring Clupea pallasi from 1936–1992: A review with comments on homing, geographic fidelity, and straying. Canadian Journal of Fisheries and Aquatic Sciences 58(7):1356–1370.

Hay, D. E., P. B. McCarter, K. S. Daniel, and J. F. Schweigert. 2009. Spatial diversity of Pacific herring (Clupea pallasi) spawning areas. ICES Journal of Marine Science 66(8):1662–1666.

Healey, M., P. Kline, and C. F. Tsai. 2001. Saving the endangered Formosa landlocked salmon. Fisheries 26(4):6–14.

Heath, M. R. 1989. Transport of larval herring (Clupea harengus L.) by the Scottish coastal current. ICES Symposium 191:85–91.

———. 1992. Field investigations of the early-life stages of marine fish. Advances in Marine Biology 28:1–174.

———. 1995. Size spectrum dynamics and the planktonic ecosystem of Loch Linnhe. ICES Journal of Marine Science 52:627–642.

———. 1996. The consequences of spawning time and dispersal patterns of larvae for spatial and temporal variability in survival to recruitment. In Survival Strategies in Early Life History Stages of Marine Resources, ed. Y. Watanabe, Y. Yamashita, and Y. Oozeki, 175–208. A. A. Balkema, Rotterdam.

Heath, M. R., P. M. MacLachlan, and J. H. A. Martin. 1987. Inshore circulation and transport of herring larvae off the north central coast of Scotland. Marine Ecology Progress Series 40:11–23.

Heath, M. R., P. A. Kunzlik, A. Gallego, S. J. Holmes, and P. J. Wright. 2008. A model of meta-population dynamics for North Sea and West of Scotland cod: The dynamic consequences of natal fidelity. Fisheries Research 93:92–116.

Hebblewhite, M., and D. T. Haydon. 2010. Distinguishing technology from biology: A critical review of the use of GPS telemetry data in ecology. Philosophical Transactions of the Royal Society B: Biological Sciences 365(1550):2303–2312.

Hedenström, A. 2002. Aerodynamics, evolution and ecology of avian flight. Trends in Ecology and Evolution 17(9):415–422.

Hedenström, A., and T. Alerstam. 1998. How fast can birds migrate? Journal of Avian Biology 29(4): 424–432.

Hedgecock, D., P. H. Barber, and S. Edmands. 2007. Genetic approaches to measuring connectivity. Oceanography 20(3):70–79.

Hedger, R. D., F. Martin, D. Hatin, F. Caron, F. G. Whoriskey, and J. J. Dodson. 2008. Active migration of wild Atlantic salmon Salmo salar smolt through a coastal embayment. Marine Ecology Progress Series 355:235–246.

Hedger, R. D., and 7 coauthors. 2010. Residency and migratory behaviour by adult Pomatomus saltatrix in a South African coastal embayment. Estuarine Coastal and Shelf Science 89(1):12–20.

Hedrick, T. L., B. W. Tobalske, and A. A. Biewener. 2002. Estimates of circulation and gait change based on a three-dimensional kinematic analysis of flight in cockatiels (Nymphicus hollandicus) and ringed turtle-doves (Streptopelia risotia). Journal of Experimental Biology 205(10):1389–1409.

Heincke, F. 1898. Naturgeschicte des Herings I: Die Lokalformen und die Wanderungen des Herings in den europaischen Meeren. Abhandlung der deutschen Seefischereivereins 2. Otto Salle, Berlin.
———. 1913. Investigation of the plaice: The plaice fishery and protective regulation. General Report, Part 1. Conseil Permanent International pour l'Exploration de la Mer 17:1–153.

Heinisch, G., and 16 coauthors. 2008. Spatial-temporal pattern of bluefin tuna (Thunnus thynnus L. 1758) gonad maturation across the Mediterranean Sea. Marine Biology 154(4):623–630.

Heino, M., and O. R. Godø. 2002. Fisheries-induced selection pressures in the context of sustainable fisheries. Bulletin of Marine Science 70:639–656.

Heithaus, M. R. 2007. Nursery areas as essential fish habitats: A theoretical perspective. American Fisheries Society Symposium 50:3–13.

Helbig, A. J. 1991. Inheritance of migratory direction in a bird species: A cross-breeding experiment with SE-migrating and SW-migrating blackcaps (Sylvia atricapilla). Behavioral Ecology and Sociobiology 28(1):9–12.

Helfman, G. S., and E. T. Schultz. 1984. Social transmission of behavioral traditions in a coral-reef fish. Animal Behaviour 32:379–384.

Helvey, M. 1982. First observations of courtship behavior in rockfish, genus Sebastes. Copeia 1982: 763–770.

Hempel, G. 1980. The Egg Stage: Early Life History of Marine Fish. Washington Sea Grant, Seattle.

Hendry, A. P. 2009. Ecological speciation! Or the lack thereof? Canadian Journal of Fisheries and Aquatic Sciences 66(8):1383–1398.

Hendry, A. P., and S. C. Stearns, eds. 2004. Appendix 1: Straying rates of anadromous salmonids. In Evolution Illuminated: Salmon and Their Relatives, 377–383. Oxford University Press, New York.

Hendry, A. P., J. K. Wenburg, P. Bentzen, E. C. Volk, and T. P. Quinn. 2000. Rapid evolution of reproductive isolation in the wild: Evidence from introduced salmon. Science 290(5491):516–518.

Hendry, A. P., V. Castric, M. T. Kinnison, and T. P. Quinn. 2004a. The evolution of philopatry and dispersal: Homing versus straying in salmonids. In Evolution Illuminated: Salmon and Their Relatives, ed. A. P. Hendry and S. C. Stearns, 52–90. Oxford University Press, New York.

Hendry, A. P., T. Bohlin, B. Jonsson, and O. E. Berg. 2004b. To sea or not to sea? Anadromy versus non-anadromy in salmonids. In Evolution Illuminated: Salmon and Their Relatives, ed. A. P. Hendry and S. C. Stearns, 92–125. Oxford University Press, New York.

Herzka, S. Z. 2005. Assessing connectivity of estuarine fishes based on stable isotope ratio analysis. Estuarine Coastal and Shelf Science 64(1):58–69.

Heupel, M. R., C. A. Simpfendorfer, and R. E. Hueter. 2003. Running before the storm: Blacktip sharks respond to falling barometric pressure associated with Tropical Storm Gabrielle. Journal of Fish Biology 63(5):1357–1363.

Heupel, M. R., J. M. Semmens, and A. J. Hobday. 2006. Automated acoustic tracking of aquatic animals: Scales, design and deployment of listening station arrays. Marine and Freshwater Research 57(1):1–13.

Heyman, W. D., R. T. Graham, B. Kjerfve, and R. E. Johannes. 2001. Whale sharks Rincodon typus aggregate to feed on fish spawn in Belize. Marine Ecology Progress Series 215:275–282.

Hilborn, R., and C. J. Walters. 1992. Quantitative Fisheries Stock Assessment: Choices, Dynamics, and Uncertainty. Chapman and Hall, New York.

Hilborn, R., T. P. Quinn, D. E. Schindler, and D. E. Rogers. 2003. Biocomplexity and fisheries sustainability. Proceedings of the National Academy of Sciences of the United States of America 100(11):6564–6568.

Hildenbrandt, H., C. Carere, and C. K. Hemelrijk. 2010. Self-organized aerial displays of thousands of starlings: A model. Behavioral Ecology 21(6):1349–1359.

Hill, M. F., A. Hastings, and L. W. Botsford. 2002. The effects of small dispersal rates on extinction times in structured metapopulation models. American Naturalist 160(3):389–402.

Hiroi, J., Y. Sakakura, M. Tagawa, T. Seikai, and M. Tanaka. 1997. Developmental changes in low-salinity tolerance and responses of prolactin, cortisol and thyroid hormones to low-salinity environment in larvae and juveniles of Japanese flounder, *Paralichthys olivaceus*. Zoological Science 14(6):987–992.

Hjort, J. 1909. A: general part II. Summary of the results of the investigations. In Rapport sur les Travaux de la Commission A, dans la Période 1902-1907. Rapports et Procès-verbaux 10, ed. J. Hjort, 20–159. Conseil International pour l'Exploration de la Mer, Copenhagen.

———. 1914. Fluctuations in the great fisheries of Northern Europe. Conseil Permanent International pour L'Exploration de la Mer 20:1–228.

Hjort, J., and E. Lea. 1914. The age of herring. Nature 94:60–61.

Hobbs, R. J., E. Higgs, and J. A. Harris. 2009. Novel ecosystems: Implications for conservation and restoration. Trends in Ecology and Evolution 24(11):599–605.

Hobson, K. A. 1999. Tracing origins and migration of wildlife using stable isotopes: A review. Oecologia 120(3):314–326.

———. 2003. Making migratory connections with stable isotopes. In Avian Migration, ed. P. Berthold, E. Gwinner, and E. Sonnenschein, 379–391. Springer-Verlag, Berlin.

Hobson, K. A., and D. R. Norris. 2008. Animal migration: A context for using new techniques and approaches. In Tracking Animal Migration with Stable Isotopes, ed. K. A. Hobson and L. I. Wassenaar, 1–18. Elsevier, New York.

Hobson, V. J., D. Righton, J. D. Metcalfe, and G. C. Hays. 2009. Link between vertical and horizontal movement patterns of cod in the North Sea. Aquatic Biology 5(2):133–142.

Hocutt, C. H., S. E. Seibold, R. M. Harrell, R. V. Jesien, and W. H. Bason. 1990. Behavioral observations of striped bass (*Morone saxatilis*) on the spawning grounds of the Choptank and Nanticoke rivers, Maryland, USA. Journal of Applied Ichthyology 6(4):211–222.

Hodgson, S., and T. P. Quinn. 2002. The timing of adult sockeye salmon migration into fresh water: Adaptations by populations to prevailing thermal regimes. Canadian Journal of Zoology 80(3):542–555.

Hodgson, S., T. P. Quinn, R. Hilborn, R. C. Francis, and D. E. Rogers. 2006. Marine and freshwater climatic factors affecting interannual variation in the timing of return migration to fresh water of sockeye salmon (*Oncorhynchus nerka*). Fisheries Oceanography 15(1):1–24.

Holland, K. N., R. W. Brill, and R. K. C. Chang. 1990. Horizontal and vertical movements of yellowfin and bigeye tuna associated with fish aggregating devices. Fishery Bulletin 88(3):493–507.

Holland, K. N., R. W. Brill, R. K. C. Chang, J. R. Sibert, and D. A. Fournier. 1992. Physiological and behavioral thermoregulation in bigeye tuna (*Thunnus obesus*). Nature 358(6385):410–412.

Holmes, J. A., G. M. W. Cronkite, H. J. Enzenhofer, and T. J. Mulligan. 2006. Accuracy and precision of fish-count data from a "dual-frequency identification sonar" (DIDSON) imaging system. ICES Journal of Marine Science 63(3):543–555.

Holt, R. D., and M. A. McPeek. 1996. Chaotic population dynamics favors the evolution of dispersal. American Naturalist 148(4):709–718.

Holyoak, M., R. Casagrandi, R. Nathan, E. Revilla, and O. Spiegel. 2008. Trends and missing parts in the study of movement ecology. Proceedings of the National Academy of Sciences of the United States of America 105(49):19,060–19,065.

Houde, E. D. 1989. Subtleties and episodes in the early life of fishes. Journal of Fish Biology 35:29–38.

———. 1996. Evaluating stage-specific survival during the early life of fishes. In Survival Strategies in Early Life History Stages of Marine Resources, ed. Y. Watanabe, Y. Yamashita, and Y. Oozeki, 51–56. A. A. Balkema, Rotterdam.

———. 2009. Recruitment variability. In Fish Reproductive Biology and Its Implications for Assessment and Management, ed. T. Jakobsen, M. J. Fogarty, B. A. Megrey, and E. Moksness, 91–170. Wiley-Blackwell, Ames, IA.

Houde, E. D., and E. S. Rutherford. 1993. Recent trends in estuarine fisheries: Predictions of fish production and yield. Estuaries 16:161–176.

Houde, E. D., and 12 coauthors. 2001. Marine Protected Areas: Tools for Sustaining Ocean Ecosystems. National Academies Press, Washington, DC.

Hourston, A. S., and H. Rosenthal. 1976. Sperm density during active spawning of Pacific herring (*Clupea harengus pallasi*). Journal of the Fisheries Research Board of Canada 33(8):1788–1790.

Hoysak, D. J., N. R. Liley, and E. B. Taylor. 2004. Raffles, roles, and the outcome of sperm competition in sockeye salmon. Canadian Journal of Zoology 82(7):1017–1026.

Hsieh, C. H., C. S. Reiss, J. R. Hunter, J. R. Beddington, R. M. May, and G. Sugihara. 2006. Fishing elevates variability in the abundance of exploited species. Nature 443(7113):859–862.

Hubbs, C. L. 1918. The flight of the California flyingfish (*Cypeselurus californicus*). Copeia 1918:85–88.

Huebert, K. B. 2008. Barokinesis and depth regulation by pelagic coral reef fish larvae. Marine Ecology Progress Series 367:261–269.

Hufnagl, M. M. A. Peck, R. D. M. Nash, T. Pohlmann, and A. D. Rijnsdorp. 2013. Changes in potential North Sea spawning grounds of plaice (*Pleuronectes platessa* L.) based on early life stage connectivity to nursery habitats. Journal of Sea Research 84:26–39.

Hughes, A. L. 2012. Evolution of adaptive phenotypic traits without positive Darwinian selection. Heredity 108(4):347–353.

Hughes, T. P., and 16 coauthors. 2003. Climate change, human impacts, and the resilience of coral reefs. Science 301(5635):929–933.

Humphries, N. E., and 15 coauthors. 2010. Environmental context explains Lévy and Brownian movement patterns of marine predators. Nature 465(7301):1066–1069.

Humphries, N. E., H. Weimerskirch, N. Queiroz, E. J. Southall, and D. W. Sims. 2012. Foraging success of biological Lévy flights recorded in situ. Proceedings of the National Academy of Sciences of the United States of America 109(19):7169–7174.

Humston, R., J. S. Ault, M. Lutcavage, and D. B. Olson. 2000. Schooling and migration of large pelagic fishes relative to environmental cues. Fisheries Oceanography 9(2):136–146.

Humston, R., D. B. Olson, and J. S. Ault. 2004. Behavioral assumptions in models of fish movement and their influence on population dynamics.

Transactions of the American Fisheries Society 133(6):1304–1328.

Hunter, E., J. D. Metcalfe, C. M. O'Brien, G. P. Arnold, and J. D. Reynolds. 2004. Vertical activity patterns of free-swimming adult plaice in the southern North Sea. Marine Ecology Progress Series 279:261–273.

Hurst, T. P., D. W. Cooper, J. S. Scheingross, E. M. Seale, B. J. Laurel, and M. L. Spencer. 2009. Effects of ontogeny, temperature, and light on vertical movements of larval Pacific cod (*Gadus macrocephalus*). Fisheries Oceanography 18(5):301–311.

Huse, G., S. Railsback, and A. Ferno. 2002. Modelling changes in migration pattern of herring: Collective behaviour and numerical domination. Journal of Fish Biology 60(3):571–582.

Huse, G., A. Ferno, and J. C. Holst. 2010. Establishment of new wintering areas in herring co-occurs with peaks in the "first time/repeat spawner" ratio. Marine Ecology Progress Series 409:189–198.

Hüssy, K., M. A. St John, and U. Bottcher. 1997. Food resource utilization by juvenile Baltic cod *Gadus morhua*: A mechanism potentially influencing recruitment success at the demersal juvenile stage? Marine Ecology Progress Series 155:199–208.

Hutchings, J. A. 1996. Spatial and temporal variation in the density of northern cod and a review of hypotheses for the stock's collapse. Canadian Journal of Fisheries and Aquatic Sciences 53(5):943–962.

———. 2005. Life history consequences of overexploitation to population recovery in Northwest Atlantic cod (*Gadus morhua*). Canadian Journal of Fisheries and Aquatic Sciences 62:824–832.

Hutchings, J. A., and J. K. Baum. 2005. Measuring marine fish biodiversity: Temporal changes in abundance, life history and demography. Philosophical Transactions of the Royal Society B: Biological Sciences 360(1454):315–338.

Hutchings, J. A., and L. Gerber. 2002. Sex-biased dispersal in a salmonid fish. Proceedings of the Royal Society B: Biological Sciences 269(1508): 2487–2493.

Hutchings, J. A., and M. E. B. Jones. 1998. Life history variation and growth rate thresholds for maturity in Atlantic salmon, *Salmo salar*. Canadian Journal of Fisheries and Aquatic Sciences 55:22–47.

Hutchings, J. A., and R. A. Myers. 1993. Effect of age on the seasonality of maturation and spawning of Atlantic cod, *Gadus morhua*, in the northwest Atlantic. Canadian Journal of Fisheries and Aquatic Sciences 50(11):2468–2474.

———. 1994a. The evolution of alternative mating strategies in variable environments. Evolutionary Ecology 8:256–268.

———. 1994b. What can be learned from the collapse of a renewable resource: Atlantic cod, *Gadus morhua*, of Newfoundland and Labrador. Canadian Journal of Fisheries and Aquatic Sciences 51(9):2126–2146.

Hutchings, J. A., and J. D. Reynolds. 2004. Marine fish population collapses: Consequences for recovery and extinction risk. Bioscience 54(4):297–309.

Hutchings, J. A., T. D. Bishop, and C. R. McGregor-Shaw. 1999. Spawning behaviour of Atlantic cod, *Gadus morhua*: Evidence of mate competition and mate choice in a broadcast spawner. Canadian Journal of Fisheries and Aquatic Sciences 56(1):97–104.

Hutchings, J. A., C. Minto, D. Ricard, J. K. Baum, and O. P. Jensen. 2010. Trends in the abundance of marine fishes. Canadian Journal of Fisheries and Aquatic Sciences 67(8):1205–1210.

Hyde, J. R., C. Kimbrell, L. Robertson, K. Clifford, E. Lynn, and R. Vetter. 2008. Multiple paternity and maintenance of genetic diversity in the live-bearing rockfishes *Sebastes* spp. Marine Ecology Progress Series 357:245–253.

Iles, T. D., and M. Sinclair. 1982. Atlantic herring: Stock discreteness and abundance. Science 215:627–633.

Inada, Y. 2001. Steering mechanism of fish schools. Complexity International 8:1–9.

Inger, R., and S. Bearhop. 2008. Applications of stable isotope analyses to avian ecology. Ibis 150(3):447–461.

Irisson, J.-O., C. B. Paris, C. Guigand, and S. Planes. 2010. Vertical distribution and ontogenetic "migration" in coral reef fish larvae. Limnology and Oceanography 55(2):909–919.

Ishikawa, S., and 14 coauthors. 2001. Spawning time and place of the Japanese eel *Anguilla japonica* in the North Equatorial Current of the western North Pacific Ocean. Fisheries Science 67(6):1097–1103.

Islam, M. S., Y. Yamashita, and M. Tanaka. 2011. A review on the early life history and ecology of Japanese sea bass and implication for recruitment. Environmental Biology of Fishes 91(4):389–405.

Istock, C. A. 1967. Evolution of complex life cycle phenomena: An ecological perspective. Evolution 21(3):592–605.

Itoh, T., S. Tjsuji, and A. Nitta. 2003. Migration patterns of young Pacific bluefin tuna (*Thunnus orientalis*) determined with archival tags. Fishery Bulletin 101:514–534.

Iverson, R. L. 1990. Control of marine fish production. Limnology and Oceanography 35:1593–1604.

Jamieson, R. E., H. P. Schwarcz, and J. Brattey. 2004. Carbon isotopic records from the otoliths of Atlantic cod (*Gadus morhua*) from eastern Newfoundland, Canada. Fisheries Research 68(1–3):83–97.

Jamon, M. 1990. A reassessment of the random hypothesis in the ocean migration of Pacific salmon. Journal of Theoretical Biology 143:197–213.

Jellyman, D., and K. Tsukamoto. 2005. Swimming depths of offshore migrating longfin eels *Anguilla dieffenbachii*. Marine Ecology Progress Series 286:261–267.

Jessop, B. M., J. C. Shiao, Y. Iizuka, and W. N. Tzeng. 2002. Migratory behaviour and habitat use by American eels *Anguilla rostrata* as revealed by otolith microchemistry. Marine Ecology Progress Series 233:217–229.

———. 2004. Variation in the annual growth, by sex and migration history, of silver American eels *Anguilla rostrata*. Marine Ecology Progress Series 272:231–244.

Johannes, R. E. 1978. Reproductive strategies of coastal marine fishes in the tropics. Environmental Biology of Fishes 3:65–84.

Johannessen, A., L. Nøttestad, A. Fernö, L. Langard, and G. Skaret. 2009. Two components of northeast Atlantic herring within the same school during spawning: Support for the existence of a metapopulation? ICES Journal of Marine Science 66(8):1740–1748.

Johnson, M. L., and M. S. Gaines. 1990. Evolution of dispersal: Theoretical models and empirical tests using birds and mammals. Annual Review of Ecology and Systematics 21:449–480.

Jones, C. M., and B. K. Wells. 1998. Age, growth, and mortality of black drum, *Pogonias cromis*, in the Chesapeake Bay region. Fishery Bulletin 96(3):451–461.

———. 2001. Yield-per-recruit analysis for black drum, *Pogonias cromis*, along the East Coast of the United States and management strategies for Chesapeake Bay. Fishery Bulletin 99(2):328–337.

Jones, F. C., C. Brown, and V. A. Braithwaite. 2008. Lack of assortative mating between incipient species of stickleback from a hybrid zone. Behaviour 145:463–484.

Jones, G. P., M. J. Milicich, M. J. Emslie, and C. Lunow. 1999. Self-recruitment in a coral reef fish population. Nature 402(6763):802–804.

Jones, G. P., and 6 coauthors. 2009. Larval retention and connectivity among populations of corals and reef fishes: History, advances and challenges. Coral Reefs 28(2):307–325.

Jonsson, B., and N. Jonsson. 1993. Partial migration: Niche shift versus sexual maturation in fishes. Reviews in Fish Biology and Fisheries 3:348–365.

———. 2001. Polymorphism and speciation in Arctic charr. Journal of Fish Biology 58(3):605–638.

Jonsson, B., N. Jonsson, and L. P. Hansen. 2003. Atlantic salmon straying from the River Imsa. Journal of Fish Biology 62(3):641–657.

Jørgensen, C., E. S. Dunlop, A. F. Opdal, and Ø. Fiksen. 2008. The evolution of spawning migrations: State dependence and fishing-induced changes. Ecology 89(12):3436–3448.

Jorgensen, S. J., and 6 coauthors. 2010. Philopatry and migration of Pacific white sharks. Proceedings of the Royal Society B: Biological Sciences 277(1682):679–688.

Jorgensen, S. J., and 9 coauthors. 2012a. Philopatry and migration of Pacific white sharks. Proceedings of the Royal Society B: Biological Sciences 277(1682):679–688.

———. 2012b. Eating or meeting? Cluster analysis reveals intricacies of white shark (*Carcharodon carcharias*) migration and offshore behavior. PLoS One 7(10):e47819. doi:10.1371/journal.pone.0047819.

Juanes, F. 2007. Role of habitat in mediating mortality during the post–settlement transition phase of temperate marine fishes. Journal of Fish Biology 70(3):661–677.

Juanes, F., and D. O. Conover. 1995. Size-structured piscivory: Advection and the linkage between predator and prey recruitment in young-of-the-year bluefish. Marine Ecology Progress Series 128(1–3):287–304.

Jung, S., and E. D. Houde. 2005. Fish biomass size spectra in Chesapeake Bay. Estuaries 28(2):226–240.

Kada, Y. 1984. The evolution of joint fisheries rights and village community structure on Lake Biwa, Japan. Senri Enthological Studies 17:137–158.

Kaitala, A., V. Kaitala, and P. Lundberg. 1993. A theory of partial migration. American Naturalist 142:59–81.

Kalland, A. 1984. Sea tenure in Tokugawa Japan: The case of Fukuoka Domain. Senri Enthological Studies 17:11–36.

Kaushal, S. S., and 8 coauthors. 2010. Rising stream and river temperatures in the United States. Frontiers in Ecology and the Environment 8(9):461–466.

Kawabe, R., Y. Naito, K. Sato, K. Miyashita, and N. Yamashita. 2004. Direct measurement of the swimming speed, tailbeat, and body angle of Japanese flounder (*Paralichthys olivaceus*). ICES Journal of Marine Science 61(7):1080–1087.

Kawasaki, T. 1980. Fundamental relations among the selections of life history in the marine teleosts. Nippon Suisan Gakkaishi 46:289–293.

Keefe, M., and K. W. Able. 1993. Patterns of metamorphosis in summer flounder, *Paralichthys dentatus*. Journal of Fish Biology 42(5):713–728.

Keefer, M. L., C. C. Caudill, C. A. Peery, and C. T. Boggs. 2008. Non-direct homing behaviours by adult Chinook salmon in a large, multi-stock river system. Journal of Fish Biology 72(1):27–44.

Keeton, W. T. 1970. Comparative orientational and homing performances of single pigeons and small flocks. Auk 87(4):797–799.

Kell, L. T., M. Dickey-Collas, N. T. Hintzen, R. D. M. Nash, G. M. Pilling, and B. A. Roel. 2009. Lumpers or splitters? Evaluating recovery and management plans for metapopulations of herring. ICES Journal of Marine Science 66(8):1776–1783.

Kemp, W. M., J. M. Testa, D. J. Conley, D. Gilbert, and J. D. Hagy. 2009. Temporal responses of coastal hypoxia to nutrient loading and physical controls. Biogeosciences 6(12):2985–3008.

Kennedy, J. S. 1985. Migration, behavioral and ecological. In Migration: Mechanisms and Adaptive Significance. Contributions in Marine Science 27, ed. M. A. Rankin, 5–27. Port Aransas, TX.

Kenyon, K. W., and D. A. Rice. 1958. Homing of the laysan albatrosses. Condor 60(1):3–6.

Kerr, L. A., and S. E. Campana. 2013. Chemical composition of fish hard parts as a natural marker of fish stocks. In Stock Identification Methods: Applications in Fisheries Science. 2nd ed., ed. S. X. Cadrin, L. A. Kerr, and S. Mariani, 205–234. Academic, London.

Kerr, L. A., and D. R. Goethel. 2013. Simulation modeling as a tool for synthesis of stock identification and information. In Stock Identification Methods: Applications in Fisheries Science. 2nd ed., ed. S. X. Cadrin, L. A. Kerr, and S. Mariani, 501–534. Academic, London.

Kerr, L. A., and D. H. Secor. 2009. Bioenergetic trajectories underlying partial migration in Patuxent River (Chesapeake Bay) white perch (*Morone americana*). Canadian Journal of Fisheries and Aquatic Sciences 66(4):602–612.

———. 2010. Latent effects of early life history on partial migration for an estuarine-dependent fish. Environmental Biology of Fishes 89(3–4):479–492.

———. 2012. Partial migration across populations of white perch (*Morone americana*): A flexible life history strategy in a variable estuarine environment. Estuaries and Coasts 35(1):227–236.

Kerr, L. A., D. H. Secor, and P. M. Piccoli. 2009. Partial migration of fishes as exemplified by the estuarine-dependent white perch. Fisheries 34(3): 114–123.

Kerr, L. A., S. X. Cadrin, and D. H. Secor. 2010a. The role of spatial dynamics in the stability, resilience, and productivity of fish populations: An evaluation based on white perch in the Chesapeake Bay. Ecological Applications 20:497–507.

———. 2010b. Simulation modelling as a tool for examining the consequences of spatial structure and connectivity on local and regional population dynamics. ICES Journal of Marine Science 67(8):1631–1639.

Kerr, S. R., and L. M. Dickie. 2001. The Biomass Spectrum: A Predator-Prey Theory of Aquatic Production. Columbia University Press, New York.

Kerstetter, D. W., J. J. Polovina, and J. E. Graves. 2004. Evidence of shark predation and scavenging on fishes equipped with pop-up satellite archival tags. Fishery Bulletin 102(4):750–756.

Kettle, A. J., and K. Haines. 2006. How does the European eel (*Anguilla anguilla*) retain its population structure during its larval migration across the North Atlantic Ocean? Canadian Journal of Fisheries and Aquatic Sciences 63(1):90–106.

Kim, H., S. Kimura, A. Shinoda, T. Kitagawa, Y. Sasai, and H. Sasaki. 2007. Effect of El Niño on migration and larval transport of the Japanese eel (*Anguilla japonica*). ICES Journal of Marine Science 64(7):1387–1395.

Kim, S., and B. Bang. 1990. Oceanic dispersion of larval fish and its implication for mortality estimates: Case study of walley pollock larvae in Shelikof Strait. Fishery Bulletin 88:303–312.

Kimura, R., D. H. Secor, E. D. Houde, and P. M. Piccoli. 2000. Up-estuary dispersal of young-of-the-year bay anchovy *Anchoa mitchilli* in the Chesapeake Bay: Inferences from microprobe analysis of strontium in otoliths. Marine Ecology Progress Series 208:217–227.

Kimura, S., and K. Tsukamoto. 2006. The salinity front in the North Equatorial Current: A landmark for the spawning migration of the Japanese eel (*Anguilla japonica*) related to the stock recruitment. Deep Sea Research Part II: Topical Studies in Oceanography 53(3–4):315–325.

Kimura, S., K. Tsukamoto, and T. Sugimoto. 1994. A model for the larval migration of the Japanese eel: Roles of the trade winds and salinity front. Marine Biology 119:185–190.

Kimura, S., T. Inoue, and T. Sugimoto. 2001. Fluctuation in the distribution of low-salinity water in the North Equatorial Current and its effect on the larval transport of the Japanese eel. Fisheries Oceanography 10(1):51–60.

King, D. M., and J. G. Sutinen. 2010. Rational non-compliance and the liquidation of northeast groundfish resources. Marine Policy 34(1):7–21.

Kingsford, M. J., J. M. Leis, A. Shanks, K. C. Lindeman, S. G. Morgan, and J. Peneda. 2002. Sensory environments, larval abilities and local self-recruitment. Bulletin of Marine Science 70(1):309–340.

Kininmonth, S., M. Drechsler, K. Johst, and H. P. Possingham. 2010. Metapopulation mean life time within complex networks. Marine Ecology Progress Series 417:139–149.

Kinlan, B. P., and S. D. Gaines. 2003. Propagule dispersal in marine and terrestrial environments: A community perspective. Ecology 84(8):2007–2020.

Kinnison, M. T., and N. G. Hairston Jr. 2007. Eco-evolutionary conservation biology: Contemporary evolution and the dynamics of persistence. Functional Ecology 21(3):444–454.

Kitamura, T., M. Kume, H. Takahashi, and A. Goto. 2006. Juvenile bimodal length distribution and sea-run migration of the lower modal group in the Pacific Ocean form of three-spined stickleback. Journal of Fish Biology 69(4):1245–1250.

Kjørsvik, E., A. Mangor-Jensen, and I. Holmefjord. 1990. Egg quality in fishes. Advances in Marine Biology 26:71–113.

Klemetsen, A. 2010. The charr problem revisited: Exceptional phenotypic plasticity promotes ecological speciation in postglacial lakes. Freshwater Reviews 3:49–74.

Klemetsen, A., and 6 coauthors. 2003. Atlantic salmon *Salmo salar* L., brown trout *Salmo trutta* L. and Arctic charr *Salvelinus alpinus* (L.): A review of aspects of their life histories. Ecology of Freshwater Fish 12(1):1–59.

Klimley, A. P. 1987. The determinants of sexual segregation in the scalloped hammerhead shark, *Sphyrna lewini*. Environmental Biology of Fishes 18(1):27–40.

———. 1993. Highly directional swimming by scalloped hammerhead sharks, *Sphyrna lewini*, and subsurface irradiance, temperature, bathymetry, and geomagnetic-field. Marine Biology 117(1):1–22.

Koenig, C. C., F. C. Coleman, and K. Kingon. 2011. Pattern of recovery of the goliath grouper *Epinephelus itajara* population in the southeastern US. Bulletin of Marine Science 87(4):891–911.

Köster, F. W., and C. Möllmann. 2000. Trophodynamic control by clupeid predators on recruitment success in Baltic cod? ICES Journal of Marine Science 57(2):310–323.

Koutsikopoulos, C., L. Fortier, and J. A. Gagne. 1991. Cross-shelf dispersion of Dover sole (*Solea solea*) eggs and larvae in Biscay Bay and recruitment to inshore nurseries. Journal of Plankton Research 13(5):923–945.

Kraus, R. T., and D. H. Secor. 2004a. Incorporation of strontium into otoliths of an estuarine fish. Journal of Experimental Marine Biology and Ecology 302(1):85–106.

———. 2004b. The dynamics of white perch (*Morone americana*) population contingents in the Patuxent River Estuary, Maryland, USA. Marine Ecology Progress Series 279:247–259.

———. 2005. Evaluation of connectivity in estuarine-dependent white perch populations of Chesapeake Bay. Estuarine and Coastal Shelf Science 64:94–107.

Kraus, R. T., R. J. D. Wells, and J. R. Rooker. 2011. Horizontal movements of Atlantic blue marlin (*Makaira nigricans*) in the Gulf of Mexico. Marine Biology 158(3):699–713.

Kristiansen, T., K. F. Drinkwater, R. G. Lough, and S. Sundby. 2011. Recruitment variability in North Atlantic cod and match-mismatch dynamics. PLoS One 6(3):e17456. doi:10.1371/journal.pone.0017456.

Kristoffersen, K., M. Halvorsen, and L. Jorgensen. 1994. Influence of parr growth, lake morphology, and freshwater parasites on the degree of anadromy in different populations of arctic char (*Salvelinus alpinus*) in northern Norway. Canadian Journal of Fisheries and Aquatic Sciences 51(6):1229–1246.

Kritzer, J. P., and O. R. Liu. 2013. Fishery management strategies for addressing complex spatial structure in marine fish stocks. In Stock Identification Methods: Applications in Fisheries Science. 2nd ed., ed. S. X. Cadrin, L. A. Kerr, and S. Mariani, 7–28. Academic, London.

Kritzer, J. P., and P. F. Sale. 2004. Metapopulation ecology in the sea: From Levins' model to marine ecology and fisheries science. Fish and Fisheries 5(2):131–140.

Kuroki, M., and 6 coauthors. 2008. Inshore migration and otolith microstructure/microchemistry of anguillid glass eels recruited to Iceland. Environmental Biology of Fishes 83(3):309–325.

Kurota, H., M. K. McAllister, G. L. Lawson, J. I. Nogueira, S. L. H. Teo, and B. A. Block. 2009. A sequential Bayesian methodology to estimate movement and exploitation rates using electronic and conventional tag data: Application to Atlantic bluefin tuna (*Thunnus thynnus*). Canadian Journal of Fisheries and Aquatic Sciences 66(2): 321–342.

Lack, D. 1944. The problem of partial migration. British Birds 37:122–130, 143–150.

Laland, K. N., F. J. Odling-Smee, and M. W. Feldman. 1999. Evolutionary consequences of niche construction and their implications for ecology. Proceedings of the National Academy of Sciences of the United States of America 96(18):10,242–10,247.

Laland, K. N., K. Sterelny, J. Odling-Smee, W. Hoppitt, and T. Uller. 2011. Cause and effect in biology revisited: Is Mayr's proximate-ultimate dichotomy still useful? Science 334(6062):1512–1516.

Lambert, T. C. 1987. Duration and intensity of spawning in herring *Clupea harengus* as related to the age structure of the mature population. Marine Ecology Progress Series 39(3):209–220.

———. 1990. The effect of population structure on recruitment in herring. Journal du Conseil pour l'Exploration de la Mer 47:249–255.

Lankford, T. E., J. M. Billerbeck, and D. O. Conover. 2001. Evolution of intrinsic growth and energy acquisition rates. 2. Trade-offs with vulnerability to predation in *Menidia menidia*. Evolution 55(9):1873–1881.

Laprise, R., and J. J. Dodson. 1989. Ontogeny and importance of tidal vertical migrations in the retention of larval smelt *Osmerus mordax* in a well-mixed estuary. Marine Ecology Progress Series 55(2–3):101–111.

Larkin, P. A. 1975. Some major problems for further study on Pacific salmon. Bulletin of the International North Pacific Fisheries Commission 32:309.

Lasker, R. 1975. Field criteria for survival of anchovy larvae: Relation between inshore chlorophyll maximum layers and successful first feeding. Fishery Bulletin 73(3):453–462.

Law, R. 2000. Fishing, selection, and phenotypic evolution. ICES Journal of Marine Science 57(3): 659–668.

Lawson, G. L., and G. A. Rose. 2000. Small-scale spatial and temporal patterns in spawning of Atlantic cod (*Gadus morhua*) in coastal Newfoundland waters. Canadian Journal of Fisheries and Aquatic Sciences 57(5):1011–1024.

Lebedev, N. V. 1969. Elementary Populations of Fish. Israel Program for Scientific Translations, Jerusalem.

Lecomte, F., W. S. Grant, J. J. Dodson, R. Rodriguez-Sanchez, and B. W. Bowen. 2004. Living with uncertainty: Genetic imprints of climate shifts in East Pacific anchovy (*Engraulis mordax*) and sardine (*Sardinops sagax*). Molecular Ecology 13(8):2169–2182.

Lecomte, V. J., and 10 coauthors. 2010. Patterns of aging in the long-lived wandering albatross. Proceedings of the National Academy of Sciences of the United States of America 107(14):6370–6375.

Leggett, W. C. 1984. Fish migrations in coastal and estuarine environments: A call for new approaches to the study of an old problem. In Mechanisms of Migration in Fishes, ed. J. D. McCleave, G. P. Arnold, and J. J. Dodson, 159–178. Plenum, New York.

———. 1985. The role of migrations in the life history evolution of fish. In Migration: Mechanisms and Adaptive Significance. Contributions to Marine Science Supplement 27, ed. M. A. R. Rankin, 258–295. Marine Science Institute, University of Texas, Austin.

Leggett, W. C., and J. E. Carscadden. 1978. Latitudinal variation in reproductive characteristics of American shad (*Alosa sapidissima*): Evidence for population specific life history strategies in fish. Journal of the Fisheries Research Board of Canada 35:1469–1478.

Leggett, W. C., and E. DeBlois. 1994. Recruitment in marine fishes: Is it regulated by starvation and predation in the egg and larval stages? Netherlands Journal of Sea Research 32:119–134.

LeGrande, A. N., and G. A. Schmidt. 2006. Global gridded data set of the oxygen isotopic composition in seawater. Geophysical Research Letters 33:L12604. doi:10.1029/2006GL026011.

Lehodey, P., M. Bertignac, J. Hampton, A. Lewis, and J. Picaut. 1997. El Niño-Southern Oscillation and tuna in the western Pacific. Nature 389(6652):715–718.

Lehodey, P., and 12 coauthors. 2006. Climate variability, fish, and fisheries. Journal of Climate 19(20):5009–5030.

Leider, S. A. 1989. Increased straying by adult steelhead trout, *Salmo gairdneri*, following the 1980 eruption of Mount St. Helens. Environmental Biology of Fishes 24(3):219–229.

Leis, J. M. 2006. Are larvae of demersal fishes plankton or nekton? Advances in Marine Biology 51:57–141.

———. 2007. Behaviour as input for modelling dispersal of fish larvae: Behaviour, biogeography, hydrodynamics, ontogeny, physiology and phylogeny meet hydrography. Marine Ecology Progress Series 347:185–193.

———. 2010. Ontogeny of behaviour in larvae of marine demersal fishes. Ichthyological Research 57(4):325–342.

Leis, J. M., and B. M. Carson-Ewart. 2001. Behaviour of pelagic larvae of four coral-reef fish species in the ocean and an atoll lagoon. Coral Reefs 19(3):247–257.

Leis, J. M., A. C. Hay, M. M. Lockett, J. P. Chen, and L. S. Fang. 2007. Ontogeny of swimming speed in larvae of pelagic-spawning, tropical, marine fishes. Marine Ecology Progress Series 349:255–267.

Levin, L. A. 2006. Recent progress in understanding larval dispersal: New directions and digressions. Integrative and Comparative Biology 46(3):282–297.

Levins, R. 1969. Some demographic and genetic consequences of environmental heterogeneity for biological control. Bulletin of the Entomological Society of America 15:237–240.

Levitan, D. R. 1993. The importance of sperm limitation to the evolution of egg size in marine invertebrates. American Naturalist 141(4):517–536.

———. 1996. Effects of gamete traits on fertilization in the sea and the evolution of sexual dimorphism. Nature 382(6587):153–155.

Li, W. M., P. W. Sorensen, and D. D. Gallaher. 1995. The olfactory system of migratory adult sea lamprey (*Petromyzon marinus*) is specifically and acutely sensitive to unique bile-acids released by conspecific larvae. Journal of General Physiology 105(5):569–587.

Liechti, F., B. Bruderer, and H. Paproth. 1995. Quantification of nocturnal bird migration by moon-watching: Comparison with radar and infrared observations. Journal of Field Ornithology 66(4):457–468.

Liechti, F., D. Peter, R. Lardelli, and B. Bruderer. 1996. The Alps, an obstacle for nocturnal broad front migration: A survey based on moon-watching. Journal für Ornithologie 137(3):337–356.

Limburg, K. E., and J. R. Waldman. 2009. Dramatic

declines in North Atlantic diadromous fishes. Bioscience 59:955–965.

Lindholm, J., P. J. Auster, and A. Knight. 2007. Site fidelity and movement of adult Atlantic cod *Gadus morhua* at deep boulder reefs in the western Gulf of Maine, USA. Marine Ecology Progress Series 342:239–247.

Lindsey, C. C. 1978. Form, function and locomotory habits in fish. In Fish Physiology, vol. 7, Locomotion, ed. W. S. Hoar and D. J. Randall, 1–100. Academic, New York.

Link, J. S. 2005. Translating ecosystem indicators into decision criteria. ICES Journal of Marine Science 62(3):569–576.

———. 2010. Ecosystem-based Fisheries Management: Confronting Tradeoffs. Cambridge University Press, New York.

Link, J. S., and L. P. Garrison. 2002. Changes in piscivory associated with fishing induced changes to the finfish community on Georges Bank. Fisheries Research 55(1–3):71–86.

Link, J. S., J. A. Nye, and J. A. Hare. 2011. Guidelines for incorporating fish distribution shifts into a fisheries management context. Fish and Fisheries 12:461–469.

Lockwood, R., J. P. Swaddle, and J. M. V. Rayner. 1998. Avian wingtip shape reconsidered: Wingtip shape indices and morphological adaptations to migration. Journal of Avian Biology 29(3):273–292.

Lohmann, K. J., N. F. Putman, and C. M. F. Lohmann. 2008a. Geomagnetic imprinting: A unifying hypothesis of long-distance natal homing in salmon and sea turtles. Proceedings of the National Academy of Sciences of the United States of America 105(49):19,096–19,101.

Lohmann, K. J., C. M. F. Lohmann, and C. S. Endres. 2008b. The sensory ecology of ocean navigation. Journal of Experimental Biology 211(11):1719–1728.

Lombardi, J. 1998. Comparative Vertebrate Reproduction. Kluwer Academic, Norwell, MA.

Long, J. A. 1995. The Rise of Fishes: 500 Million Years of Evolution. Johns Hopkins University Press, Baltimore, MD.

Longhurst, A. 2002. Murphy's law revisited: Longevity as a factor in recruitment to fish populations. Fisheries Research 56(2):125–131.

Loreau, M., and 11 coauthors. 2001. Biodiversity and ecosystem functioning: Current knowledge and future challenges. Science 294(5543):804–808.

Lough, R. G. 2010. Juvenile cod (*Gadus morhua*) mortality and the importance of bottom sediment type to recruitment on Georges Bank. Fisheries Oceanography 19(2):159–181.

Lough, R. G., and 6 coauthors. 1989. Ecology and distribution of juvenile cod and haddock in relation to sediment type and bottom currents on eastern Georges Bank. Marine Ecology Progress Series 56(1–2):1–12.

Love, M. S., P. Morris, M. McCrae, and R. Collins. 1990. Life history aspects of 19 rockfish species (Scorpaenidae: *Sebastes*) from the Southern California Bight. NOAA Technical Report NMFS 87, 38 pp. National Oceanic and Atmospheric Association, Silver Spring, MD.

Lowe, W. H., and F. W. Allendorf. 2010. What can genetics tell us about population connectivity? Molecular Ecology 19(23):3038–3051.

Lozano, C., E. D. Houde, R. L. Wingate, and D. H. Secor. 2012. Age, growth and hatch dates of ingressing larvae and surviving juveniles of Atlantic menhaden *Brevoortia tyrannus*. Journal of Fish Biology 81(5):1665–1685.

Lubchenco, J., S. R. Palumbi, S. D. Gaines, and S. Andelman. 2003. Plugging a hole in the ocean: The emerging science of marine reserves. Ecological Applications 13(1):S3–S7.

Lucas, M. C., and E. Baras. 2001. Migration of Freshwater Fishes. Blackwell Science, London.

Luczkovich, J. J., D. A. Mann, and R. A. Rountree. 2008. Passive acoustics as a tool in fisheries science. Transactions of the American Fisheries Society 137(2):533–541.

Ludwig, A., and 8 coauthors. 2002. When the American sea sturgeon swam east. Nature 419:447–448.

Ludwig, A., and 6 coauthors. 2008. Tracing the first steps of American sturgeon pioneers in Europe. BMC Evolutionary Biology 8:221. doi:10.1186/1471-2148-8-221.

Lukeman, R., Y.-X. Li, and L. Edelstein-Keshet. 2010. Inferring individual rules from collective behavior. Proceedings of the National Academy of Sciences of the United States of America 107(28): 12,576–12,580.

Lundberg, P. 1987. Partial bird migration and evolutionarily stable strategies. Journal of Theoretical Biology 125(3):351–360.

Luo, J., and J. A. Musick. 1991. Reproductive biology of the bay anchovy in Chesapeake Bay. Transactions of the American Fisheries Society 120: 701–710.

MacCall, A. D. 1990. Dynamic Geography of Marine Fish Populations. Washington Sea Grant, Seattle.

———. 2012. Data-limited management reference points to avoid collapse of stocks dependent on learned migration behaviour. ICES Journal of Marine Science 69:267–270.

Mace, P. M. 2004. In defense of fisheries scientists, single-species models and other scapegoats: Confronting the real problems. Marine Ecology Progress Series 274:285–291.

Machut, L. S., K. E. Limburg, R. E. Schmidt, and D. Dittman. 2007. Anthropogenic impacts on American eel demographics in Hudson River tributaries, New York. Transactions of the American Fisheries Society 136(6):1699–1713.

Magnuson, J. J. 1978. Locomotion by scombrid fishes. In Fish Physiology, vol. 7, Locomotion, ed. W. S. Hoar and D. J. Randall, 239–313. Academic, New York.

Mäkinen, H. S., J. M. Cano, and J. Merilä. 2006. Genetic relationships among marine and freshwater populations of the European three-spined stickleback (*Gasterosteus aculeatus*) revealed by microsatellites. Molecular Ecology 15(6):1519–1534.

Makris, N. C., P. Ratilal, D. T. Symonds, S. Janannathan, S. Lee, and R. W. Nero. 2006. Fish population and behavior revealed by instantaneous continental shelf-scale imaging. Science 311(5761):660–663.

Makris, N. C., and 8 coauthors. 2009. Critical population density triggers rapid formation of vast oceanic fish shoals. Science 323(5922):1734–1737.

Malone, T. C., L. H. Crocker, S. E. Pike, and B. W. Wendler. 1988. Influences of river flow on the dynamics of phytoplankton production in a partially stratified estuary. Marine Ecology Progress Series 48(3):235–249.

Manderson, J. 2008. The spatial scale of phase synchrony in winter flounder (*Pseudopleuronectes americanus*) production increased among southern New England nurseries in the 1990s. Canadian Journal of Fisheries and Aquatic Sciences 65:340–351.

Manderson, J., L. Palamara, J. Kohut, and M. J. Oliver. 2010. Ocean observatory data are useful for regional habitat modeling of species with different vertical habitat preferences. Marine Ecology Progress Series 438:1–23.

Mansueti, R. 1958. Eggs, larvae and young of the striped bass, *Roccus saxatilis*. Volume Contribution 112. Chesapeake Biological Laboratory, Maryland Department of Research and Education, Solomons.

———. 1961. Movements, reproduction, and mortality of the White Perch, *Roccus americanus*, in the Patuxent Estuary, Maryland. Chesapeake Science 2:142–205.

Marconato, A., and D. Y. Shapiro. 1996. Sperm allocation, sperm production and fertilization rates in the bucktooth parrotfish. Animal Behaviour 52:971–980.

Mariani, S., and D. Bekkevold. 2013. The nuclear genome: Natural and adaptive markers in fisheries science. In Stock Identification Methods: Applications in Fisheries Science, 2nd ed., ed. S. X. Cadrin, L. A. Kerr, and S. Mariani, 297–328. Academic, London.

Marriott, R. J., B. D. Mapstone, and G. A. Begg. 2007. Age-specific demographic parameters, and their implications for management of the red bass, *Lutjanus bohar* (Forsskal 1775): A large, long-lived reef fish. Fisheries Research 83(2–3):204–215.

Marshall, C. T., and 12 coauthors. 2003. Developing alternative indices of reproductive potential for use in fisheries management: Case studies for stocks spanning an information gradient. Journal of Northwest Atlantic Fisheries Science 33:161–190.

Marshall, D. J., S. S. Heppell, S. B. Munch, and R. R. Warner. 2010. The relationship between maternal phenotype and offspring quality: Do older mothers really produce the best offspring? Ecology 91(10):2862–2873.

Marteinsdottir, G., and G. A. Begg. 2002. Essential relationships incorporating the influence of age, size and condition on variables required for estimation of reproductive potential in Atlantic cod *Gadus morhua*. Marine Ecology Progress Series 235:235–256.

Marteinsdottir, G., and K. Thorarinsson. 1998. Improving the stock–recruitment relationship in Icelandic cod (*Gadus morhua*) by including age diversity of spawners. Canadian Journal of Fisheries and Aquatic Sciences 55(6):1372–1377.

Martino, E. J., and E. D. Houde. 2010. Recruitment of striped bass in Chesapeake Bay: Spatial and temporal environmental variability and availability of zooplankton prey. Marine Ecology Progress Series 409:213–228.

Mather, M. E., J. T. Finn, K. H. Ferry, L. A. Deegan, and G. A. Nelson. 2009. Use of non-natal estuaries by migratory striped bass (*Morone saxatilis*) in summer. Fishery Bulletin 107(3):329–338.

Matsumiya, Y., T. Mitani, and M. Tanaka. 1982. Studies on the juvenile Japanese sea bass in the

Chikugo Estuary of Ariake Bay. 2. Changes in distribution pattern and condition coefficient of the juvenile Japanese sea bass with the Chikugo River ascending. Bulletin of the Japanese Society of Scientific Fisheries 48(2):129–138.

May, R. M. 1973. Stability in randomly fluctuating versus deterministic environments. American Naturalist 107(957):621–650.

Mayr, E. 1961. Cause and effect in biology: Kinds of causes, predictability, and teleology are viewed by a practicing biologist. Science 134(348):1501–1506.

McBride, R. S., and K. W. Able. 1998. Ecology and fate of butterflyfishes, *Chaetodon* spp., in the temperate, western North Atlantic. Bulletin of Marine Science 63(2):401–416.

McBride, R. S., and K. A. McKown. 2000. Consequences of dispersal of subtropically spawned crevalle jacks, *Caranx hippos*, to temperate estuaries. Fishery Bulletin 98:528–538.

McCann, K. S. 2000. The diversity-stability debate. Nature 405(6783):228–233.

McCleave, J. D., and R. C. Kleckner. 1982. Selective tidal stream transport in the estuarine migration of glass eels of the American eel (*Anguilla rostrata*). Journal du Conseil pour l'Exploration de la Mer 40(3):262–271.

McCleave, J. D., G. P. Arnold, J. J. Dodson, and W. H. Neil, eds. 1984. Mechanisms of Migration of Fishes. NATO Conference Series. Plenum, New York.

McCleave, J. D., and 6 coauthors. 1998. Do leptocephali of the European eel swim to reach continental waters? Status of the question. Journal of the Marine Biological Association of the United Kingdom 78(1):285–306.

McCulloch, M., M. Cappo, J. Aumend, and W. Muller. 2005. Tracing the life history of individual barramundi using laser ablation MC-ICP-MS Sr-isotopic and Sr/Ba ratios in otoliths. Marine and Freshwater Research 56(5):637–644.

McDowall, R. M. 1988. Diadromy in Fishes: Migrations between Freshwater and Marine Environments. Timber Press, Portland, OR.

———. 2001. The origin of the salmonid fishes: Marine, freshwater . . . or neither? Reviews in Fish Biology and Fisheries 11(3).171–179.

———. 2004. Ancestry and amphidromy in island freshwater fish faunas. Fish and Fisheries 5(1):75–85.

———. 2007. On amphidromy, a distinct form of diadromy in aquatic organisms. Fish and Fisheries 8(1):1–13.

———. 2008. Why are so many boreal freshwater fishes anadromous? Confronting "conventional wisdom." Fish and Fisheries 9(2):208–213.

———. 2009. Making the best of two worlds: Diadromy in the evolution, ecology, and conservation of aquatic organisms. American Fisheries Society Symposium 69:1–22.

———. 2010. Why be amphidromous: Expatrial dispersal and the place of source and sink population dynamics? Reviews in Fish Biology and Fisheries 20(1):87–100.

McFarlane, G. A., and J. R. King. 2003. Migration patterns of spiny dogfish (*Squalus acanthias*) in the North Pacific Ocean. Fishery Bulletin 101(2):358–367.

McGilliard, C. R., A. E. Punt, and R. Hilborn. 2011. Spatial structure induced by marine reserves shapes population responses to catastrophes in mathematical models. Ecological Applications 21(4):1399–1409.

McGuigan, K., and C. M. Sgró. 2009. Evolutionary consequences of cryptic genetic variation. Trends in Ecology and Evolution 24(6):305–311.

McGurk, M. D. 1987. The spatial patchiness of Pacific herring larvae. Environmental Biology of Fishes 20(2):81–89.

———. 1989. Advection, diffusion and mortality of Pacific herring larvae *Clupea harengus pallasi* in Bamfield Inlet, British Columbia. Marine Ecology Progress Series 51:1–18.

McIntyre, T. M., and J. A. Hutchings. 2003. Small-scale temporal and spatial variation in Atlantic cod (*Gadus morhua*) life history. Canadian Journal of Fisheries and Aquatic Sciences 60(9):1111–1121.

McKeown, B. A. 1985. Fish Migration. Timber Press, Portland, OR.

McKinnon, J. S., and H. D. Rundle. 2002. Speciation in nature: The threespine stickleback model systems. Trends in Ecology and Evolution 17(10):480–488.

McKinnon, J. S., and 7 coauthors. 2004. Evidence for ecology's role in speciation. Nature 429(6989): 294–298.

McMahon, K. W., L. H. Hamady, and S. R. Thorrold. 2013. A review of ecogeochemistry approaches to estimating the movements of marine animals. Limnology Oceanography 58(2):697–714.

McMahon, K. W., M. L. Berumen, and S. R. Thorrold. 2012. Linking habitat mosaics and connectivity in a coral reef seascape. Proceedings of the National Academy of Sciences of the United States of America 109(38):15,372–15,376.

McMahon, T. E., and W. J. Matter. 2006. Linking

habitat selection, emigration and population dynamics of freshwater fishes: A synthesis of ideas and approaches. Ecology of Freshwater Fish 15(2): 200–210.

McPhee, M. V., D. L. G. Noakes, and F. W. Allendorf. 2012. Developmental rate: A unifying mechanism for sympatric divergence in postglacial fishes? Current Zoology 58(1):21–34.

McQuinn, I. H. 1997a. Metapopulations and the Atlantic herring. Reviews in Fish Biology and Fisheries 7:297–329.

———. 1997b. Year-class twinning in sympatric seasonal spawning populations of Atlantic herring, *Clupea harengus*. Fishery Bulletin 95(1):126–136.

Mehner, T., and P. Kasprzak. 2011. Partial diel vertical migrations in pelagic fish. Journal of Animal Ecology 80(4):761–770.

Metcalfe, J. D. 2006. Fish population structuring in the North Sea: Understanding processes and mechanisms from studies of the movements of adults. Journal of Fish Biology 69:48–65.

———, ed. 2012. Fish migration in the 21st century: Opportunities and challenges. Special issue, Journal of Fish Biology 81(2).

Metcalfe, J. D., and G. P. Arnold. 1997. Tracking fish with electronic tags. Nature 387(6634):665–666.

Metcalfe, J. D., G. P. Arnold, and P. W. McDowall. 2002. Migration. In Handbook of Fish Biology and Fisheries, ed. P. J. B. Hart and J. D. Reynolds, 175–199. Blackwell Scientific, Oxford.

Metcalfe, N. B., and P. Monaghan. 2001. Compensation for a bad start: Grow now, pay later? Trends in Ecology and Evolution 16(5):254–260.

Metcalfe, N. B., F. A. Huntingford, J. E. Thorpe, and C. E. Adams. 1990. The effects of social status on life-history variation in juvenile salmon. Canadian Journal of Zoology 68(12):2630–2636.

Metcalfe, N. B., A. C. Taylor, and J. E. Thorpe. 1995. Metabolic rate, social status and life-history strategies in Atlantic salmon. Animal Behaviour 49(2):431–436.

Meyer, C. G., and K. N. Holland. 2005. Movement patterns, home range size and habitat utilization of the bluespine unicornfish, *Naso unicornis* (Acanthuridae) in a Hawaiian marine reserve. Environmental Biology of Fishes 73(2):201–210.

Meyer, C. G., K. N. Holland, and Y. P. Papastamatiou. 2005. Sharks can detect changes in the geomagnetic field. Journal of the Royal Society Interface 2:129–130.

Meyer, C. G., T. B. Clark, Y. P. Papastamatiou, N. M. Whitney, and K. N. Holland. 2009. Long-term movement patterns of tiger sharks *Galeocerdo cuvier* in Hawaii. Marine Ecology Progress Series 381:223–235.

Miller, B. A., and S. Sadro. 2003. Residence time and seasonal movements of juvenile coho salmon in the ecotone and lower estuary of Winchester Creek, South Slough, Oregon. Transactions of the American Fisheries Society 132(3):546–559.

Miller, C. B., B. W. Frost, P. A. Wheeler, M. R. Landry, N. Welschmeyer, and T. M. Powell. 1991. Ecological dynamics in the sub-arctic Pacific, a possibly iron-limited ecosystem. Limnology and Oceanography 36(8):1600–1615.

Miller, K. M., and 14 coauthors. 2011. Genomic signatures predict migration and spawning failure in wild Canadian salmon. Science 331(6014): 214–217.

Miller, M. J. 2009. Ecology of Anguilliform Leptocephali: Remarkable transparent fish larvae of the ocean surface layer. Aqua-BioSci Monographs 2(4):1–94.

Miller, M. J., and K. Tsukamoto. 2004. An introduction to leptocephali: Biology and identification. Ocean Research Institute, University of Tokyo, Tokyo.

Miller, R. J., and E. L. Brannon. 1982. The origin and development of life history patterns in Pacific salmonids. In Proceedings of the Salmon and Trout Migratory Behavior Symposium, ed. E. L. Brannon and E. O. Salo, 296–309. University of Washington Press, Seattle.

Miller, T. J. 2007. Contribution of individual-based coupled physical-biological models to understanding recruitment in marine fish populations. Marine Ecology Progress Series 347:127–138.

Miller, T. J., L. B. Crowder, J. A. Rice, and E. A. Marschall. 1988. Larval size and recruitment mechanisms in fishes: Toward a conceptual framework. Canadian Journal of Fisheries and Aquatic Science 45(9):1657–1670.

Miller, T. J., T. Herra, and W. C. Leggett. 1995. An individual-based analysis of the variability of eggs and their newly hatched larvae of Atlantic cod (*Gadus morhua*) on the Scotian Shelf. Canadian Journal of Fisheries and Aquatic Sciences 52(5):1083–1093.

Miller, T. J., J. Blair, T. H. Ihde, R. M. Jones, D. H. Secor, and M. J. Wilberg. 2010. FishSmart: An innovative role for science in stakeholder-centered approaches to fisheries management. Fisheries 35:424–433.

Milton, D. A., and S. R. Chenery. 2005. Movement

patterns of barramundi *Lates calcarifer*, inferred from ^{87}Sr/^{86}Sr and Sr/Ca ratios in otoliths, indicate non-participation in spawning. Marine Ecology Progress Series 301:279-291.

Mims, M. C., J. D. Olden, Z. R. Shattuck, and N. L. Poff. 2010. Life history trait diversity of native freshwater fishes in North America. Ecology of Freshwater Fish 19(3):390-400.

Minami, H., and H. Ogi. 1997. Determination of migratory dynamics of the sooty shearwater in the Pacific using stable carbon and nitrogen isotope analysis. Marine Ecology Progress Series 158:249-256.

Mochioka, N., M. Iwamizu, and T. Kanda. 1993. Leptocephalus eel larvae will feed in aquariums. Environmental Biology of Fishes 36(4):381-384.

Morgan, M. J., E. M. DeBlois, and G. A. Rose. 1997. An observation on the reaction of Atlantic cod (*Gadus morhua*) in a spawning shoal to bottom trawling. Canadian Journal of Fisheries and Aquatic Sciences 54:217-223.

Morinville, G. R., and J. B. Rasmussen. 2003. Early juvenile bioenergetic differences between anadromous and resident brook trout (*Salvelinus fontinalis*). Canadian Journal of Fisheries and Aquatic Sciences 60(4):401-410.

Morita, K., and T. Nagasawa. 2010. Latitudinal variation in the growth and maturation of masu salmon (*Oncorhynchus masou*) parr. Canadian Journal of Fisheries and Aquatic Sciences 67(6):955-965.

Morita, K., S. Yamamoto, and N. Hoshino. 2000. Extreme life history change of white-spotted char (*Salvelinus leucomaenis*) after damming. Canadian Journal of Fisheries and Aquatic Sciences 57(6):1300-1306.

Morrison, W. E., and D. H. Secor. 2003. Demographic attributes of yellow-phase American eels (*Anguilla rostrata*) in the Hudson River estuary. Canadian Journal of Fisheries and Aquatic Sciences 60(12):1487-1501.

Morrison, W. E., D. H. Secor, and P. M. Piccoli. 2003. Estuarine habitat use by Hudson River American eels. American Fisheries Society Symposium 33:87-99.

Mucientes, G. R., N. Queiroz, L. L. Sousa, P. Tarroso, and D. W. Sims. 2009. Sexual segregation of pelagic sharks and the potential threat from fisheries. Biology Letters 5(2):156-159.

Mueller, J. C., F. Pulido, and B. Kempanaers. 2011. Identification of a gene associated with avian migratory behavior. Proceedings of the Royal Society B: Biological Sciences 278:2848-2856.

Mueller, T., R. B. O'Hara, S. J. Converse, R. P. Urbanek, and W. F. Fagan. 2013. Social learning of migratory performance. Science 341(6149):999-1002.

Müller, U. K., E. J. Stamhuis, and J. J. Videler. 2000. Hydrodynamics of unsteady fish swimming and the effects of body size: Comparing the flow fields of fish larvae and adults. Journal of Experimental Biology 203(2):193-206.

Munch, S. B., and D. O. Conover. 2000. Recruitment dynamics of bluefish (*Pomatomus saltatrix*) from Cape Hatteras to Cape Cod, 1973-1995. ICES Journal of Marine Science 57:393-402.

Munk, K. M. 2001. Maximum ages of groundfishes in waters off Alaska and British Columbia and considerations of age determination. Alaska Fisheries Research Bulletin 8:12-21.

Murawski, S. A. 1993. Climate change and marine fish distributions: Forecasting from historical analogy. Transactions of the American Fisheries Society 122(5):647-658.

———. 2010. Rebuilding depleted fish stocks: The good, the bad, and, mostly, the ugly. ICES Journal of Marine Science 67:1830-1840.

Murdy, E. O., R. S. Birdsong, and J. A. Musick. 1997. Fishes of Chesapeake Bay. Smithsonian Institution, Washington, DC.

Murphy, C. A., and J. A. Cowan. 2007. Production, marine larval retention or dispersal, and recruitment of amphidromous Hawaiian gobioids: Issues and implications. In Biology of Hawaiian Streams and Estuaries. Bulletin of Cultural Environmental Studies 3, ed. L. Evenhuis and J. M. Fitzsimons, 63-74. Bishop Museum, Honolulu, HI.

Musick, J. A., and 17 coauthors. 2000. Marine, estuarine, and diadromous fish stocks at risk of extinction in North America (exclusive of Pacific salmonids). Fisheries 25(11):6-30.

Musyl, M. K., R. W. Brill, C. H. Boggs, D. S. Curran, T. K. Kazama, and M. P. Seki. 2003. Vertical movements of bigeye tuna (*Thunnus obesus*) associated with islands, buoys, and seamounts near the main Hawaiian Islands from archival tagging data. Fisheries Oceanography 12(3):152-169.

Myers, G. S. 1949. Usage of anadromous, catadromous and allied terms for migratory fishes. Copeia 1949:89-97.

Myers, R. A., and B. Worm. 2003. Rapid worldwide depletion of predatory fish communities. Nature 423(6937):280-283.

Myers, R. A., J. A. Hutchings, and N. J. Barrowman. 1996. Hypotheses for the decline of cod in the

North Atlantic. Marine Ecology Progress Series 138(1–3):293–308.

——. 1997. Why do fish stocks collapse? The example of cod in Atlantic Canada. Ecological Applications 7(1):91–106.

Naeem, S. 1998. Species redundancy and ecosystem reliability. Conservation Biology 12(1):39–45.

Nagel, L., and D. Schluter. 1998. Body size, natural selection, and speciation in sticklebacks. Evolution 52(1):209–218.

Naish, K. A., and J. J. Hard. 2008. Bridging the gap between the genotype and the phenotype: Linking genetic variation, selection and adaptation in fishes. Fish and Fisheries 9(4):396–422.

Nakamura, I., Y. Y. Watanabe, Y. P. Papastamatiou, K. Sato, and C. G. Meyer. 2011. Yo-yo vertical movements suggest a foraging strategy for tiger sharks *Galeocerdo cuvier*. Marine Ecology Progress Series 424:237–246.

Nakayama, S., J. L. Harcourt, R. A. Johnstone, and A. Manica. 2012. Initiative, personality and leadership in pairs of foraging fish. PLoS One 7(5):e36606. doi:10.1371/journal.pone.0036606.

Nammack, M. F., J. A. Musick, and J. A. Colvocoresses. 1985. Life history of spiny dogfish off the northeastern United States. Transactions of the American Fisheries Society 114(3):367–376.

Näslund, I. 1990. The development of regular seasonal habitat shifts in a landlocked Arctic charr, *Salvelinus alpinus* L., population. Journal of Fish Biology 36:401–414.

Näslund, I., G. Milbrink, L. O. Eriksson, and S. Holmgren. 1993. Importance of habitat productivity differences, competition and predation for the migratory behavior of Arctic charr. Oikos 66(3): 538–546.

Nathan, R. 2008. An emerging movement ecology paradigm. Proceedings of the National Academy of Sciences of the United States of America 105(49):19,050–19,051.

Nathan, R., and 6 coauthors. 2008. A movement ecology paradigm for unifying organismal movement research. Proceedings of the National Academy of Sciences of the United States of America 105(49):19,052–19,059.

National Research Council. 1994. An Assessment of Atlantic Bluefin Tuna. National Academy Press, Washington, DC.

Neat, F. C., and 7 coauthors. 2006. Residency and depth movements of a coastal group of Atlantic cod (*Gadus morhua* L.). Marine Biology 148(3): 643–654.

Neave, F. 1964. Ocean migrations of Pacific Salmon. Journal of the Fisheries Research Board of Canada 21(5):1227–1244.

Neeson, T. M., M. J. Wiley, S. A. Adlerstein, and R. L. Riolo. 2012. How river network structure and habitat availability shape the spatial dynamics of larval sea lampreys. Ecological Modelling 226:62–70.

Nehlson, W., J. E. Williams, and J. A. Lichatowich. 1991. Pacific salmon at the crossroads: Stocks at risk from California, Oregon, Idaho, and Washington. Fisheries 16:4–21.

Neilson, J. D., and R. I. Perry. 1990. Diel vertical migrations of marine fishes: An obligate or facultative process? Advances in Marine Biology 26:115–168.

Neilson, J. D., R. I. Perry, P. Valerio, and K. G. Waiwood. 1986. Condition of Atlantic cod *Gadus morhua* larvae after the transition to exogenous feeding: Morphometrics, buoyancy and predator avoidance. Marine Ecology Progress Series 32(2–3):229–235.

Nelson, D. R., and 6 coauthors. 1997. An acoustic tracking of a megamouth shark, *Megachasma pelagios*: A crepuscular vertical migrator. Environmental Biology of Fishes 49(4):389–399.

Nemerson, D., S. Berkeley, and C. Safina. 2000. Spawning site fidelity in Atlantic bluefin tuna, *Thunnus thynnus*: The use of size-frequency analysis to test for the presence of migrant east Atlantic bluefin tuna on Gulf of Mexico spawning grounds. Fishery Bulletin 98(1):118–126.

Neubauer, P., O. P. Jensen, J. A. Hutchings, and J. K. Baum. 2013. Resilience and recovery of overexploited marine populations. Science 6130:347–349.

Newlands, N. K., M. E. Lutcavage, and T. J. Pitcher. 2006. Atlantic bluefin tuna in the Gulf of Maine. 1. Estimation of seasonal abundance accounting for movement, school and school-aggregation behaviour. Environmental Biology of Fishes 77(2):177–195.

Newton, I. 2008. The Migration Ecology of Birds. Elsevier Academic, New York.

Nichol, D. G., and E. K. Pikitch. 1994. Reproduction of darkblotched rockfish off the Oregon coast. Transactions of the American Fisheries Society 123(4):469–481.

Nichols, J. D. 1992. Capture-recapture models. Bioscience 42(2):94–102.

Nielsen, E. E., M. M. Hansen, and D. Meldrup. 2006. Evidence of microsatellite hitch-hiking selection in Atlantic cod (*Gadus morhua* L.): Implications for inferring population structure in nonmodel organisms. Molecular Ecology 15(11):3219-3229.

Nielsen, E. E., J. Hemmer-Hansen, P. F. Larsen, and D. Bekkevold. 2009. Population genomics of marine fishes: Identifying adaptive variation in space and time. Molecular Ecology 18(15): 3128-3150.

Niklitschek, E. J., D. H. Secor, P. Toledo, A. Lafon, and M. George-Nascimento. 2010. Segregation of SE Pacific and SW Atlantic southern blue whiting stocks: Integrating evidence from complementary otolith microchemistry and parasite assemblage approaches. Environmental Biology of Fishes 89(3-4):399-413.

Nishi, T., G. Kawamura, and K. Matsumoto. 2004. Magnetic sense in the Japanese eel, *Anguilla japonica*, as determined by conditioning and electrocardiography. Journal of Experimental Biology 207(17):2965-2970.

Nixon, S. W. 1988. Physical energy inputs and the comparative ecology of lake and marine ecosystems. Limnology and Oceanography 33(4):1005-1025.

NOAA. National Oceanic and Atmospheric Administration. 2007. Magnuson-Stevens Fisheries Conservation and Management Act. January 2007 Amendment. National Marine Fisheries Service, Silver Spring, MD.

———. 2012. 2011 Report to Congress: The Status of U.S. Fisheries. National Marine Fisheries Service, Silver Spring, MD.

Norberg, U. M. L. 2002. Structure, form, and function of flight in engineering and the living world. Journal of Morphology 252(1):52-81.

Norcross, B. L., and R. F. Shaw. 1984. Oceanic and estuarine transport of fish eggs and larvae: A review. Transactions of the American Fisheries Society 113:153-165.

Norcross, J. J., S. L. Richardson, W. H. Massmann, and E. B. Joseph. 1974. Development of young bluefish (*Pomatomus saltatrix*) and distribution of eggs and young in Virginian coastal waters. Transactions of the American Fisheries Society 103(3):477-497.

Nordeng, H. 1971. Is the local orientation of anadromous fishes determined by pheromones? Nature 233(5319):411-413.

———. 1983. Solution to the "char problem" based on Arctic char (*Salvelinus alpinus*) in Norway. Canadian Journal of Fisheries and Aquatic Science 40:1372-1387.

North, E. W., and E. D. Houde. 2001. Retention of white perch and striped bass larvae: Biological-physical interactions in Chesapeake Bay estuarine turbidity maximum. Estuaries 24:756-769.

———. 2006. Retention mechanisms of white perch (*Morone americana*) and striped bass (*Morone saxatilis*) early-life stages in an estuarine turbidity maximum: An integrative fixed-location and mapping approach. Fisheries Oceanography 15(6): 429-450.

North, E. W., R. R. Hood, S. Y. Chao, and L. P. Sanford. 2005. The influence of episodic events on transport of striped bass eggs to the estuarine turbidity maximum nursery area. Estuaries 28(1):108-123.

Nyström, M. 2006. Redundancy and response diversity of functional groups: Implications for the resilience of coral reefs. Ambio 35(1):30-35.

Odling-Smee, F. J., K. N. Laland, and M. W. Feldman. 2003. Niche Construction: The Neglected Process in Evolution. Monographs in Population Biology 37. Princeton University Press, Princeton, NJ.

Odling-Smee, L., and V. A. Braithwaite. 2003. The role of learning in fish orientation. Fish and Fisheries 4(3):235-246.

Okamura, A., and 6 coauthors. 2002. Development of lateral line organs in leptocephali of the freshwater eel *Anguilla japonica* (Teleostei, Anguilliformes). Journal of Morphology 254(1):81-91.

———. 2007. Effects of water temperature on early development of Japanese eel *Anguilla japonica*. Fisheries Science 73(6):1241-1248.

Olivieri, I., and P.-H. Gouyan. 1997. Evolution of migration rate and other traits: The metapopulation effect. In Metapopulation Biology: Ecology, Genetics, and Evolution, ed. I. A. Hanski and M. E. Gilpin, 293-323. Academic, New York.

Olla, B. L., and M. W. Davis. 1990. Effects of physical factors on the vertical distribution of larval walleye pollock *Theragra chalcogramma* under controlled laboratory conditions. Marine Ecology Progress Series 63(2-3):105-112.

Olney, J. E., and J. M. Hoenig. 2001. Managing a fishery under moratorium: Assessment opportunities for Virginia's stocks of American shad. Fisheries 26(2):6-12.

Olsen, G. H. 2001. Of cranes and men: Reintroduction of cranes to a migratory pathway. Part 1. Journal of Avian Medicine and Surgery 15(2):133-137.

Olsson, I. C., L. A. Greenberg, E. Bergman, and K. Wysujack. 2006. Environmentally induced migration: The importance of food. Ecology Letters 9(6):645-651.

O'Neal, S. L., and J. A. Stanford. 2011. Partial migra-

tion in a robust brown trout population of a Pata-
gonian river. Transactions of the American Fisheries Society 140(3):623–635.

Ortiz, M., E., and 7 coauthors. 2003. Global overview of the major constituent-based billfish tagging programs and their results since 1954. Marine and Freshwater Research 54(4):489–507.

Óskarsson, G. J., A. Gudmundsdottir, and T. Sig-
urdsson. 2009. Variation in spatial distribution and migration of Icelandic summer-spawn-
ing herring. ICES Journal of Marine Science
66(8):1762–1767.

Otake, T. 1996. Fine structure and function of the alimentary canal in leptocephali of the Japanese eel
Anguilla japonica. Fisheries Science 62(1):28–34.

Otake, T., K. Nogami, and K. Maruyama. 1993. Dissolved and particulate organic matter as possible
food sources for eel leptocephali. Marine Ecology
Progress Series 92:27–34.

Otake, T., T. Inagaki, H. Hasumoto, N. Mochioka,
and K. Tsukamoto. 1998. Diel vertical distribution
of Anguilla japonica leptocephali. Ichthyological
Research 45(2):208–211.

Ottersen, G., D. O. Hjermann, and N. C. Stenseth.
2006. Changes in spawning stock structure
strengthen the link between climate and recruitment in a heavily fished cod (Gadus morhua) stock.
Fisheries Oceanography 15(3):230–243.

Ottersen, G., S. Kim, G. Huse, J. J. Polovina, and
N. C. Stenseth. 2010. Major pathways by which climate may force marine fish populations. Journal
of Marine Systems 79(3–4):343–360.

Overholtz, W. J., and K. D. Friedland. 2002. Recovery of the Gulf of Maine-Georges Bank Atlantic
herring (Clupea harengus) complex: Perspectives
based on bottom trawl survey data. Fishery Bulletin 100(3):593–608.

Overholtz, W. J., J. A. Hare, and C. M. Keith. 2011.
Impacts of interannual environmental forcing
and climate change on the distribution of Atlantic
mackerel on the US Northeast Continental Shelf.
Marine and Coastal Fisheries 3(1):219–232.

Páez, D. J., C. Brisson-Bonenfant, O. Rossignol,
H. E. Guderley, L. Bernatchez, and J. J. Dodson.
2011. Alternative developmental pathways and
the propensity to migrate: A case study in the Atlantic salmon. Journal of Evolutionary Biology
24(2):245–255.

Page, F. H., and 6 coauthors. 1999. Cod and haddock
spawning on Georges Bank in relation to water
residence times. Fisheries Oceanography 8(3):
212–226.

Palkovacs, E. P., K. B. Dion, D. M. Post, and A. Cac-
cone. 2008. Independent evolutionary origins of
landlocked alewife populations and rapid parallel
evolution of phenotypic traits. Molecular Ecology
17(2):582–597.

Palmer, A. R., and R. R. Strathmann. 1981. Scale of
dispersal in varying environments and its implications for life histories of marine invertebrates.
Oecolgia 48:308–318.

Palumbi, S. R. 2003. Population genetics, demographic connectivity, and the design of marine reserves. Ecological Applications 13(1):S146–S158.

Pampoulie, C., M. O. Stefánsson, T. D. Jörundsdóttir,
B. S. Danilowicz, and A. K. Daníelsdóttir. 2008.
Recolonization history and large-scale dispersal
in the open sea: The case study of the North Atlan-
tic cod, Gadus morhua L. Biological Journal of the
Linnean Society 94(2):315–329.

Pampoulie, C., A. K. Daníelsdóttir, M. Storr-Pulsen,
H. Hovgård, E. Hjörleifsson, and B. E. Steinarsson.
2011. Neutral and nonneutral genetic markers revealed the presence of inshore and offshore stock
components of Atlantic cod in Greenland waters.
Transactions of the American Fisheries Society
140(2):307–319.

Paradis, E., S. R. Baillie, W. J. Sutherland, and
R. D. Gregory. 1998. Patterns of natal and breeding dispersal in birds. Journal of Animal Ecology
67(4):518–536.

Paragamian, V. L., and M. J. Hansen. 2008. Evaluation of recovery goals for endangered white sturgeon in the Kootenai River, Idaho. North American
Journal of Fisheries Management 28(2):463–470.

Paris, C. B., L. M. Cherubin, and R. K. Cowen. 2007.
Surfing, spinning, or diving from reef to reef: Ef-
fects on population connectivity. Marine Ecology
Progress Series 347:285–300.

Park, H., and H. Choi. 2010. Aerodynamic characteristics of flying fish in gliding flight. Journal of
Experimental Biology 213(19):3269–3279.

Parker, G. A., and M. E. Begon. 1993. Sperm competition games: Sperm size and number under gamet-
ic control. Proceedings of the Royal Society B: Biological Sciences 253(1338):255–262.

Parker, G. A., and T. Pizzari. 2010. Sperm competition and ejaculate economics. Biological Reviews
85(4):897–934.

Parker, N. C., A. E. Giorgi, R. C. Heidinger, D. B.
Jester, E. D. Prince, and G. A. Winans. 1990.
Fish-Marking Techniques. American Fisheries
Society Symposium 7:1–879.

Parker, S. J., and J. D. McCleave. 1997. Selective tidal

stream transport by American eels during homing movements and estuarine migration. Journal of the Marine Biological Association of the United Kingdom 77:871–889.

Parker, S. J., J. M. Olson, P. S. Rankin, and J. S. Malvitch. 2008. Patterns in vertical movements of black rockfish *Sebastes melanops*. Aquatic Biology 2(1):57–65.

Parrish, J. K., and L. Edelstein-Keshet. 1999. Complexity, pattern, and evolutionary trade-offs in animal aggregation. Science 284(5411):99–101.

Parrish, J. K., S. V. Viscido, and D. Grunbaum. 2002. Self-organized fish schools: An examination of emergent properties. Biological Bulletin 202(3):296–305.

Parrish, R. H., C. S. Nelson, and A. Bakun. 1981. Transport mechanisms and reproductive success of fishes in the California Current. Biological Oceanography 1:175–203.

Parsons, D. M., M. A. Morrison, J. R. McKenzie, B. W. Hartill, R. Bain and R. I. C. C. Francis. 2011. A fisheries perspective of behavioural variability: Differences in movement behaviour and extraction rate of an exploited sparid, snapper (*Pagrus auratus*). Canadian Journal of Fisheries and Aquatic Sciences 68(4):632–642.

Partridge, B. L. 1980. Effect of school size on the structure and dynamics of minnow schools. Animal Behaviour 28:68–77.

———. 1981. Internal dynamics and the interrelations of fish in schools. Journal of Comparative Physiology 144(3):313–325.

Partridge, B. L., and T. J. Pitcher. 1980. The sensory basis of fish schools: Relative roles of lateral line and vision. Journal of Comparative Physiology 135(4):315–325.

Partridge, B. L., T. Pitcher, J. M. Cullen, and J. Wilson. 1980. The three-dimensional structure of fish schools. Behavioral Ecology and Sociobiology 6(4):277–288.

Pascual, M. A., and T. P. Quinn. 1991. Evaluation of alternative models of the coastal migration of adult Fraser River sockeye salmon (*Oncorhynchus nerka*). Canadian Journal of Fisheries and Aquatic Sciences 48(5):799–810.

Patterson, H. M., M. J. Kingsford, and M. T. McCulloch. 2005. Resolution of the early life history of a reef fish using otolith chemistry. Coral Reefs 24(2):222–229.

Patterson, T. A., L. Thomas, C. Wilcox, O. Ovaskainen, and J. Matthiopoulos. 2008. State-space models of individual animal movement. Trends in Ecology and Evolution 23(2):87–94.

Pauly, D. 1981. The relationship between gill surface area and growth performance in fish: A generalization of von Bertanffy's theory of growth. Meereforschung 28:251–282.

———. 1995. Anecdotes and the shifting base-line syndrome of fisheries. Trends in Ecology and Evolution 10(10):430.

Pauly, D., and R. S. V. Pullin. 1988. Hatching time in spherical, pelagic, marine fish eggs in response to temperature and egg size. Environmental Biology of Fishes 22:261–271.

Pautzke, S. M., M. E. Mather, J. T. Finn, L. A. Deegan, and R. M. Muth. 2010. Seasonal use of a New England estuary by foraging contingents of migratory striped bass. Transactions of the American Fisheries Society 139(1):257–269.

Pawson, M. G., D. F. Kelley, and G. D. Pickett. 1987. The distribution and migrations of bass, *Dicentrarchus labrax* L., in waters around England and Wales as shown by tagging. Journal of the Marine Biological Association of the United Kingdom 67:183–217.

Pearcy, W. G. 1992. Ocean Ecology of North Pacific Salmonids. University of Washington Press, Seattle.

Pearson, J. C. 1941. The young of some marine fishes taken in lower Chesapeake Bay, Virginia, with special reference to gray sea trout, *Cynoscion regalis* (Bloch). Fisheries Bulletin 50:79–102.

Pechenik, J. A. 2006. Larval experience and latent effects: Metamorphosis is not a new beginning. Integrative and Comparative Biology 46(3):323–333.

Pedersen, M. W., D. Righton, U. H. Thygesen, K. H. Andersen, and H. Madsen. 2008. Geolocation of North Sea cod (*Gadus morhua*) using hidden Markov models and behavioural switching. Canadian Journal of Fisheries and Aquatic Sciences 65(11):2367–2377.

Pepin, P. 1991. Effect of temperature and size on development, mortality, and survival rates of the pelagic early life history stages of marine fish. Canadian Journal of Fisheries and Aquatic Science 48(3):503–518.

Pepin, P., and J. A. Helbig. 1997. Distribution and drift of Atlantic cod (*Gadus morhua*) eggs and larvae on the northeast Newfoundland Shelf. Canadian Journal of Fisheries and Aquatic Sciences 54(3):670–685.

Pepin, P., and T. J. Miller. 1993. Potential use and abuse of general empirical-models of early-life history processes in fish. Canadian Journal of Fisheries and Aquatic Sciences 50(6):1343–1345.

Pepin, P., E. Colbourne, and G. Maillet. 2011. Seasonal patterns in zooplankton community structure on the Newfoundland and Labrador Shelf. Progress in Oceanography 91:273–285.

Perry, A. L., P. J. Low, J. R. Ellis, and J. D. Reynolds. 2005. Climate change and distribution shifts in marine fishes. Science 308(5730):1912–1915.

Perry, R. I., P. Cury, K. Brander, S. Jennings, C. Moellmann, and B. Planque. 2010. Sensitivity of marine systems to climate and fishing: Concepts, issues and management responses. Journal of Marine Systems 79(3–4):427–435.

Petersen, C. H., P. T. Drake, C. A. Edwards, and S. Ralston. 2010. A numerical study of inferred rockfish (*Sebastes* spp.) larval dispersal along the central California coast. Fisheries Oceanography 19(1):21–41.

Petersen, C. W., R. R. Warner, S. Cohen, H. C. Hess, and A. T. Sewell. 1992. Variable pelagic fertilization success: Implications for mate choice and spatial patterns of mating. Ecology 73(2):391–401.

Petersen, C. W., R. R. Warner, D. Y. Shapiro, and A. Marconato. 2001. Components of fertilization success in the bluehead wrasse, *Thalassoma bifasciatum*. Behavioral Ecology 12(2):237–245.

Peterson, G., C. R. Allen, and C. S. Holling. 1998. Ecological resilience, biodiversity, and scale. Ecosystems 1(1):6–18.

Petitgas, P. 1998. Biomass-dependent dynamics of fish spatial distributions characterized by geostatistical aggregation curves. ICES Journal of Marine Science 55(3):443–453.

Petitgas, P., and J. J. Levenez. 1996. Spatial organization of pelagic fish: Echogram structure, spatio-temporal condition, and biomass in Senegalese waters. ICES Journal of Marine Science 53(2):147–153.

Petitgas, P., and 8 coauthors. 2007. Report of the Workshop on Testing the Entrainment Hypothesis (WKTEST). ICES CM2007/LRC:10. Living Resource Committee, International Council for the Exploration of the Sea, Denmark.

Petitgas, P., D. H. Secor, I. McQuinn, G. Huse, and N. Lo. 2010. Stock collapses and their recovery: Mechanisms that establish and maintain life-cycle closure in space and time. ICES Journal of Marine Science 67(9):1841–1848.

Petitgas, P., and 10 coauthors. 2012. Anchovy population expansion in the North Sea. Marine Ecology Progress Series 444:1–13.

Petitgas, P., and 8 coauthors. 2013. Impacts of climate change on the complex life cycles of fish. Fisheries Oceanography 22(2):121–139.

Pfeiler, E. 1986. Towards an explanation of the developmental strategy in leptocephalus larvae of marine teleost fishes. Environmental Biology of Fishes 15(1):3–13.

Pfennig, D. W., M. A. Wund, E. C. Snell-Rood, T. Cruickshank, C. D. Schlichting, and A. P. Moczek. 2010. Phenotypic plasticity's impacts on diversification and speciation. Trends in Ecology and Evolution 25(8):459–467.

Phalan, B., and 7 coauthors. 2007. Foraging behaviour of four albatross species by night and day. Marine Ecology Progress Series 340:271–286.

Pianka, E. R. 1970. *r*-selection and *K*-selection. American Naturalist 104(940):592–597.

Piche, J., J. A. Hutchings, and W. Blanchard. 2008. Genetic variation in threshold reaction norms for alternative reproductive tactics in male Atlantic salmon, *Salmo salar*. Proceedings of the Royal Society B: Biological Sciences 275(1642):1571–1575.

Pigliucci, M. 2001. Phenotypic Plasticity: Beyond Nature and Nurture. Johns Hopkins University Press, Baltimore.

Pimm, S. L., and J. C. Rice. 1987. The dynamics of multispecies, multi-life-stage models of aquatic food webs. Theoretical Population Biology 32(3):303–325.

Pine, W. E., K. H. Pollock, J. E. Hightower, T. J. Kwak, and J. A. Rice. 2003. A review of tagging methods for estimating fish population size and components of mortality. Fisheries 28(10):10–23.

Pineda, J., J. A. Hare, and S. Sponaugle. 2007. Larval transport and dispersal in the coastal ocean and consequences for population connectivity. Oceanography 20(3):22–39.

Pinsky, M. L., B. Worm, M. J. Fogarty, J. L. Sarmiento, and S. A. Levin. 2013. Marine taxa track local climate velocities. Science 341(6151):1239–1242.

Pitcher, R. J., and J. K. Parrish. 1993. Function of shoaling behavior in teleosts. In Behaviour of Teleosts, ed. T. J. Pitcher, 363–439. Chapman and Hall, London.

Pitcher, T. J., and B. L. Partridge. 1979. Fish school density and volume. Marine Biology 54(4):383–394.

Pitcher, T. J., O. A. Misund, A. Fernö, B. Totland, and V. Melle. 1996. Adaptive behaviour of herring schools in the Norwegian Sea as revealed by high-resolution sonar. ICES Journal of Marine Science 53(2):449–452.

Planque, B., and 6 coauthors. 2010. How does fishing alter marine populations and ecosystems sensitivity to climate? Journal of Marine Systems 79
(3–4):403–417.

Plaza, G., S. Katayama, and M. Omori. 2004. Spatio-temporal patterns of parturition of the black
rockfish *Sebastes inermis* in Sendai Bay, northern
Japan. Fisheries Science 70(2):256–263.

Podolsky, R. D., and A. L. Moran. 2006. Integrating
function across marine life cycles. Integrative and
Comparative Biology 46(5):577–586.

Polovina, J. J. 1996. Decadal variation in the trans-
Pacific migrations of northern bluefin tuna
(*Thunnus thynnus*) coherent with climate-induced
change in prey abundance. Fisheries Oceanography 5:114–119.

Polunin, N. V. C. 2002. Marine protected areas, fish
and fisheries. In Handbook of Fish Biology and
Fisheries, vol. 2, ed. P. J. B. Hart and J. D. Reynolds,
293–341. Wiley-Blackwell, Ames, IA.

Pope, J. G., J. G. Shepherd, and J. Webb. 1994. Successful surf-riding on size spectra: The secret
of survival in the sea. Philosophical Transactions of the Royal Society B: Biological Sciences
343(1303):41–49.

Post, D. M. 2002. Using stable isotopes to estimate
trophic position: Models, methods, and assumptions. Ecology 83(3):703–718.

Post, D. M., and E. P. Palkovacs. 2009. Eco-evolutionary feedbacks in community and ecosystem ecology: Interactions between the ecological
theatre and the evolutionary play. Philosophical
Transactions of the Royal Society B: Biological Sciences 364(1523):1629–1640.

Poulsen, N. A., J. Hemmer-Hansen, V. Loeschcke,
G. R. Carvalho, and E. E. Nielsen. 2011. Microgeo-
graphical population structure and adaptation in
Atlantic cod *Gadus morhua*: Spatio-temporal insights from gene-associated DNA markers. Marine Ecology Progress Series 436:231–243.

Price, T. D., A. Qvarnström, and D. E. Irwin. 2003.
The role of phenotypic plasticity in driving genet-
ic evolution. Proceedings of the Royal Society B:
Biological Sciences 270(1523):1433–1440.

Psenicka, M., and 7 coauthors. 2008. Morphology,
chemical contents and physiology of Chondroste-
an fish sperm: A comparative study between Sibe-
rian sturgeon (*Acipenser baerii*) and sterlet (*Acipenser ruthenus*). Journal of Applied Ichthyology
24(4):371–377.

Pulido, F. 2011. Evolutionary genetics of partial migration: The threshold model of migration revisited. Oikos 120(12):1776–1783.

Pulido, F., and P. Berthold. 2010. Current selection for lower migratory activity will drive the
evolution of residency in a migratory bird population. Proceedings of the National Academy of
Sciences of the United States of America 107(16):
7341–7346.

Pulliam, H. R. 1988. Sources, sinks, and population
regulation. American Naturalist 132(5):652–661.

Putman, N. F., K. J. Lohmann, E. M. Putnam, T. P.
Quinn, A. P. Klimley, and D. L. G. Noakes. 2013.
Evidence for geomagnetic imprinting as a homing mechanism in Pacific salmon. Current Biology 23(4):312–316.

Quammen, D. 2010. Great migrations. National Geographic 218(5):28–51.

Quinn, T. P. 1980. Evidence for celestial and magnetic compass orientation in lake migrating sockeye salmon fry. Journal of Comparative Physiology
137(3):243–248.

———. 1983. A review of homing and straying of
wild and hatchery-produced salmon. Fisheries
Research 18:29–44.

———. 2005. The Behavior and Ecology of Pacific Salmon and Trout. University of Washington
Press, Seattle.

Quinn, T. P., and E. L. Brannon. 1982. The use of celestial and magnetic cues by orienting sockeye
salmon smolts. Journal of Comparative Physiology
147(4):547–552.

Quinn, T. P., and R. D. Brodeur. 1991. Intra-specif-
ic variations in the movement patterns of marine
animals. American Zoologist 31(1):231–241.

Quinn, T. P., and K. Fresh. 1984. Homing and straying in Chinook salmon (*Oncorhynchus tshawytscha*)
from Cowlitz River Hatchery, Washington. Canadian Journal of Fisheries and Aquatic Sciences
41(7):1078–1082.

Quinn, T. P., and K. W. Myers. 2004. Anadromy and
the marine migrations of Pacific salmon and trout:
Rounsefell revisited. Reviews in Fish Biology and
Fisheries 14(4):421–442.

Quinn, T. P., R. S. Nemeth, and D. O. McIsaac. 1991.
Homing and straying patterns of fall Chinook
salmon in the Lower Columbia River. Transactions
of the American Fisheries Society 120:150–156.

Quinn, T. P., J. A. Peterson, V. F. Gallucci, W. K. Her-
shberger, and E. L. Brannon. 2002. Artificial selection and environmental change: Countervailing factors affecting the timing of spawning by

Coho and Chinook salmon. Transactions of the American Fisheries Society 131:591–598.

Rabøl, J. 1969. Reversed migration as the cause of westward vagrancy by four *Phyllocscopus* warblers. British Birds 62:89–92.

Radtke, R., and R. A. Kinzie. 1996. Evidence of a marine larval stage in endemic Hawaiian stream gobies from isolated high-elevation locations. Transactions of the American Fisheries Society 125:613–621.

Radtke, R., M. Svenning, D. Malone, A. Klementsen, J. Ruzicka, and D. Fey. 1996. Migrations in an extreme northern population of Arctic charr *Salvelinus alpinus*: Insights from otolith microchemistry. Marine Ecology Progress Series 136:13–23.

Rago, P. J., K. A. Sosebee, J. K. T. Brodziak, S. A. Murawski, and E. D. Anderson. 1998. Implications of recent increases in catches on the dynamics of northwest Atlantic spiny dogfish (*Squalus acanthias*). Fisheries Research 39(2):165–181.

Rahn, H., C. V. Paganelli, and A. Ar. 1975. Relation of avian egg weight to body weight. Auk 92 (4):750–765.

Rakitin, A., M. F. Ferguson, and E. A. Trippel. 1999. Sperm competition and fertilization success in Atlantic cod (*Gadus morhua*): Effect of sire size and condition factor on gamete quality. Canadian Journal of Fisheries and Aquatic Science 56(12):2315–2323.

Ramenofsky, M., and J. C. Wingfield. 2007. Regulation of migration. Bioscience 57(2):135–143.

Ravier, C., and J. M. Fromentin. 2001. Long-term fluctuations in the eastern Atlantic and Mediterranean bluefin tuna population. ICES Journal of Marine Science 58:1299–1317.

Ray, G. C. 2005. Connectivities of estuarine fishes to the coastal realm. Estuarine Coastal and Shelf Science 64(1):18–32.

Rayner, J. M. V. 1988. The evolution of vertebrate flight. Biological Journal of the Linnaean Society 34(3):269–287.

Reed, R. K., L. S. Incze, and J. D. Schumacher. 1989. Estimation of the effects of flow on dispersion of larval pollock, *Theragra chalcogramma*, in Shelikov Strait, Alaska. Canadian Journal of Fisheries and Aquatic Sciences Special Publication108:280–294.

Reese, D. C., R. T. O'Malley, R. D. Brodeur, and J. H. Churnside. 2011. Epipelagic fish distributions in relation to thermal fronts in a coastal upwelling system using high-resolution remote-sensing techniques. ICES Journal of Marine Science 68(9):1865–1874.

Reiss, C. S., G. Panteleev, C. T. Taggart, J. Sheng, and B. deYoung. 2000. Observations on larval fish transport and retention on the Scotian Shelf in relation to geostrophic circulation. Fisheries Oceanography 9(3):195–213.

Reiss, H., G. Hoarau, M. Dickey-Collas, and W. J. Wolff. 2009. Genetic population structure of marine fish: Mismatch between biological and fisheries management units. Fish and Fisheries 10(4):361–395.

Restrepo, V. R., and 10 coauthors. 1998. Technical guidance on the use of precautionary approaches to implementing National Standard 1 of the Magnuson-Stevens Fishery Conservation and Management Act. NOAA Technical Memorandum NMFS-F/SPO-31. National Marine Fisheries Service, Silver Spring, MD.

Reynolds, A. M., and C. J. Rhodes. 2009. The Lévy flight paradigm: Random search patterns and mechanisms. Ecology 90(4):877–887.

Reznick, D. 1985. Costs of reproduction: An evaluation of the empirical evidence. Oikos 44:257–267.

Reznick, D., H. Bryga, and J. A. Endler. 1990. Experimentally induced life-history evolution in a natural population. Nature 346:357–359.

Rice, J., and H. Gislason. 1996. Patterns of change in the size spectra of numbers and diversity of the North Sea fish assemblage, as reflected in surveys and models. ICES Journal of Marine Science 53(6):1214–1225.

Rice, J. A., L. B. Crowder, and E. A. Marschall. 1997. Predation on juvenile fishes: Dynamic interactions between size-structured predators and prey. In Early Life History and Recruitment in Fish Populations, ed. R. C. Chambers and E. A. Trippel, 333–356. Chapman and Hall, New York.

Richards, R. A., and P. J. Rago. 1999. A case history of effective fishery management: Chesapeake Bay striped bass. North American Journal of Fisheries Management 19:356–375.

Ricker, W. E. 1958. Maximum sustainable yield from fluctuating environments and mixed stocks. Journal of the Fisheries Research Board of Canada 15:991–1006.

Ricklefs, R. E. 1979. Ecology. 2nd ed. Chiron, Asheville, NC.

Rideout, R. M., G. A. Rose, and M. P. M. Burton. 2005. Skipped spawning in female iteroparous fishes. Fish and Fisheries 6:50–72.

Rieman, B. E., and J. B. Dunham. 2000. Metapopulations and salmonids: A synthesis of life history patterns and empirical observations. Ecology of Freshwater Fish 9(1–2):51–64.

Riffell, J. A., and R. K. Zimmer. 2007. Sex and flow: The consequences of fluid shear for sperm-egg interactions. Journal of Experimental Biology 210(20):3644–3660.

Righton, D., A. Aarestrup, D. Jellyman, P. Sébert, G. van den Thillart, and K. Tsukamoto. 2012. The *Anguilla* spp. migration problem: 40 million years of evolution and two millennia of speculation. Journal of Fish Biology 81(2):365–386.

Rijnsdorp, A. D. 1993. Fisheries as a large-scale experiment on life-history evolution: Disentangling phenotypic and genetic effects in changes in maturation and reproduction of North Sea plaice, *Pleuronectes platessa* L. Oecologia, 96:391–401.

———. 1994. Population-regulating processes during the adult phase in flatfish. Netherlands Journal of Sea Research 32(2):207–223.

Rijnsdorp, A. D., and M. A. Pastoors. 1995. Modeling the spatial dynamics and fisheries of North Sea plaice (*Pleuronectes platessa* l) based on tagging data. ICES Journal of Marine Science 52(6): 963–980.

Rijnsdorp, A. D., M. Vanstralen, and H. W. Vanderveer. 1985. Selective tidal transport of North Sea plaice larvae *Pleuronectes platessa* in coastal nursery areas. Transactions of the American Fisheries Society 114(4):461–470.

Robards, M. D., and T. P. Quinn. 2002. The migratory timing of adult summer-run steelhead in the Columbia River over six decades of environmental change. Transactions of the American Fisheries Society 131(3):523–536.

Roberts, C. M. 1997. Connectivity and management of Caribbean coral reefs. Science 278(5342): 1454–1457.

Robichaud, D., and G. A. Rose. 2001. Multiyear homing of Atlantic cod to a spawning ground. Canadian Journal of Fisheries and Aquatic Sciences 58(12):2325–2329.

———. 2004. Migratory behaviour and range in Atlantic cod: Inference from a century of tagging. Fish and Fisheries 5(3):185–214.

Robichaud-leBlanc, K. A., S. C. Courtenay, and A. Locke. 1996. Spawning and early life history of a northern population of striped bass (*Morone saxatilis*) in the Miramichi River estuary, Gulf of St. Lawrence. Canadian Journal of Zoology 74(9):1645–1655.

Robinson, W. D., and 8 coauthors. 2010. Integrating concepts and technologies to advance the study of

bird migration. Frontiers in Ecology and the Environment 8(7):354–361.

Rochet, M. J., and V. M. Trenkel. 2003. Which community indicators can measure the impact of fishing? A review and proposals. Canadian Journal of Fisheries and Aquatic Sciences 60(1):86–99.

Rodriguez-Munoz, R., J. R. Waldman, C. Grunwald, N. K. Roy, and I. Wirgin. 2004. Absence of shared mitochondrial DNA haplotypes between sea lamprey from North American and Spanish rivers. Journal of Fish Biology 64(3):783–787.

Roff, D. A. 1978. Size and survival in a stochastic environment. Oecologia 36(2):163–172.

———. 1988. The evolution of migration and some life history parameters in marine fishes. Environmental Biology of Fishes 22:133–146.

———. 1992. The Evolution of Life Histories. Chapman and Hall, New York.

———. 1996. The evolution of threshold traits in animals. Quarterly Review of Biology 71(1):3–35.

Rogers, D. 1983. Patterns and process in large-scale animal movement. In The Ecology of Animal Movement, ed. I. R. Swingland and P. J. Greenwood, 160–180. Clarendon, Oxford.

Rooker, J. R., and 9 coauthors. 2007. Life history and stock structure of Atlantic bluefin tuna (*Thunnus thynnus*). Reviews in Fisheries Science 15(4):265–310.

Rooker, J. R., D. H. Secor, G. D. DeMetrio, R. Schloesser, B. A. Block, and J. D. Neilson. 2008. Natal homing and connectivity in Atlantic bluefin tuna populations. Science 322:742–744.

Rose, G. A. 1993. Cod spawning on a migration highway in the northwest Atlantic. Nature 366 (6454):458–461.

Rose, G. A., and D. W. Kulka. 1999. Hyperaggregation of fish and fisheries: How catch-per-unit-effort increased as the northern cod (*Gadus morhua*) declined. Canadian Journal of Fisheries and Aquatic Sciences 56:118–127.

Rose, G. A., B. deYoung, D. W. Kulka, S. V. Goddard, and G. L. Fletcher. 2000. Distribution shifts and overfishing the northern cod (*Gadus morhua*): A view from the ocean. Canadian Journal of Fisheries and Aquatic Sciences 57:644–663.

Rothschild, B. J. 1986. Dynamics of Marine Fish Populations. Harvard University Press, Cambridge, MA.

Rothschild, B. J., and T. R. Osborn. 1988. Small-scale turbulence and plankton contact rates. Journal of Plankton Research 10(3):465–474.

Rounsefell, G. A. 1957. Fecundity of North American Salmonidae. Fisheries Bulletin 57:449–468.

Royer, F., and J. M. Fromentin. 2006. Recurrent and density-dependent patterns in long-term fluctuations of Atlantic bluefin tuna trap catches. Marine Ecology Progress Series 319:237–249.

Royer, F., J. M. Fromentin, and P. Gaspar. 2005. A state-space model to derive bluefin tuna movement and habitat from archival tags. Oikos 109(3):473–484.

Rozhok, A. 2008. Orientation and Navigation in Vertebrates. Springer, Berlin.

Rubenstein, D. R., and 6 coauthors. 2002. Linking breeding and wintering ranges of a migratory songbird using stable isotopes. Science 295 (5557):1062–1065.

Rulifson, R. A., and K. A. Tull. 1999. Striped bass spawning in a tidal bore river: The Shubenacadie Estuary, Atlantic Canada. Transactions of the American Fisheries Society 128(4):613–624.

Rundle, H. D., L. Nagel, J. W. Boughman, and D. Schluter. 2000. Natural selection and parallel speciation in sympatric sticklebacks. Science 287(5451):306–308.

Russell, D. J., and A. J. McDougall. 2005. Movement and juvenile recruitment of mangrove jack, Lutjanus argentimaculatus (Forsskal), in northern Australia. Marine and Freshwater Research 56(4):465–475.

Russell, E. S. 1931. Some theoretical considerations on the "overfishing" problem. Journal du Conseil International pour l'Exploration de la Mer 6:3–20.

Russell, F. S. 1976. The Eggs and Planktonic Stages of British Marine Fishes. Academic, London.

Rutherford, E. S., and E. D. Houde. 1995. The influence of temperature on cohort-specific growth, survival, and recruitment of striped bass, Morone saxatilis, larvae in Chesapeake Bay. Fishery Bulletin 93(2):315–332.

Ruttenberg, B. I., and 7 coauthors. 2012. Rapid invasion of Indo-Pacific lionfishes (Pterois volitans and Pterois miles) in the Florida Keys, USA: Evidence from multiple pre- and post-invasion data sets. Bulletin of Marine Science 88(4):1051–1059.

Ruzzante, D. E., C. T. Taggart, C. Cook, and S. Goddard. 1996. Genetic differentiation between inshore and offshore Atlantic cod (Gadus morhua) off Newfoundland: Microsatellite DNA variation and antifreeze level. Canadian Journal of Fisheries and Aquatic Sciences 53(3):634–645.

Ruzzante, D. E., J. S. Wroblewski, C. T. Taggart, R. K. Smedbol, D. Cook, and S. V. Goddard. 2000. Bay-scale population structure in coastal Atlantic cod in Labrador and Newfoundland, Canada. Journal of Fish Biology 56(2):431–447.

Ruzzante, D. E., and 16 coauthors. 2006. Biocomplexity in a highly migratory pelagic marine fish, Atlantic herring. Proceedings of the Royal Society B: Biological Sciences 273(1593):1459–1464.

Sackett, D. K., K. W. Able, and T. M. Grothues. 2007. Dynamics of summer flounder, Paralichthys dentatus, seasonal migrations based on ultrasonic telemetry. Estuarine Coastal and Shelf Science 74 (1–2):119–130.

Saenz-Agudelo, P., G. P. Jones, S. R. Thorrold, and S. Planes. 2011. Connectivity dominates larval replenishment in a coastal reef fish metapopulation. Proceedings of the Royal Society B: Biological Sciences 278(1720):2954–2961.

Sagarese, S. R., and M. G. Frisk. 2011. Movement patterns and residence of adult winter flounder within a Long Island estuary. Marine and Coastal Fisheries 3(1):295–306.

Sale, P. F. 1977. Maintenance of high diversity in coral-reef fish communities. American Naturalist 111(978):337–359.

———. 1982. Stock-recruit relationships and regional coexistence in a lottery competitive system: A simulation study. American Naturalist 120(2):139–159.

Sale, P. F., and 10 coauthors. 2005. Critical science gaps impede use of no-take fishery reserves. Trends in Ecology and Evolution 20(2):74–80.

Salt, G. W. 1979. A comment on the use of the term emergent properties. American Naturalist 113(1):145–148.

Salvanes, A. G. V., J. E. Skjaeraasen, and T. Nilsen. 2004. Sub-populations of coastal cod with different behaviour and life-history strategies. Marine Ecology Progress Series 267:241–251.

Sato, K., and 17 coauthors. 2007. Stroke frequency, but not swimming speed, is related to body size in free-ranging seabirds, pinnipeds and cetaceans. Proceedings of the Royal Society B: Biological Sciences 274(1609):471–477.

Sato, K., K. Shiomi, Y. Watanabe, Y. Watanuki, A. Takahashi, and P. J. Ponganis. 2010. Scaling of swim speed and stroke frequency in geometrically similar penguins: They swim optimally to minimize cost of transport. Proceedings of the Royal Society B: Biological Sciences 277:707–714.

Scharf, F. S., J. A. Buckel, and F. Juanes. 2009. Contrasting patterns of resource utilization between juvenile estuarine predators: The influence of relative prey size and foraging ability on the ontogeny of piscivory. Canadian Journal of Fisheries and Aquatic Sciences 66(5):790–801.

Scheffer, M. 2009. Alternative stable states and regime shifts in ecosystems. In The Princeton Guide to Ecology, ed. S. A. Levin, 395–406. Princeton University Press, Princeton, NJ.

Scheffer, M., and S. R. Carpenter. 2003. Catastrophic regime shifts in ecosystems: Linking theory to observation. Trends in Ecology and Evolution 18(12):648–656.

Scheffer, M., and 9 coauthors. 2009. Early-warning signals for critical transitions. Nature 461(7260): 53–59.

Schellinck, J., and T. White. 2011. A review of attraction and repulsion models of aggregation: Methods, findings and a discussion of model validation. Ecological Modelling 222(11):1897–1911.

Schick, R. S., and 9 coauthors. 2008. Understanding movement data and movement processes: Current and emerging directions. Ecology Letters 11(12):1338–1350.

Schindler, D. E., and 6 coauthors. 2010. Population diversity and the portfolio effect in an exploited species. Nature 465(7298):609–612.

Schluter, D., and L. M. Nagel. 1995. Parallel speciation by natural selection. American Naturalist 146(2):292–301.

Schmidt, J. 1909. The distribution of the pelagic fry and the spawning regions of the gadoids in the North Atlantic from Iceland to Spain. Rapports et Procès-verbaux des Réunions du Conseil International pour l'Exploration de la Mer 10(4):1–229.

———. 1922. The breeding places of the eel. Philosophical Transactions of the Royal Society B: Biological Sciences 211:179–208.

Schmidt-Nielsen, K. 1972. Locomotion: Energy cost of swimming, flying, and running. Science 177(4045):222–228.

———. 1984. Scaling: Why Is Animal Size So Important? Cambridge University Press, Cambridge.

Schwartzlose, R. A., and 20 coauthors. 1999. Worldwide large-scale fluctuations of sardine and anchovy populations. South African Journal of Marine Science 21:289–347.

Sclafani, M., C. T. Taggart, and K. R. Thompson. 1993. Condition, buoyancy and the distribution of larval fish: Implications for vertical migration and retention. Journal of Plankton Research 15:413–435.

Secor, D. H. 1999. Specifying divergent migrations in the concept of stock: The contingent hypothesis. Fisheries Research 43(1–3):13–34.

———. 2000a. Spawning in the nick of time? Effect of adult demographics on spawning behavior and recruitment of Chesapeake Bay striped bass. ICES Journal of Marine Science 57:403–411.

———. 2000b. Longevity and resilience of Chesapeake Bay striped bass. ICES Journal of Marine Science 57:808–815.

———. 2002a. Historical roots of the migration triangle. ICES Journal of Marine Science 215:329–335.

———. 2002b. Estuarine dependency and life history evolution in temperate sea basses. Fisheries Science 68:178–181.

———. 2004. Fish migration and the unit stock: Three formative debates. In Stock Identification Methods, ed. S. X. Cadrin, K. D. Friedland, and J. R. Waldman, 17–44. Elsevier, Burlington, VT.

———. 2007. The year-class phenomenon and the storage effect in marine fishes. Journal of Sea Research 57:91–103.

———. 2008. Influence of skipped spawning and misspecified reproductive schedules on biological reference points in sustainable fisheries. Transactions of the American Fisheries Society 137:782–789.

———. 2010. Is otolith science transformative? New views on fish migration. Environmental Biology of Fishes 89(3–4):209–220.

———. 2013. The unit stock concept: Bounded fish and fisheries. In Stock Identification Methods: Applications in Fisheries Science, 2nd ed., ed. S. X. Cadrin, L. A. Kerr, and S. Mariani, 7–28. Academic, London.

Secor, D. H., and E. D. Houde. 1995a. Temperature effects on the timing of striped bass egg production, larval viability, and recruitment potential in the Patuxent River (Chesapeake Bay). Estuaries 18:527–544.

———. 1995b. Larval mark-release experiments: Potential for research on dynamics and recruitment in fish stocks. In Recent Developments in Fish Otolith Research, ed. D. H. Secor, S. E. Campana, and J. M. Dean, 423–444. University of South Carolina Press, Columbia.

———. 1998. Use of larval stocking in restoration of Chesapeake Bay striped bass. ICES Journal of Marine Science 55:228–239.

Secor, D. H., and L. A. Kerr. 2009. A lexicon of life cycle diversity in diadromous and other fishes. American Fisheries Society Symposium 69:537–556.

Secor, D. H., and P. M. Piccoli. 1996. Age- and sex-dependent migrations of striped bass in the Hudson

River as determined by chemical microanalysis of otoliths. Estuaries 19:778–793.

———. 2007. Oceanic migration rates of Upper Chesapeake Bay striped bass (*Morone saxatilis*), determined by otolith microchemical analysis. Fishery Bulletin 105(1):62–73.

Secor, D. H., and J. R. Rooker. 2000. Is otolith strontium a useful scalar of life cycles in estuarine fishes? Fisheries Research 46:359–371.

Secor, D. H., J. M. Dean, T. A. Curtis, and F. W. Sessions. 1992. Effect of female size and propagation methods on larval production at a South Carolina striped bass (*Morone saxatilis*) hatchery. Canadian Journal of Fisheries and Aquatic Science 49:1778–1787.

Secor, D. H., E. D. Houde, and D. M. Monteleone. 1995. A mark-release experiment on larval striped bass *Morone saxatilis* in a Chesapeake Bay tributary. ICES Journal of Marine Science 52:87–101.

Secor, D. H., T. Ohta, K. Nakayama, and M. Tanaka. 1998. Use of otolith microanalysis to determine estuarine migrations of Japanese sea bass *Lateolabrax japonicus* distributed in Ariake Sea. Fisheries Science 64(5):740–743.

Secor, D. H., V. Arefjev, A. Nikolaev, and A. Sharov. 2000. Restoration of sturgeons: Lessons from the Caspian Sea Sturgeon Ranching Programme. Fish and Fisheries 1:215–230.

Secor, D. H., L. A. Kerr, and S. X. Cadrin. 2009. Connectivity effects on productivity, stability, and persistence in an Atlantic herring metapopulation. ICES Journal of Maine Science 66:1726–1732.

Secor, D. H., B. I. Gahagan, and J. R. Rooker 2013. Atlantic bluefin tuna stock mixing within the U.S. North Carolina recreational fishery, 2011–2012. SCRS/2013/88. International Commission for the Conservation of Atlantic Tunas, Madrid.

Seel, D. C. 1977. Migration of northwestern European population of cuckoo *Cuculus canorus*, as shown by ringing. Ibis 119(3):309–322.

Seitz, A. C., T. Loher, B. L. Norcross, and J. L. Nielsen. 2011. Dispersal and behavior of Pacific halibut *Hippoglossus stenolepis* in the Bering Sea and Aleutian Islands region. Aquatic Biology 12(3): 225–239.

Sentchev, A., and K. Korotenko. 2007. Modelling distribution of flounder larvae in the eastern English Channel: Sensitivity to physical forcing and biological behaviour. Marine Ecology Progress Series 347:233–245.

Setzler, E. M., and 8 coauthors. 1980. Synopsis of biological data on striped bass, *Morone saxatilis*

(Walbaum). NOAA Technical Report NMFS 433. US Department of Commerce, Washington, DC.

Sfakiotakis, M., D. M. Lane, and J. B. C. Davies. 1999. Review of fish swimming modes for aquatic locomotion. IEEE Journal of Oceanic Engineering 24:237–252.

Shackell, N. L., K. T. Frank, and D. W. Brickman. 2005. Range contraction may not always predict core areas: An example from marine fish. Ecological Applications 15(4):1440–1449.

Shaffer, S. A., and 11 coauthors. 2006. Migratory shearwaters integrate oceanic resources across the Pacific Ocean in an endless summer. Proceedings of the National Academy of Sciences of the United States of America 103(34):12,799–12,802.

Shanks, A. L. 1995. Mechanisms of cross-shelf dispersal of larval invertebrates and fish. In Ecology of Marine Invertebrate Larvae, ed. L. McEdward, 323–367. CRC Press, New York.

Shapiro, D. Y., A. Marconato, and T. Yoshikawa. 1994. Sperm economy in a coral-reef fish, *Thalassemia bifasciatum*. Ecology 75(5):1334–1344.

Shaw, A. K., and S. A. Levin. 2011. To breed or not to breed: A model of partial migration. Oikos 120(12):1871–1879.

Shaw, K. A., M. L. Scotti, and S. A. Foster. 2007. Ancestral plasticity and the evolutionary diversification of courtship behaviour in threespine sticklebacks. Animal Behaviour 73:415–422.

Sheldon, R. W., and T. R. Parsons. 1967. A continuous size spectrum for particulate matter in sea. Journal of the Fisheries Research Board of Canada 24(5):909–915.

Shepherd, G. R., J. Moser, D. Deuel, and P. Carlsen. 2006. The migration patterns of bluefish (*Pomatomus saltatrix*) along the Atlantic coast determined from tag recoveries. Fishery Bulletin 104(4):559–570.

Shepherd, J. G., and D. H. Cushing. 1990. Regulation in fish populations: Myth or mirage? Philosophical Transactions of the Royal Society B: Biological Sciences 330(1257):151–164.

Shepherd, T. D., and M. K. Litvak. 2004. Density-dependent habitat selection and the ideal free distribution in marine fish spatial dynamics: Consideration and cautions. Fish and Fisheries 5:141–152.

Shepherd, T. D., K. E. Costain, and M. K. Litvak. 2000. Effect of development rate on the swimming, escape responses, and morphology of yolk-sac stage larval American plaice, *Hippoglossoides platessoides*. Marine Biology 137(4):737–745.

Sherwood, G. D., and J. H. Grabowski. 2010. Exploring the life-history implications of colour variation in offshore Gulf of Maine cod (*Gadus morhua*). ICES Journal of Marine Science 67(8): 1640–1649.

Shinomiya, A., and O. Ezaki. 1991. Mating habits of the rockfish *Sebastes inermis*. Environmental Biology of Fishes 30(1–2):15–22.

Shoji, J., and M. Tanaka. 2007. Growth and mortality of larval and juvenile Japanese seaperch *Lateolabrax japonicus* in relation to seasonal changes in temperature and prey abundance in the Chikugo estuary. Estuarine Coastal and Shelf Science 73(3–4):423–430.

Shuozeng, D. 1995. Life history cycles of flatfish species in the Bohai Sea, China. Netherlands Journal of Sea Research 34:195–210.

Sibert, J., and J. Hampton. 2003. Mobility of tropical tunas and the implications for fisheries management. Marine Policy 27(1):87–95.

Sibert, J., J. Hampton, D. A. Fournier, and P. J. Bills. 1999. An advection-diffusion-reaction model for the estimation of fish movement parameters from tagging data, with application to skipjack tuna (*Katsuwonus pelamis*). Canadian Journal of Fisheries and Aquatic Sciences 56(6):925–938.

Sih, A., and A. M. Bell. 2008. Insights for behavioral ecology from behavioral syndromes. Advances in the Study of Behavior 38:227–281.

Simons, A. M. 2004. Many wrongs: The advantage of group navigation. Trends in Ecology and Evolution 19(9):453–455.

Simpson, M. R., and S. J. Walsh. 2004. Changes in the spatial structure of Grand Bank yellowtail flounder: Testing MacCall's basin hypothesis. Journal of Sea Research 51(3–4):199–210.

Sims, D. W., E. J. Southall, A. J. Richardson, P. C. Reid, and J. D. Metcalfe. 2003. Seasonal movements and behaviour of basking sharks from archival tagging: No evidence of winter hibernation. Marine Ecology Progress Series 248:187–196.

Sims, D. W., E. J. Southall, G. A. Tarling, and J. D. Metcalfe. 2005. Habitat-specific normal and reverse diel vertical migration in the plankton-feeding basking shark. Journal of Animal Ecology 74(4):755–761.

Sims, D. W., and 17 coauthors. 2008. Scaling laws of marine predator search behaviour. Nature 451(7182):1098–1102.

Sinclair, M. 1988. Marine Populations: An Essay on Population Regulation and Speciation. University of Washington Press, Seattle.

Skomal, G. B., and 6 coauthors. 2009. Transequatorial migrations by basking sharks in the western Atlantic Ocean. Current Biology 19(12):1019–1022.

Slade, J. W., and 9 coauthors. 2003. Techniques and methods for estimating abundance of larval and metamorphosed sea lampreys in Great Lakes tributaries, 1995 to 2001. Journal of Great Lakes Research 29:137–151.

Smedbol, R. K., and R. Stephenson. 2001. The importance of managing within-species diversity in cod and herring fisheries of the north-western Atlantic. Journal of Fish Biology 59:109–128.

Smedbol, R. K., and J. S. Wroblewski. 2002. Metapopulation theory and northern cod population structure: Interdependency of subpopulations in recovery of a groundfish population. Fisheries Research 55:161–174.

Smith, C. C., and S. D. Fretwell. 1974. Optimal balance between size and number of offspring. American Naturalist 108(962):499–506.

Smith, R. J. F. 1984. The Control of Fish Migration. Springer, New York.

Smith, T. B., and S. Skúlason. 1996. Evolutionary significance of resource polymorphisms in fishes, amphibians, and birds. Annual Review of Ecology and Systematics 27:111–133.

Smith, T. D. 1994. Scaling Fisheries: The Science of Measuring the Effects of Fishing, 1855–1955. Cambridge University Press, Cambridge.

Smogor, R. A., P. L. Angermeier, and C. K. Gaylord. 1995. Distribution and abundance of American eels in Virginia streams: Tests of null models across spatial scales. Transactions of the American Fisheries Society 124:789–803.

Snelgrove, P. V. R. 2010. Discoveries of the Census of Marine Life: Making Ocean Life Count. Cambridge University Press, Cambridge.

Snyder, J. O. 1931. Salmon of the Klamath River, California. Fish Bulletin 34. Department of Fish and Game, State of California, Sacramento.

Sogard, S. M. 1997. Size-selective mortality in the juvenile stage of teleost fishes: A review. Bulletin of Marine Science 60(3):1129–1157.

Sogard, S. M., E. Gilbert-Hovarth, E. C. Anderson, R. Fishee, S. A. Berkeley, and J. C. Garza. 2008. Multiple paternity in viviparous kelp rockfish, *Sebastes atrovirens*. Environmental Biology of Fishes 81(1):7–13.

Solemdal, P., E. Dahl, D. S. Danielssen, and E. Moksness. 1984. The cod hatchery in Flodevigen: Background and realities. In The Propagation of Cod, 17–45. Arendal, Norway.

Solmundsson, J., E. Jonsson, and H. Bjornsson. 2010.

Phase transition in recruitment and distribution of monkfish (*Lophius piscatorius*) in Icelandic waters. Marine Biology 157(2):295-305.

Sparre, P., and S. C. Venema. 1998. Introduction to Tropical Fish Stock Assessment, Part 1: Manual. FAO Fisheries Technical Paper 306/Rev. 2. Food and Agriculture Organization of the United Nations, Rome.

Speed, C. W., I. C. Field, M. G. Meekan, and C. J. A. Bradshaw. 2010. Complexities of coastal shark movements and their implications for management. Marine Ecology Progress Series 408:275-293.

Spice, E. K., D. H. Goodman, S. B. Reid, and M. F. Docker. 2012. Neither philopatric nor panmictic: Microsatellite and mtDNA evidence suggests lack of natal homing but limits to dispersal in Pacific lamprey. Molecular Ecology 21(12):2916-2930.

Sprules, W. G., and A. P. Goyke. 1994. Size-based structure and production in the pelagia of Lakes Ontario and Michigan. Canadian Journal of Fisheries and Aquatic Sciences 51(11):2603-2611.

Stabell, O. B. 1984. Homing and olfaction in salmonids: A critical review with special reference to Atlantic salmon. Biological Review 59:333-388.

Stabeno, P. J., J. D. Schumacher, K. M. Bailey, R. D. Brodeur, and E. D. Cokelet. 1996. Observed patches of walleye pollock eggs and larvae in Shelikof Strait, Alaska: Their characteristics, formation and persistence. Fisheries Oceanography 5:81-91.

Stacey, P. B., V. A. Johnson, and M. L. Taper. 1997. Migration with metapopulations: The impact of local population demographics. In Metapopulation Biology: Ecology, Genetics, and Evolution, ed. I. A. Hanski and M. E. Gilpin, 267-291. Academic, New York.

Standing Committee on Research and Statistics. 2013. Report of the 2012 Atlantic Bluefin Tuna Stock Assessment Session. International Commission for the Conservation of Atlantic Tunas, Madrid. http://www.iccat.es/en/assess.htm.

Stearns, S. C. 1977. The evolution of life history traits: A critique of the theory and a review of the data. Annual Review of Ecological Systems 8:145-171.

Steele, J. H. 1991. Can ecological theory cross the land-sea boundary? Journal of Theoretical Biology 153(3):425-436.

Stehfest, K. M., T. A. Patterson, L. Dagorn, K. N. Holland, D. Itano, and J. M. Semmens. 2013. Network analysis of acoustic tracking data reveals the structure and stability of fish aggregations in the ocean. Animal Behaviour 85(4):839-848.

Stehlik, L. L. 2009. Effects of seasonal change on activity rhythms and swimming behavior of age-0 bluefish (*Pomatomus saltatrix*) and a description of gliding behavior. Fishery Bulletin 107(1):1-12.

Steneck, R. S., and 8 coauthors. 2009. Thinking and managing outside the box: Coalescing connectivity networks to build region-wide resilience in coral reef ecosystems. Coral Reefs 28(2):367-378.

Stephenson, R. L., G. D. Melvin, and M. J. Power. 2009. Population integrity and connectivity in northwest Atlantic herring: A review of assumptions and evidence. ICES Journal of Marine Science 66:1733-1739.

Stewart, I. J., T. P. Quinn, and P. Bentzen. 2003. Evidence for fine-scale natal homing among island beach spawning sockeye salmon, *Oncorhynchus nerka*. Environmental Biology of Fishes 67(1):77-85.

Stockley, P., M. J. G. Gage, G. A. Parker, and A. P. Moller. 1997. Sperm competition in fishes: The evolution of testis size and ejaculate characteristics. American Naturalist 149(5):933-954.

Stockwell, C. A., A. P. Hendry, and M. T. Kinnison. 2003. Contemporary evolution meets conservation. Trends in Ecology and Evolution 18(2):94-101.

Strathmann, R. R. 1990. Why life histories evolve differently in the sea. American Zoologist 30(1): 197-207.

Sullivan, C. V., B. L. Berlinsky, and R. G. Hodson. 1997. Reproduction. In Striped Bass and other *Morone* Culture, ed. R. M. Harrell, 11-65. Elsevier, New York.

Sullivan, M. C., R. K. Cowen, K. W. Able, and M. P. Fahay. 2006. Applying the basin model: Assessing habitat suitability of young-of-the-year demersal fishes on the New York Bight continental shelf. Continental Shelf Research 26(14):1551-1570.

Sumpter, D. J. T. 2006. The principles of collective animal behaviour. Philosophical Transactions of the Royal Society B: Biological Sciences 361(1465):5-22.

Sundby, S., H. Bjoerke, A. Soldal, and S. Olsen. 1989. Mortality rates during the early life stages and year-class strength of Arctic cod (*Gadus morhua* L.). Rapports et Procès-verbaux des Réunions du Conseil International pour l'Exploration de la Mer 191:351-358.

Suzuki, K. W., A. Kasai, T. Ohta, K. Nakayama, and M. Tanaka. 2008. Migration of Japanese temperate bass *Lateolabrax japonicus* juveniles within the Chikugo River estuary revealed by $\delta^{13}C$ analysis. Marine Ecology Progress Series 358:245-256.

Svedäng, H., D. Righton, and P. Jonsson. 2007. Migratory behaviour of Atlantic cod *Gadus morhua*: Natal homing is the prime stock-separating mechanism. Marine Ecology Progress Series 345:1–12.

Swaney, W., J. Kendal, H. Capon, C. Brown, and K. N. Laland. 2001. Familiarity facilitates social learning of foraging behaviour in the guppy. Animal Behaviour 62:591–598.

Swearer, S. E., J. E. Caselle, D. W. Lea, and R. R. Warner. 1999. Larval retention and recruitment in an island population of a coral-reef fish. Nature 402(6763):799–802.

Taborsky, M. 1998. Sperm competition in fish: "Bourgeois" males and parasitic spawning. Trends in Ecology and Evolution 13(6):222–227.

Taggart, C. T., and K. F. Frank. 1990. Perspectives on larval fish ecology and recruitment processes: Probing the scales of relationships. In Patterns, Processes, and Yields of Large Marine Ecosystems. AAAS Selected Symposium Series, ed. K. Sherman and L.M. Alexander, 151–164. American Association for the Advancement of Science, Washington, DC.

Taggart, C. T., and 10 coauthors. 1994. Overview of cod stocks, biology, and environment in the northwest Atlantic region of Newfoundland with emphasis on northern cod. ICES Journal of Marine Science Symposium 198:140–157.

Tallman, R. F., and M. C. Healey 1994. Homing, straying, and gene flow among seasonally separated populations of chum salmon (*Oncorhynchus keta*). Canadian Journal of Fisheries and Aquatic Sciences 51(3):577–588.

Tamm, S. 1980. Bird orientation: Single homing pigeons compared with small flocks. Behavioral Ecology and Sociobiology 7(4):319–322.

Tanaka, H., H. Kagawa, and H. Ohta. 2001. Production of leptocephali of Japanese eel (*Anguilla japonica*) in captivity. Aquaculture 201(1–2):51–60.

Tanaka, M., T. Goto, M. Tomiyama, H. Sudo, and M. Azuma. 1989. Lunar-phased immigration and settlement of metamorphosing Japanese flounder larvae into the nearshore nursery ground. Rapports et Procès-verbaux des Réunions du Conseil International pour l'Exploration de la Mer 191.303–310.

Tarazona, J., D. Guitiérrez, C. Paredes, and A. Inacochea. 2003. Overview and challenges of marine biodiversity research in Peru. Gayana (Concepción) 67(2):206–231.

Taylor, E. B. 1990. Environmental correlates of life-history variation in juvenile chinook salmon, *Oncorhynchus tshawytscha* (Walbaum). Journal of Fish Biology 37(1):1–17.

Taylor, N. G., M. K. McAllister, G. L. Lawson, T. Carruthers, and B. A. Block. 2011. Atlantic bluefin tuna: A novel multistock spatial model for assessing population biomass. PLoS One 6(12):e27693. doi:10.1371/journal.pone.0027693.

Tedesco, P. A., B. Hugueny, T. Oberdorff, H. H. Durr, S. Merigoux, and B. Merona. 2008. River hydrological seasonality influences life history strategies of tropical riverine fishes. Oecologia 156(3):691–702.

ten Cate, C. 2000. How learning mechanisms might affect evolutionary processes. Trends in Ecology and Evolution 15(5):179–181.

Teo, S. L. H., A. Boustany, S. Blackwell, A. Walli, K. C. Weng, and B. A. Block. 2004. Validation of geolocation estimates based on light level and sea surface temperature from electronic tags. Marine Ecology Progress Series 283:81–98.

Teo, S. L. H., A. M. Boustany, and B. A. Block. 2007. Oceanographic preferences of Atlantic bluefin tuna, *Thunnus thynnus*, on their Gulf of Mexico breeding grounds. Marine Biology 152(5):1105–1119.

Terrill, S. B., and K. P. Able. 1988. Bird migration terminology. Auk 105(1):205–206.

Thériault, V., and J. J. Dodson. 2003. Body size and the adoption of a migratory tactic in brook charr. Journal of Fish Biology 63(5):1144–1159.

Thornton, K. R., J. D. Jensen, C. Becquet, and P. Andolfatto. 2007. Progress and prospects in mapping recent selection in the genome. Heredity 98(6):340–348.

Thorpe, J. E. 1987. Smolting versus residency: Developmental conflict in salmonids. American Fisheries Society Symposium 1:244–252.

———. 1994. Salmonid flexibility: Responses to environmental extremes. Transactions of the American Fisheries Society 123:606–612.

Thorrold, S. R., C. Latkoczy, P. K. Swart, and C. M. Jones. 2001. Natal homing in a marine fish metapopulation. Science 291:297–299.

Thorup, K., and 6 coauthors. 2007. Evidence for a navigational map stretching across the continental US in a migratory songbird. Proceedings of the National Academy of Sciences of the United States of America 104(46):18,115–18,119.

Thresher, R. E. 1984. Patterns in the reproduction of reef fishes. In Reproduction in Reef Fishes, 343–388. T. F. H. Publications, Neptune, NJ.

———. 1988. Latitudinal variation in egg sizes of

tropical and sub-tropical North Atlantic shore fishes. Environmental Biology of Fishes 21:17–25.

Thünken, T., T. C. M. Bakker, and H. Kullmann. 2007. Extraordinarily long sperm in the socially monogamous cichlid fish *Pelvicachromis taeniatus*. Naturwissenschaften 94(6):489–491.

Tilman, D., C. L. Lehman, and C. E. Brostow. 1998. Diversity-stability relationships: Statistical inevitability or ecological consequence. American Naturalist 151:277–282.

Tobin, D., P. J. Wright, F. M. Gibb, and I. M. Gibb. 2010. The importance of life stage to population connectivity in whiting (*Merlangius merlangus*) from the northern European shelf. Marine Biology 157(5):1063–1073.

Toole, C. L., D. F. Markle, and P. M. Harris. 1993. Relationships between otolith microstructure, microchemistry, and early life history events in Dover sole, *Microstomus pacificus*. Fishery Bulletin 91:732–753.

Townsend, D. W., R. L. Radtke, M. A. Morrison, and S. D. Folsom. 1989. Recruitment implications of larval herring overwintering distributions in the Gulf of Maine, inferred using a new otolith technique. Marine Ecology Progress Series 55:1–13.

Trippel, E. A., and M. J. Morgan. 1994. Sperm longevity in Atlantic cod (*Gadus morhua*). Copeia 1994:1025–1029.

Trippel, E. A., and J. D. Neilson. 1992. Fertility and sperm quality of virgin and repeat-spawning Atlantic cod (*Gadus morhua*) and associated hatching success. Canadian Journal of Fisheries and Aquatic Sciences 49(10):2118–2127.

Tsukamoto, K. 2009. Oceanic migration and spawning of anguillid eels. Journal of Fish Biology 74(9): 1833–1852.

Tsukamoto, K., and J. Aoyama. 1998. Evolution of freshwater eels of the genus *Anguilla*: A probable scenario. Environmental Biology of Fishes 52:139–148.

Tsukamoto, K., and T. Arai. 2001. Facultative catadromy of the eel *Anguilla japonica* between freshwater and seawater habitats. Marine Ecology Progress Series 220:265–276.

Tsukamoto, K., R. Ishida, K. Naka, and T. Kajihara. 1987. Switching of size and migratory patterns in successive generation of landlocked ayu. American Fisheries Society Symposium 1:492–506.

Tsukamoto, K., I. Nakai, and W. V. Tesch. 1998. Do all freshwater eels migrate? Nature 396:635–636.

Tsukamoto, K., and 12 coauthors. 2003. Seamounts, new moon and eel spawning: The search for the spawning site of the Japanese eel. Environmental Biology of Fishes 66(3):221–229.

Tsukamoto, K., and 8 coauthors. 2009. Positive buoyancy in eel leptocephali: An adaptation for life in the ocean surface layer. Marine Biology 156(5):835–846.

Tsukamoto, K., and 20 coauthors. 2011. Oceanic spawning ecology of freshwater eels in the western North Pacific. Nature Communications 2:179. doi:10.1038/ncomms1174.

Tsunagawa, T., and T. Arai. 2009. Migration diversity of the freshwater goby *Rhinogobius* sp. BI, as revealed by otolith Sr:Ca ratios. Aquatic Biology 5(2):187–194.

Tucker, D. W. 1959. A new solution to the Atlantic eel problem. Nature 183:495–501.

Tupper, M., and R. G. Boutilier. 1995. Effects of habitat on settlement, growth, and postsettlement survival of Atlantic cod (*Gadus morhua*). Canadian Journal of Fisheries and Aquatic Sciences 52(9):1834–1841.

Tuset, V. M., E. A. Trippel, and J. de Monserrat. 2008. Sperm morphology and its influence on swimming speed in Atlantic cod. Journal of Applied Ichthyology 24(4):398–405.

Tyler, C. R., and J. P. Sumpter. 1996. Oocyte growth and development in teleosts. Reviews in Fish Biology and Fisheries 6(3):287–318.

Tyler, C. R., J. J. Nagler, T. G. Pottinger, and M. A. Turner. 1994. Effects of unilateral ovariectomy on recruitment and growth of follicles in the rainbow trout, *Oncorhynchus mykiss*. Fish Physiology and Biochemistry 13(4):309–316.

Uchida, S., M. Toda, K. Teshima, and K. Yano. 1996. Pregnant white sharks and full-term embryos from Japan. In Great White Sharks: The Biology of *Carcharodon carcharias*, ed. A. P. Klimley and D. G. Ainley, 139–155. Academic, San Diego.

Ulanowicz, R. E. 1997. Ecology: The Ascendant Perspective. Columbia University Press, New York.

Unwin, M. J., and T. P. Quinn. 1993. Homing and straying patterns of chinook salmon (*Oncorhynchus tshawytscha*) from a New Zealand hatchery: Spatial distribution of strays and effects of release date. Canadian Journal of Fisheries and Aquatic Sciences 50(6):1168–1175.

Uphoff, J. H. 1989. Environmental effects on survival of eggs, larvae, and juveniles of striped bass in the Choptank River, Maryland. Transactions of the American Fisheries Society 118:251–263.

US Fish and Wildlife Service. 2007. Endangered and

threatened wildlife and plants: 12-month finding on a petition to list the American eel as threatened or endangered. Federal Register 72(22):4967–4997.

US Fish and Wildlife Service and National Marine Fisheries Service. 1996. Policy regarding the recognition of distinct vertebrate population segments under the Endangered Species Act. Federal Register 61(26):4722–4725.

Vabø, R., and G. Skaret. 2008. Emerging school structures and collective dynamics in spawning herring: A simulation study. Ecological Modelling 214(2–4):125–140.

Valiela, I. 1995. Marine Ecological Processes. 2nd ed. Springer, New York.

van der Veer, H. W. 1985. Impact of coelenterate predation on larval plaice *Pleuronectes platessa* and flounder *Platichthys flesus* stock in the western Wadden Sea. Marine Ecology Progress Series 25:229–238.

Van der Veer, H. W., and M. J. N. Bergman. 1987. Predation by crustaceans on a newly settled 0-group plaice *Pleuronectes platessa* population in the western Wadden Sea. Marine Ecology Progress Series 35:203–215.

van der Veer, H. W., T. Ellis, J. M. Miller, L. Pihl, and A. D. Rijnsdorp. 1997. Size-selective predation on juvenile North Sea flatfish and possible interactions for recruitment. In Early Life History and Recruitment in Fish Populations, ed. R. C. Chambers and E. A. Trippel, 279–304. Chapman and Hall, New York.

van Wilgenburg, S. L., and K. A. Hobson. 2011. Combining stable-isotope (δD) and band recovery data to improve probabilistic assignment of migratory birds to origin. Ecological Applications 21(4):1340–1351.

Varsamos, S., C. Nebel, and G. Charmantier. 2005. Ontogeny of osmoregulation in postembryonic fish: A review. Comparative Biochemistry and Physiology A: Molecular and Integrative Physiology 141(4):401–429.

Vélez-Espino, L. A., and M. A. Koops. 2010. A synthesis of the ecological processes influencing variation in life history and movement patterns of American eel: Towards a global assessment. Reviews in Fish Biology and Fisheries 20(2):163–186.

Vianna, G. M. S., M. G. Meekan, J. J. Meeuwig, and C. W. Speed. 2013. Environmental influences on patterns of vertical movement and site fidelity of grey reef sharks (*Carcharhinus amblyrhynchos*) at aggregation sites. PLoS One 8(4):e60331. doi:10.1371/journal.pone.0060331.

Victor, B. C. 1986. Larval settlement and juvenile mortality in a recruitment-limited coral reef fish population. Ecological Monographs 56:145–160.

Videler, J. J., and C. S. Wardle. 1991. Fish swimming stride by stride: Speed limits and endurance. Reviews in Fish Biology and Fisheries 1:23–40.

Vila-Gispert, A., R. Moreno-Amich, and E. Garcia-Berthou. 2002. Gradients of life-history variation: An intercontinental comparison of fishes. Reviews in Fish Biology and Fisheries 12(4):417–427.

Viscido, S. V., J. K. Parrish, and D. Grunbaum. 2004. Individual behavior and emergent properties of fish schools: A comparison of observation and theory. Marine Ecology Progress Series 273:239–249.

———. 2005. The effect of population size and number of influential neighbors on the emergent properties of fish schools. Ecological Modelling 183(2–3): 347–363.

Viswanathan, G. M., V. Afanasyev, S. V. Buldyrev, E. J. Murphy, P. A. Prince, and H. E. Stanley. 1996. Lévy flight search patterns of wandering albatrosses. Nature 381(6581):413–415.

von Herbing, I. H. 2002. Effects of temperature on larval fish swimming performance: The importance of physics to physiology. Journal of Fish Biology 61(4):865–876.

von Herbing, I. H., and R. G. Boutilier. 1996. Activity and metabolism of larval Atlantic cod (*Gadus morhua*) from Scotian shelf and Newfoundland source populations. Marine Biology 124(4): 607–617.

von Herbing, I. H., R. G. Boutilier, T. Miyake, and B. K. Hall. 1996. Effects of temperature on morphological landmarks critical to growth and survival in larval Atlantic cod (*Gadus morhua*). Marine Biology 124:593–606.

von Herbing, I. H., S. M. Gallager, and W. Halteman. 2001. Metabolic costs of pursuit and attack in early larval Atlantic cod. Marine Ecology Progress Series 216:201–212.

Vrieze, L. A., and P. W. Sorensen. 2001. Laboratory assessment of the role of a larval pheromone and natural stream odor in spawning stream localization by migratory sea lamprey (*Petromyzon marinus*). Canadian Journal of Fisheries and Aquatic Sciences 58(12):2374–2385.

Waldman, J. 2013. Running Silver: Restoring Atlantic Rivers and Their Great Fish Migrations. Lyons Press, Guilford, CT.

Waldman, J., D. J. Dunning, Q. E. Ross, and M. T. Mattson. 1990. Range dynamics of Hudson River striped bass along the Atlantic Coast. Transactions of the American Fisheries Society 119:910–919.

Waldman, J., C. Grunwald, and I. Wirgin. 2008. Sea lamprey *Petromyzon marinus*: An exception to the rule of homing in anadromous fishes. Biology Letters 4(6):659–662.

Walker, B. 1992. Biological diversity and ecological redundancy. Conservation Biology 6:18–23.

Walker, M. M., C. E. Diebel, C. V. Haugh, P. M. Pankhurst, J. C. Montgomery, and C. R. Green. 1997. Structure and function of the vertebrate magnetic sense. Nature 390(6658):371–376.

Walker, M. M., J. L. Kirschvink, S. B. R. Chang, and A. E. Dizon. 1984. A candidate magnetic sense organ in the yellowfin tuna, *Thunnus albacares*. Science 224(4650):751–753.

Walli, A., and 7 coauthors. 2009. Seasonal movements, aggregations and diving behavior of Atlantic bluefin tuna (*Thunnus thynnus*) revealed with archival tags. PLoS One 4(7):e6151. doi:10.1371/journal.pone.0006151.

Walters, C., and A. M. Parma. 1996. Fixed exploitation rate strategies for coping with effects of climate change. Canadian Journal of Fisheries and Aquatic Sciences 53:148–158.

Walther, B. D., S. R. Thorrold, and J. E. Olney. 2008. Geochemical signatures in otoliths record natal origins of American shad. Transactions of the American Fisheries Society 137(1):57–69.

Walther, B. D., T. Dempster, M. Letnic, and M. T. McCulloch. 2011. Movements of diadromous fish in large unregulated tropical rivers inferred from geochemical tracers. PLoS One 6(4):e18351. doi:10.1371/journal.pone.0018351

Waples, R. S., A. E. Punt, and J. M. Cope. 2008. Integrating genetic data into management of marine resources: How can we do it better? Fish and Fisheries 9:423–449.

Ward, A. J. W., J. Krause, and D. J. T. Sumpter. 2012. Quorum decision-making in foraging fish shoals. PLoS One 7(3):e32411. doi:10.1371/journal.pone.0032411.

Ward, D. L., 6 coauthors. 2012. Translocating adult Pacific lamprey within the Columbia River Basin: State of the science. Fisheries 37(8):351–361.

Ware, D. M., and R. W. Tanasichuk. 1989. Biological basis of maturation and spawning waves in Pacific herring (*Clupea harengus pallasi*). Canadian Journal of Fisheries and Aquatic Sciences 46(10):1776–1784.

Warner, R. R. 1990. Resource assessment versus tradition in mating-site determination. American Naturalist 135:205–217.

———. 1995. Large mating aggregations and daily long-distance spawning migrations in the bluehead wrasse, *Thalassoma bifasciatum*. Environmental Biology of Fishes 44(4):337–345.

Warner, R. R., and P. L. Chesson. 1985. Coexistence mediated by recruitment fluctuations: A field guide to the storage effect. American Naturalist 125:769–787.

Watanabe, Y. 2002. Resurgence and decline of the Japanese sardine population. In Fisheries Science: The Unique Contributions of Early Life Stages, ed. L. A. Fuiman and R. G. Werner, 243–257. Blackwell, Oxford.

Watanabe, Y., and 6 coauthors. 1998. Naupliar copepod concentrations in the spawning grounds of Japanese sardine, *Sardinops melanostictus*, along the Kuroshio Current. Fisheries Oceanography 7(2):101–109.

Watanabe, Y., and 8 coauthors. 2008. Swimming behavior in relation to buoyancy in an open swimbladder fish, the Chinese sturgeon. Journal of Zoology 275(4):381–390.

Watanuki, Y., Y. Niizuma, G. W. Gabrielsen, K. Sato, and Y. Naito. 2003. Stroke and glide of wing-propelled divers: Deep diving seabirds adjust surge frequency to buoyancy change with depth. Proceedings of the Royal Society B: Biological Sciences 270(1514):483–488.

Wearmouth, V. J., and D. W. Sims. 2008. Sexual segregation in marine fish, reptiles, birds and mammals: Behaviour patterns, mechanisms and conservation implications. Advances in Marine Biology 54:107–170.

Wearmouth, V. J., and 6 coauthors. 2012. Year-round sexual harassment as a behavioral mediator of vertebrate population dynamics. Ecological Monographs 82(3):351–366.

Webb, P. W. 1978. Hydromechanics of swimming. Science 199(4329):678.

———. 1982. Locomotor patterns in the evolution of Actinopterygian fishes. American Zoologist 22(2):329–342.

———. 1988. Simple physical principles and vertebrate aquatic locomotion. American Zoologist 28(2):709–725.

Webb, P. W., and R. T. Corolla. 1981. Burst swimming performance of northern anchovy, *Engraulis mordax*, larvae. Fishery Bulletin 79(1):143–150.

Webb, P. W., and D. Weihs. 1986. Functional locomotor morphology of early life-history stages of fishes. Transactions of the American Fisheries Society 115(1):115–127.

Webb, P. W., P. T. Kostecki, and E. D. Stevens. 1984. The effect of size and swimming speed on locomotor kinematics of rainbow trout. Journal of Experimental Biology 109:77–95.

Webster, M. S., P. P. Marra, S. M. Haig, S. Bensch, and R. T. Holmes. 2002. Links between worlds: Unraveling migratory connectivity. Trends in Ecology and Evolution 17(2):76–83.

Weihs, D. 1980. Respiration and depth control as possible reasons for swimming of northern anchovy, *Engraulis mordax*, yolk-sac larvae. Fishery Bulletin 78(1):109–117.

Weimerskirch, H., M. Le Corre, F. Marsac, C. Barbraud, O. Tostain, and O. Chastel. 2006. Postbreeding movements of frigatebirds tracked with satellite telemetry. Condor 108(1):220–225.

Welch, D. W., B. R. Ward, and S. D. Batten. 2004. Early ocean survival and marine movements of hatchery and wild steelhead trout (*Oncorhynchus mykiss*) determined by an acoustic array: Queen Charlotte Strait, British Columbia. Deep Sea Research Part II: Topical Studies in Oceanography 51(6–9):897–909.

Welch, D. W., and 8 coauthors. 2008. Survival of migrating salmon smolts in large rivers with and without dams. PLoS Biology 6(10):e314. doi:10.1371/journal.pbio.0060314.

Wells, R. J. D., J. R. Rooker, and E. D. Prince. 2010. Regional variation in the otolith chemistry of blue marlin (*Makaira nigricans*) and white marlin (*Tetrapturus albidus*) from the western North Atlantic Ocean. Fisheries Research 106(3):430–435.

Werner, E. E. 1986. Amphibian metamorphosis: Growth rate, predation risk, and the optimal size at transformation. American Naturalist 128(3): 319–341.

Werner, E. E., and J. F. Gilliam. 1984. The ontogenetic niche and species interactions in size-structured populations. Annual Review of Ecological Systems 15:393–425.

Werner, E. E., and D. J. Hall. 1988. Ontogenetic habitat shifts in bluegill: The foraging rate-predation risk trade-off. Ecology 69(5):1352–1366.

Werner, E. E., J. F. Gilliam, D. J. Hall, and G. G. Mittelbach. 1983a. An experimental test of the effects of predation risk on habitat use in fish. Ecology 64:1540–1548.

Werner, E. E., G. G. Mittelbach, D. J. Hall, and J. F. Gilliam. 1983b. Experimental tests of optimal habitat use in fish: The role of relative habitat profitability. Ecology 64:1525–1539.

West-Eberhard, M. J. 2005. Developmental plasticity and the origin of species differences. Proceedings of the National Academy of Sciences of the United States of America 102:6543–6549.

Whitfield, A. K. 1999. Ichthyofaunal assemblages in estuaries: A South African case study. Reviews in Fish Biology and Fisheries 9:151–186.

Whitledge, G. W., B. M. Johnson, and P. J. Martinez. 2006. Stable hydrogen isotopic composition of fishes reflects that of their environment. Canadian Journal of Fisheries and Aquatic Sciences 63(8):1746–1751.

Wiens, J. A. 1997. Metapopulation dynamics and landscape ecology. In Metapopulation Biology: Ecology, Genetics, and Evolution, ed. I. A. Hanski and M. E. Gilpin, 43–62. Academic, San Diego.

Wilberg, M. J., M. E. Livings, J. S. Barkman, B. T. Morris, and J. M. Robinson. 2011. Overfishing, disease, habitat loss, and potential extirpation of oysters in upper Chesapeake Bay. Marine Ecology Progress Series 436:131–144.

Williams, E. P., A. C. Peer, T. J. Miller, D. H. Secor, and A. R. Place. 2012. A phylogeny of the temperate seabasses (Moronidae) characterized by a translocation of the mt-nd6 gene. Journal of Fish Biology 80(1):110–130.

Williams, L. M., and M. F. Oleksiak. 2011. Ecologically and evolutionarily important SNPs identified in natural populations. Molecular Biology and Evolution 28(6):1817–1826.

Williams, T. C., J. M. Williams, P. G. Williams, and P. Stokstad. 2001. Bird migration through a mountain pass studied with high resolution radar, ceilometers, and census. Auk 118(2):389–403.

Willis, J., J. Phillips, R. Muheim, F. Javier Diego-Rasilla, and A. J. Hobday. 2009. Spike dives of juvenile southern bluefin tuna (*Thunnus maccoyii*): A navigational role? Behavioral Ecology and Sociobiology 64(1):57–68.

Wiltschko, W., and R. Wiltschko. 1972. Magnetic compass of European robins. Science 176(4030): 62–64.

———. 1996. Magnetic orientation in birds. Journal of Experimental Biology 199(1):29–38.

Windle, M. J. S., and G. A. Rose. 2007. Do cod form spawning leks? Evidence from a Newfoundland spawning ground. Marine Biology 150(4):671–680.

Winemiller, K. O. 2005. Life history strategies, population regulation, and implications for fisheries management. Canadian Journal of Fisheries and Aquatic Science 62:872–885.

Winemiller, K. O., and K. A. Rose. 1992. Patterns of life-history diversification in North American fishes: Implications for population regulation. Canadian Journal of Fisheries and Aquatic Science 49:2196–2218.

———. 1993. Why do most fish produce so many tiny offspring? American Naturalist 142:585–603.

Wingate, R. L., and D. H. Secor. 2007. Intercept telemetry of the Hudson River striped bass resident contingent: Migration and homing patterns. Transactions of the American Fisheries Society 136(1):95–104.

Wingate, R. L., D. H. Secor, and R. T. Kraus. 2009. Seasonal patterns of movement and residency by striped bass within a subestuary of the Chesapeake Bay. Transactions of the American Fisheries Society 140(6):1441–1450.

Winkler, G., J. J. Dodson, N. Bertrand, D. Thivierge, and W. F. Vincent. 2003. Trophic coupling across the St. Lawrence River estuarine transition zone. Marine Ecology Progress Series 251:59–73.

Wirgin, I., A. I. Kovach, L. Maceda, N. K. Roy, J. Waldman, and D. L. Berlinsky. 2007. Stock identification of Atlantic cod in US waters using microsatellite and single nucleotide polymorphism DNA analyses. Transactions of the American Fisheries Society 136(2):375–391.

Woodland, R. J., M. A. Rodriguez, P. Magnan, H. Glemet, and G. Cabana. 2012. Incorporating temporally dynamic baselines in isotopic mixing models. Ecology 93(1):131–144.

Wooton, R. J. 1984. Introduction: Strategies and tactics in fish reproduction. In Fish Reproduction: Strategies and Tactics, ed. G. W. Potts and R. J. Wooton, 1–12. Academic, London.

Worm, B., and 20 coauthors. 2009. Rebuilding global fisheries. Science 325:578–585.

Wourms, J. P. 1981. Viviparity: The maternal-fetal relationship in fishes. American Zoologist 21(2):473–515.

———. 1991. Reproduction and development of Sebastes in the context of the evolution of Piscine viviparity. Environmental Biology of Fishes 30 (1–2):111–126.

Wray, G. A. 1995. Evolution of larvae and developmental modes. In Ecology of Marine Invertebrate Larvae, ed. L. McEdward, 413–447. CRC Press, New York.

Wright, P. J., F. C. Neat, F. M. Gibb, I. M. Gibb, and H. Thordarson. 2006. Evidence for metapopulation structuring in cod from the west of Scotland and North Sea. Journal of Fish Biology 69:181–199.

Wu, L.-Q., and J. D. Dickman. 2012. Neural correlates of a magnetic sense. Science 336(6084):1054–1057.

Wunder, M. B., and D. R. Norris. 2008. Analysis and design for isotope-based studies of migratory animals. In Tracking Animal Migration with Stable Isotopes, ed. K. A. Hobson and L. I. Wassenaar, 107–128. Elsevier, New York.

Yamada, J., and M. Kusakari. 1991. Staging and the time course of embryonic development in kurosoi, Sebastes schlegeli. Environmental Biology of Fishes 30(1–2):103–110.

Yano, A., and 6 coauthors. 1997. Effect of modified magnetic field on the ocean migration of maturing chum salmon, Oncorhynchus keta. Marine Biology 129(3):523–530.

Yates, P. M., M. R. Heupel, A. J. Tobin, and C. A. Simpfendorfer. 2012. Diversity in young shark habitats provides the potential for portfolio effects. Marine Ecology Progress Series 458:269–281.

Yin, M. C., and J. H. S. Blaxter. 1987. Feeding ability and survival during starvation of marine fish larvae reared in the laboratory. Journal of Experimental Marine Biology and Ecology 105:73–83.

Yoklavich, M. M., V. J. Loeb, M. Nishimoto, and B. Daly. 1996. Nearshore assemblages of larval rockfishes and their physical environment off central California during an extended El Niño event, 1991–1993. Fishery Bulletin 94(4):766–782.

Youngson, N. A., and E. Whitelaw. 2008. Transgenerational epigenetic effects. Annual Review of Genomics and Human Genetics 9:233–257.

Yund, P. O. 2000. How severe is sperm limitation in natural populations of marine free-spawners? Trends in Ecology and Evolution 15(1):10–13.

Zastrow, C. E., E. D. Houde, and E. H. Saunders. 1989. Quality of striped bass (Morone saxatilis) eggs in relation to river source and female weight. Rapports et Procès-verbaux des Réunions du Conseil International pour l'Exploration de la Mer 191:34–42.

Zhang, J. Z. 2003. Evolution by gene duplication: An update. Trends in Ecology and Evolution 18(6):292–298.

Zlokovitz, E. R., D. H. Secor, and P. M. Piccoli. 2003. Patterns of migration in Hudson River striped bass as determined by otolith microchemistry. Fisheries Research 63(2):245–259.

Index

The letter *t* following a page number denotes a table. The letter *f* denotes a figure.

acceleration reaction, 24
acoustic-lateral line system, 45, 52, 181
acoustic receiver arrays, 6, 37, 41
adopted migration, 4t; and conservatism, 142; and density dependence, 141, 166; and entrainment, 49, 139, 141, 153, 166, 216; and learning, 56–57, 140–41; and lost traditions, 142; and personality modes, 141; in reef fish, 141; theory of, 137, 139–42
Aegean Sea, 132
age resonance, 216
age structure, 124–26, 140, 142, 162, 165, 216–17, 224–26, 226f
Alaska, 194
Alaska Gyre, 129–30, 130f
albatross, Laysan (*Diomedea immutabilis*), 34, 56
albatross, wandering (*Diomedea exulans*), 42–43, 53
alevin, 129
alewife (*Alosa pseudoharengus*), 17, 202
Allee effect, 221
ammocoetes larvae, 137, 143, 168
amphidromy, 203, 205
anadromy: 22t; evolution of, 195–96
anchovy, bay (*Anchoa mitchilli*), 69, 82, 106, 116, 118
anchovy, northern (*Engraulis mordax*), 76, 147
anchovy, Peruvian (*Engraulis ringens*), 121, 138, 214f, 219–20, 230
Antarctic Peninsula, 39, 155
Appalachian Mountains, 39
Arctic Ocean watersheds, 124, 129, 171
area-restricted search, 42, 150
Argentina, 40, 195
Ariake Sea, 196
arrival time hypothesis, 198
artificial selection: effect on behaviors, 94; effect on life history traits, 93–95; effect on mating systems, 92; in partial migration experiments, 17, 171–73, 175f
Asian temperate bass family (Lateolabracidae), 196
aspect ratio, 27–28f, 159
assemblage, marine, 30, 160, 219
Atlantic Multidecadal Oscillation, 220

Atlantic temperate bass family (Moronidae), 174t, 196–97, 203
Australia, 133, 135, 149, 157, 161, 193, 197, 202
averaging effect, 221–22, 227, 231
avian ecology, 6, 205
ayu (*Plecoglussus altivelis*), 176, 178f, 205

Baja California, 193
Balearic Sea, 132
Barents Sea, 33, 225
barramundi (*Lates calcarifer*), 149, 197–98, 202
basin model, 145–48, 147f
bass, European sea (*Dicentrarchus labrax*), 65, 196
bass, Japanese sea (*Lateolabrax japonicus*), 90–91, 196, 204
bass, red (*Lutjanus bohar*), 149
bass, striped (*Morone saxatilis*), 57–58, 62, 69–70, 73, 88–94, 90f, 103–4, 104f, 113, 114f, 119f, 134–35, 134t, 149, 197–200, 200f, 212, 214f, 216–17, 225–26, 226f
bass, white (*Morone chrysops*), 197
bass, yellow (*Morone mississippiensis*), 197
Bayesian parameter estimation, 41, 60
Beaufort Sea, 190t
behavioral syndromes, 150–51
Bering Sea, 53
bet hedging, 9f, 50, 165; and spawning behavior, 113
biocomplexity, 21, 205; and functional genetic markers, 188–89; as networks of interacting genes, 188f
biological hierarchy, 10
biological population, 5t, 162
birds of two worlds, 55, 135
Biscay Bay, 85
bluefish (*Pomatomus saltatrix*), 28, 92, 105–7f, 198
bluegill (*Lepomis macrochirus*), 104–5, 115
bocaccio (*Sebastes paucispinis*), 119f
boldness, 150
bonefishes (*Albula* spp.), 77
Brazil, 142, 215, 215f
Bristol Bay, 227
British Columbia, 58, 64, 67, 125f, 130f, 151, 152f, 202
British Islands, 58, 166, 196
buoyancy, 24, 28, 70–71, 73–74, 76–80, 83, 95; in diving birds, 29